STATISTICAL and THERMAL PHYSICS

Part II

Quantum Statistical Mechanics and Simple Applications

by
Shigeji Fujita
State University of New York at Buffalo

ROBERT E. KRIEGER PUBLISHING COMPANY
MALABAR, FLORIDA
1986

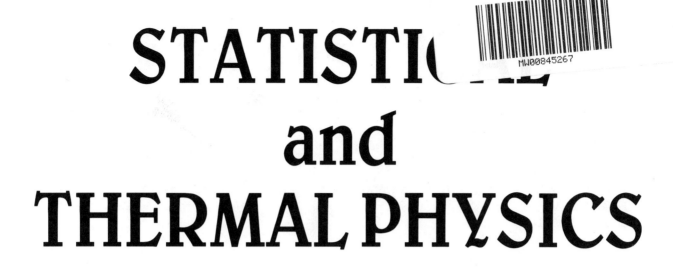

Original Edition 1986

Printed and Published by
ROBERT E. KRIEGER PUBLISHING COMPANY, INC.
KRIEGER DRIVE
MALABAR, FLORIDA 32950

Printed in the United States of America.

Library of Congress Cataloging in Publication Data

Fujita, Shigeji

Statistical and thermal physics.

Bibliography: p.
Includes index.
Contents: -- pt. 2. Quantum statistical mechanics and simple applications.
1. Statistical physics. 2. Thermodynamics. 3. Statistical mechanics. I. Title.
QC174.8.F85 1986 530.1'3 83-22250
ISBN 0-89874-689-2 (pt. 1)
ISBN 0-89874-866-6 (pt. 2)

10 9 8 7 6 5 4 3 2

To Sachiko

Acknowledgments

I wish to thank all those colleagues and students who have suggested improvements, pointed out numerous errors, and given valuable comments in the various stages of the present work. I am especially indebted to Evan Glaser, Sergio Ulloa, Michio, Isao, Yoshiko and Eriko (my children) for their careful readings of the manuscript and valuable suggestions about stylistic and technical revisions. The book is dedicated to my wife, Sachiko, who typed most of the manuscript, and who has given encouragement and support. I wish to express my thanks to authors and publishers for their permission to reproduce figures used in the text.

CONTENT OF VOLUME 2

Chapter 13. TRANSPORT PHENOMENA

CONTENTS OF VOLUME 1

APPENDICES

PREFACE TO SECOND VOLUME

This volume contains fundamentals of quantum statistical mechanics and simple applications. Standard topics include the derivation of the Bose and Fermi distribution functions, discussions of conduction electrons, liquid helium, black body radiation and lattice vibrations, spin and magnetism, and the Boltzmann theory of the electrical conductivity. For pedagogical reasons and convenience, a number of preparatory materials are included here. The Dirac formulation of quantum mechanics by using the bra and ket notations, the permutation group, the normal-mode description of the electromagnetic waves and the elastic waves in solids, the quantum formulation of orbital and spin angular momentum are presented in detail. If they are read outside the classroom, they should contribute greatly a deeper understanding of quantum statistical mechanics proper. Advanced special topics include Ising model and phase transition, conformation of polymers in solution and helix-coil transition, Kubo's formula and simple applications, a theory of the diffusion based on the correlated walk model. They represent somewhat newer developments of the topics in statistical mechanics.

Chapter 8. QUANTUM MECHANICS. FUNDAMENTALS REVIEWED.

Fundamental concepts and elementary applications of the quantum mechanics of a single particle are developed and discussed, following Dirac's quantization scheme. The chapter begins with basic experimental facts in § 8.1. The following four sections, §§ 8.2-8.5, deal with the mathematical preliminaries. The rest of the Chapter, §§ 8.6-8.15, is dedicated to the quantum mechanical description of a particle moving in one dimension and in three dimensions.

Readers who wish to study the statistical mechanics proper in a minimum time may skip the entire chapter for their first reading and may come back to pertinent material in this chapter when need arises. Those readers who have learned quantum mechanics primarily by means of the wave mechanics, will learn more about the fundamental structures of quantum mechanics by looking at them from Dirac's angle presented in this chapter. All serious future physicists are urged to refer to Dirac's classic book, <u>Priciples of Quantum Mechanics</u>, [1] for a deeper understanding of quantum mechanics.

8.1 Basic Experimental Facts

a. Waves and particles

In 1900, in order to explain the distribution of energy for black body radiation [see §11.3], Planck (Max Planck, 1858-1947) assumed that a heated body does not radiate or absorb energy continuously, as dictated by classical electrodynamics, but only in integral multiples of a fundamental unit called a <u>quantum</u> . If ν is the frequency of radiation, the energy of one quantum is given by

$$\boxed{E = h\nu ,} \tag{1.1}$$

where

$$h = 6.626 \times 10^{-27} \text{ erg sec.} = 6.626 \times 10^{-34} \text{ J·s} \tag{1.2}$$

is a fundamental constant known as Planck's constant. Note that this quantity has the dimensions of angular momentum.

In 1905, Einstein, in his theory of the photo-electric effect, proposed that light (or electromagnetic radiation in general) can sometimes be regarded as having a particle-like nature. He postulated that the light-quantum called the photon has the energy given by (1.1) and the momentum (magnitude) given by

$$\boxed{p = \frac{h}{\lambda} = \frac{h\nu}{c} = \frac{E}{c},} \tag{1.3}$$

where c is the speed of light, and λ is the wavelength. Note that the momentum of a photon is inversely proportional to the wavelength λ.

According to the theory of relativity, the energy E and momentum p of a particle with the rest mass m_o are related by

$$E^2 = c^2p^2 + m_o^2 c^4. \tag{1.4}$$

The energy-momentum relation for a zero-rest-mass particle ($m_o = 0$) is then

$$E = pc , \tag{1.5}$$

which is consistent with the relations (1.1) and (1.3). That is, the photon can be regarded as a relativistic particle with zero rest mass.

Although light had been ascribed a particle-like nature by Newton, this concept had been abandoned because of the sucesses of the wave theory. Einstein's theory explained very well the photo-electric effect: when a metallic surface is irradiated by ultra-violet light, electrons are emitted with energies which do not depend on the intensity of the light but only on the frequency.

In summary, light can be regarded as a wave with frequencies and wavelengths in some conditions, or as zero-mass particles with energies and momenta in others.

In 1924, de Broglie extended this dual nature of wave and particles to electrons. In some experiments, an electron with mass m and momentum p will display wave-like properties with frequency ν and wavelength λ given by

$$\nu = \frac{E}{h} \tag{1.6}$$

$$\lambda = \frac{h}{p}, \tag{1.7}$$

which can be obtained by solving eqs.(1.1) and (1.3) for ν and λ.

This prediction was confirmed in 1929, when Davisson and Germer observed the diffraction of a beam of electrons by a crystal of nickel. Electrons emitted by a hot filament and accelerated through an electric field to a known velocity, were allowed to fall on the surface of nickel crystal. The reflected beam exhibited direction-dependent maxima and minima in intensity just as in the case of X-ray diffraction. This electron diffraction could be accounted for by assuming that the electron beam consisted of waves with frequency ν and wave length λ as given in (1.6) and (1.7). Diffraction phenomena have since been observed with beams of other microscopic particles.

b. Line spectra and stationary states

According to classical electrodynamics, an accelerated electric charge will emit radiation. In Rutherford's model of the hydogen atom, an electron moves in an orbit around a proton. Since the electron in circular motion is in an accelerated state, it should continuously lose energy by emitting radiation and should spiral in smaller and smaller orbits and eventually collide with the proton. This does not happen in reality since we know that the atom is stable.

In addition to this difficulty, we have another kind of difficulty. If a system of classical particles is subjected to some force, then the energy of the system may vary in a continuous manner. That is, a classical system can change its energy in a continuous manner. If this were true of atomic systems, we should expect a gas of hydrogen atoms, for example, to absorb light of all frequencies, and thus, should observe a continuous spectrum of absorption. In reality, part of the hydrogen spectrum consists of a series of discrete spectral lines.

In 1913, Bohr proposed that an atom may exist in any one of a set of discrete states, each with a definite energy (stationary state rule). When an atom is in a discrete state, it does not radiate; if, however, it passes from one discrete state with energy E_1 to another with lower energy E_0, a radiation of frequency ν will be emitted, where E_1, E_0 and ν are related by

$$E_1 - E_0 = h\nu \tag{1.8}$$

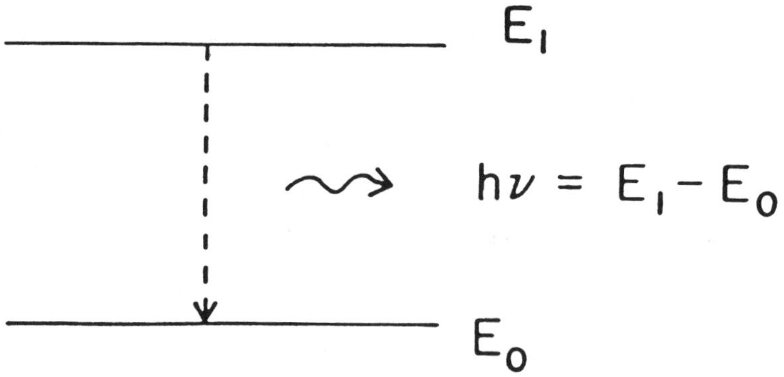

Fig. 8.1 A radiation of energy $h\nu = E_1 - E_0$ is emitted when an atom passes from the state of energy E_1 to that of energy E_0.

This is called Bohr's frequency rule. Conversely, if the atom is in the lower energy state, absorption of a photon with energy $h\nu$ will raise the atom to the higher energy state.

c. Heat capacity paradox

Experiments show that most solids obey Dulong-Petit's law, that is, the molar heat capacity of a solid equals 3R. This is understandable for a non-metallic solid. In fact, by applying the equipartition theorem to the kinetic and potential energies of N vibrating atoms, we obtain $3Nk_B T = 3RT$ for the thermal energy. Differentiation with respect to T yields 3R, which is in agreement with Dulong-Petit's law. For a metal, this argument fails. We know that the electric charge in a metal is transported by mobile electrons. Therefore, the kinetic energy of the conduction electrons should contribute $\frac{3}{2}$ R to the molar heat capacity. But such contribution appears to be missing.

This difficulty, called the heat capacity paradox, can be resolved by applying the quantum statistics to electrons. By now we know that electrons are quantum particles which obey the Fermi-Dirac statistics. For quantum particles, the equipartition theorem of classical statistical mechanics does not apply. We will discuss the electronic heat capacity in §§10.5 and 10.6 .

8.2° Generalized Vectors. Matrices

We wish to construct a class of mathematical quantities which have the same algebraic properties as vectors. Such quantities, called generalized vectors, can be generated by matrix algebra, which will be discussed in the present section.

A vector may be defined geometrically as a quantity having magnitude and direction. Various properties of vectors are enumerated in Appendix C. Among the basic properties of vectors are : (a) any two vectors $\vec{A}$ and $\vec{B}$ can be added together, yielding the vector $\vec{A} + \vec{B}$; (b) any vector $\vec{A}$ can be multiplied by a real number p, yielding the vector $p\vec{A}$; (c) from two vectors $\vec{A}$ and $\vec{B}$, a scalar product can be constructed, yielding a real number $\vec{A}\cdot\vec{B}$; (d) every vector $\vec{A}$ has a non-negative magnitude (or length) equal to $\sqrt{\vec{A}\cdot\vec{A}} \equiv A$. In addition, the following algebraic properties are satisfied by vectors.

$$\vec{A} + \vec{B} = \vec{B} + \vec{A} \qquad \text{(commutative)}$$

$$(\vec{A} + \vec{B}) + \vec{C} = \vec{A} + (\vec{B} + \vec{C}) \equiv \vec{A} + \vec{B} + \vec{C} \qquad \text{(associative)} \qquad (2.1)$$

$$p\vec{A} = \vec{A}p$$

$$p(q\vec{A}) = q(p\vec{A}) = (pq)\vec{A}$$

$$p(\vec{A} + \vec{B}) = p\vec{A} + p\vec{B} \qquad \text{(distributive)}$$

$$(p + q)\vec{A} = p\vec{A} + q\vec{A} \qquad (2.2)$$

$$\vec{A}\cdot(p\vec{B}) = p(\vec{A}\cdot\vec{B}) = (p\vec{A})\cdot\vec{B}$$

$$\vec{A}\cdot(\vec{B} + \vec{C}) = \vec{A}\cdot\vec{B} + \vec{A}\cdot\vec{C} \qquad (2.3)$$

$$\vec{A}\cdot\vec{A} \equiv A^2 \geqq 0. \qquad (2.4)$$

In general, a rectangular array of numbers with m rows and n columns will be called an $m \times n$ matrix, e.g.

$$\begin{pmatrix} a_{11} & a_{12} & \cdots & a_{1n} \\ a_{21} & a_{22} & \cdots & a_{2n} \\ \cdots & \cdots & \cdots & \cdots \\ a_{m1} & a_{m2} & & a_{mn} \end{pmatrix} \equiv (a_{ij})_{(m,n)} = A ,$$

where a_{ij} is a general complex number. Two matrices A and B are equal if and only if they have the same elements:

$$A = B \quad \text{if } a_{ij}=b_{ij} \; . \tag{2.6}$$

The following operations are assumed:

(a) The sum of two matrices $(a_{ij})_{(m,n)}$ and $(b_{ij})_{(m,n)}$ is a matrix whose elements are the sum of the corresponding elements. For example, for m=n=2,

$$\begin{pmatrix} a_{11} & a_{12} \\ a_{21} & a_{22} \end{pmatrix} + \begin{pmatrix} b_{11} & b_{12} \\ b_{21} & b_{22} \end{pmatrix} \equiv \begin{pmatrix} a_{11}+b_{11} & a_{12}+b_{12} \\ a_{21}+b_{21} & a_{22}+b_{22} \end{pmatrix} . \tag{2.7}$$

(b) The product of a matrix A and a complex number c is a matrix whose elements are c times the corresponding elements of A. For example,

$$cA \equiv c \begin{pmatrix} a_{11} & a_{12} \\ a_{21} & a_{22} \end{pmatrix} \equiv \begin{pmatrix} ca_{11} & ca_{12} \\ ca_{21} & ca_{22} \end{pmatrix} . \tag{2.8}$$

(c) The ordered product of a m x n matrix $(a_{ij})_{(m,n)}$ and a n x p matrix $(b_{ij})_{(n,p)}$ is a m x p matrix whose (i,j) elements are given by

$$\sum_{k=1}^{n} a_{ik}\, b_{kj} \; . \tag{2.9}$$

In particular, the product of two m x m matrices A and B is a m x m matrix. For example, for m=2,

$$\begin{pmatrix} a_{11} & a_{12} \\ a_{21} & a_{22} \end{pmatrix} \begin{pmatrix} b_{11} & b_{12} \\ b_{21} & b_{22} \end{pmatrix}$$

$$= \begin{pmatrix} a_{11}b_{11}+a_{12}b_{21} & a_{11}b_{12}+a_{12}b_{22} \\ a_{21}b_{11}+a_{22}b_{21} & a_{21}b_{12}+a_{22}b_{22} \end{pmatrix} . \tag{2.9a}$$

The ordered product of a 1 x n matrix and a n x 1 matrix is a number. For example,

$$\begin{pmatrix} a_1 & a_2 \end{pmatrix} \begin{pmatrix} b_1 \\ b_2 \end{pmatrix} = a_1b_1 + a_2b_2 . \tag{2.9b}$$

Consider the m x n matrix $(a_{ij})_{(m,n)}$ in (2.5). Interchanging rows and columns, we obtain a n x m matrix

$$\begin{pmatrix} a_{11} & a_{21} & \cdots & a_{m1} \\ a_{12} & a_{22} & \cdots & a_{m2} \\ a_{1n} & a_{2n} & \cdots & a_{mn} \end{pmatrix} \equiv A^T . \tag{2.10}$$

This matrix is called the transpose of matrix A, and is denoted by A^T.

A complex number c can in general be written as the sum of real and imaginary numbers:

$$c = p + iq \tag{2.11}$$

where

$$i \equiv \sqrt{-1} , \tag{2.12}$$

and p and q are real numbers. The complex conjugate of c is defined by p - iq, and will be denoted by c^*:

$$c^* = (p + iq)^* = p - iq . \tag{2.13}$$

The Hermitean conjugate of a matrix A is obtained by applying transposition and complex-conjugation, and will be denoted by $A^\dagger$. Thus, for the matrix A in (2.5),

$$A^\dagger \equiv \begin{pmatrix} a_{11}^* & a_{21}^* & \cdots & a_{m1}^* \\ a_{12}^* & a_{22}^* & \cdots & a_{m2}^* \\ \cdots & \cdots & \cdots & \cdots \\ a_{1n}^* & a_{2n}^* & \cdots & a_{mn}^* \end{pmatrix} . \tag{2.14}$$

The symbol $\dagger$ is read dagger.

From the definition, we can show that

$$(c^*)^* = c , \tag{2.15}$$

$$[A^T]^T = A , \tag{2.16}$$

$$[A^\dagger]^\dagger = A . \tag{2.17}$$

A matrix having the same number of rows and columns, i.e., a n x n matrix, is called a square matrix. A 1 x n matrix, i.e., a matrix with a single row, is called a row matrix. A n x 1 matrix, i.e., a matrix with a single column, is called a column matrix. The transpose (and the Hermitean conjugate) of a square matrix is a square matrix. The transpose of a row matrix is a column matrix, and vice versa.

A matrix whose elements are all real: $a_{ij}^* = a_{ij}$, is called a real matrix. For real matrices, $A^\dagger = A^T$. That is, the distinction between transposition and Hermitean conjugation disappears for real matrices.

We can easily see from (2.7) and (2.8) that addition of matrices is commutative and associative, just as for ordinary vectors [see (2.1)]. Multiplication of matrices by a number is also commutative, associative and distributive.

Let us take two column matrices,

$$\begin{pmatrix} a_1 \\ a_2 \\ a_3 \end{pmatrix} \equiv |A>, \qquad \begin{pmatrix} b_1 \\ b_2 \\ b_3 \end{pmatrix} \equiv |B> . \tag{2.18}$$

Their Hermitean conjugates are row matrices given by

$$(|A>)^\dagger = \left(a_1^*, a_2^*, a_3^* \right) \equiv <A| ,$$

$$(|B>)^\dagger = \left(b_1^*, b_2^*, b_3^* \right) \equiv <B| . \tag{2.19}$$

Column and row matrices are quite different mathematical entities. For example, they cannot be added together. In the above, we distinguished column and row matrices by "ket" and "bra" symbols, $|\ >$ and $<\ |$, respectively. The names "bra" and "ket" (due to Dirac) are derived from the first and second halves of the word "bracket".

The scalar product $<B||A>$, the row matrix on the left and the column matrix on the right, can be calculated as follows:

$$<B||A> \equiv <B|A>$$

$$\equiv \left(b_1^* \ b_2^* \ b_3^* \right) \begin{pmatrix} a_1 \\ a_2 \\ a_3 \end{pmatrix} = b_1^* a_1 + b_2^* a_2 + b_3^* a_3 . \tag{2.20a}$$

That is, the product written in the form of a complete bracket, $<B|A>$, is a (complex) number.

Since

$$<A|B> = \left(a_1^* \ a_2^* \ a_3^* \right) \begin{pmatrix} b_1 \\ b_2 \\ b_3 \end{pmatrix} = a_1^* b_1 + a_2^* b_2 + a_3^* b_3 , \tag{2.20b}$$

the numbers $\langle B|A\rangle$ and $\langle A|B\rangle$ are complex conjugates to each other:

$$\langle B|A\rangle = [\langle A|B\rangle]^* . \tag{2.21}$$

In particular, if $|B\rangle = |A\rangle$, then

$$\langle A|A\rangle = a_1^* a_1 + a_2^* a_2 + a_3^* a_3 \quad (\geq 0) , \tag{2.22}$$

which is non-negative.

It is noted that ordered products such as $|A\rangle\langle B|$ and $|A\rangle\langle A|$ do not generate numbers, but instead are more complicated mathematical quantities (called linear operators, see §8.3).

Let us now consider an ordinary vector in three dimensions. In the Cartesian representation, such a vector $\vec{A}$ can be specified by the set of three components, (A_x, A_y, A_z), which are all real. We can represent the vector $\vec{A}$ by a real column matrix:

$$\vec{A} = \begin{pmatrix} A_x \\ A_y \\ A_z \end{pmatrix} \equiv |A\rangle , \tag{2.23}$$

with the provisions that (a) the magnitude of the vector is given by the square root of

$$\begin{aligned} \langle A|A\rangle &\equiv A_x^* A_x + A_y^* A_y + A_z^* A_z \\ &= A_x^2 + A_y^2 + A_z^2 \quad (= \text{ squared magnitude}) , \end{aligned} \tag{2.24}$$

and (b) the scalar product $\vec{B}\cdot\vec{A}$ with a real vector $\vec{B} = (B_x, B_y, B_z)$ is given by

$$\begin{aligned} \langle B|A\rangle &\equiv B_x^* A_x + B_y^* A_y + B_z^* A_z \\ &= B_x A_x + B_y A_y + B_z A_z \\ &= \vec{B}\cdot\vec{A} . \end{aligned} \tag{2.25}$$

We can alternatively represent ordinary vectors in terms of real row matrices. The resulting algebras for row matrices are not much different from those for column matrices. Essentially, the two algebras are Hermitean conjugates to each other.

The above representation of vectors in terms of column or row matrices suggests a natural generalization. Let us consider n x 1 column matrices with complex elements. Clearly the collection of such n x 1 column matrices satisfies the same algebraic laws as ordinary vectors: [compare with (2.1) - (2.4)]

$$|A\rangle + |B\rangle = |B\rangle + |A\rangle$$

$$(|A\rangle + |B\rangle) + |C\rangle = |A\rangle + (|B\rangle + |C\rangle) \equiv |A\rangle + |B\rangle + |C\rangle \tag{2.26}$$

$$c|A\rangle = |A\rangle c$$

$$c(d|A\rangle) = d(c|A\rangle) = (dc)|A\rangle$$

$$c(|A\rangle + |B\rangle) = c|A\rangle + c|B\rangle$$

$$(c + d)|A\rangle = c|A\rangle + d|A\rangle \tag{2.27}$$

$$\langle B|(c|A\rangle) = c\langle B|A\rangle = (\langle B|c)|A\rangle = \langle B|c|A\rangle$$

$$\langle B|(|A\rangle + |C\rangle) = \langle B|A\rangle + \langle B|C\rangle \tag{2.28}$$

$$\langle B|A\rangle^* = \langle A|B\rangle$$

$$\langle A|A\rangle \geqq 0 . \tag{2.29}$$

where c and d are complex numbers.

With the rule that the column matrix $|A\rangle$ has the magnitude given by

$$\sqrt{\langle A|A\rangle} = \text{magnitude of } |A\rangle , \tag{2.30}$$

we may say that n x 1 column matrices represent generalized vectors. The generalization is two-fold. First, n can be any positive integer; second, the elements may be complex numbers. Because of the latter condition, we have introduced minor changes in algebraic properties between eqs.(2.1) - (2.4) and eqs.(2.26) - (2.29). In order to assure that $<A|A> \geqq 0$, we chose $<A|$ to be the Hermitean conjugate (rather than the transpose) of $|A>$ [see (2.24) and (2.30)]

In summary, the column matrices (and again row matrices) satisfy the same algebraic properties as ordinary vectors. Because of this, these column and row matrices are also referred to as <u>column and row vectors</u>.

Problem 2.1 Show that

(a) $[(x+iy)(a+ib)]^* = (x-iy)(a-ib)$

(b) $[(x+iy)^n]^* = (x-iy)^n$

(c) $[e^{(a+ib)}]^* = e^{(a-ib)}$

(d) $[e^{(a+ib)(x+iy)}]^* = e^{(a-ib)(x-iy)}$

where a, b, x and y are real numbers.

Problem 2.2 Calculate the lengths of the following column vectors

(a) $\begin{pmatrix} 1 \\ 2 \\ -3 \end{pmatrix}$ (b) $\begin{pmatrix} 1 \\ i \\ -1 \\ -i \end{pmatrix}$

Problem 2.3 Show that

(a) $<A|(c|B>) = c<A|B> = (<A|c)|B>$

(b) $<A|B>^* = <B|A>$

8.3° Linear Operators

Let us consider a n x 1 column vector $|A>$. We multiply it from the left by a n x n matrix α. According to the multiplication rule, we then obtain a n x 1 column vector, which will be denoted by $|B>$. In mathematical terms,

$$|B> \;=\; \alpha|A> \;. \tag{3.1}$$

We may look at this equation from a few different angles. First, we may regard it as the definition equation for the vector $|B>$ in terms of the vector $|A>$ and the matrix α. Second, we may say that the two vectors $|B>$ and $|A>$ are related by the *mapping* operation α. Third, we may say that the matrix α, upon *acting* on the vector $|A>$ *from the left*, generates the vector $|B>$ with the action being the operation of matrix multiplication. In general, a quantity capable of performing a mathematical operation is called an *operator*. Thus, in the last view, the n x n matrix α, which upon acting on a column vector generates another column vector, is an operator.

An operator is said to be known (or defined) if all of the results of the operation are given. In the case of (3.1), if the results $\alpha|A>$ for all possible vectors $|A>$ are known, then the operator α is defined.

We now take the special case of (3.1):

$$|A> \;=\; I|A> \quad \text{for all } |A> \;. \tag{3.2}$$

The operator I is well defined, and clearly is given by

$$I \;=\; \begin{pmatrix} 1 & 0 & 0 \\ 0 & 1 & 0 \\ 0 & 0 & 1 \end{pmatrix} \qquad \text{for } n=3 \;. \tag{3.3}$$

A matrix having non-zero elements only along the diagonal is called a *diagonal matrix*. If these diagonal elements are all equal to unity, the diagonal matrix is called a *unit matrix*. The matrix I in (3.3) is a unit matrix.

Using the operational rules of matrices, we can easily show that

$$\alpha(|A\rangle + |B\rangle) = \alpha|A\rangle + \alpha|B\rangle$$

$$\alpha(c|A\rangle) = c(\alpha|A\rangle) \; . \qquad (3.4)$$

An operator which satisfies the above algebraic relations is called a linear operator. Thus the matrix α is a linear operator.

Consider now a complex number a. We can multiply any n x 1 column vector by this number and obtain a column vector. We can also obtain relations like (3.4) with a replacing α throughout. The number a can therefore be regarded as a linear operator. Let us denote this operator by $\hat{a}$. This operator may be defined by

$$\hat{a}|A\rangle = a|A\rangle \qquad \text{for all } |A\rangle \; . \qquad (3.5)$$

It is clear that $\hat{a}$ can be represented by the matrix

$$\begin{pmatrix} a & 0 & 0 \\ 0 & a & 0 \\ 0 & 0 & a \end{pmatrix} = a \begin{pmatrix} 1 & 0 & 0 \\ 0 & 1 & 0 \\ 0 & 0 & 1 \end{pmatrix} = aI. \qquad (3.6)$$

In other words, the number a can be represented by the matrix aI.

In general, linear operators relating any two n x 1 column vectors in the form (3.1) can be represented by n x n matrices (see below). Conversely, as we have said before, n x n matrices are linear operators. Because of this close connection, the same symbol will be used to denote both a linear operator and its corresponding n x n matrix. An advantage of referring to such a quantity as a linear operator is that the name reminds us of its principal operational properties. The term "matrix" can be used to describe a quantity less restricted in operational characters. A tensor (whose definition and properties are given in Volume 1, Appendix C) can be regarded as a linear operator connecting two vectors in ordinary space.

Any two linear operators α and γ can be added together, yielding a third linear operator, denoted by $\alpha + \gamma$, which is defined by

$$(\alpha + \gamma)|A> \;\equiv\; \alpha|A> + \gamma|A> \quad \text{for any } |A> \,. \tag{3.7}$$

From this definition, we can obtain

$$\alpha + \gamma \;=\; \gamma + \alpha \qquad [\text{commutative}]$$

$$(\alpha + \gamma) + \delta \;=\; \alpha + (\gamma + \delta) \;\equiv\; \alpha + \gamma + \delta \,. \qquad [\text{associative}] \tag{3.8}$$

The product of two linear operators α and γ is defined by

$$(\alpha\gamma)\,|A> \;\equiv\; \alpha(\gamma|A>) \quad \text{for any } |A> \,. \tag{3.9}$$

From this, we may deduce that

$$(\alpha\gamma)\delta \;=\; \alpha(\gamma\delta) \;\equiv\; \alpha\gamma\delta \,. \qquad [\text{associative}] \tag{3.10}$$

However, the product $\alpha\gamma$ and the product of the reversed order, $\gamma\alpha$, are not in general equal to each other:

$$\boxed{\alpha\gamma \;\neq\; \gamma\alpha \quad \text{in general} \,.} \tag{3.11}$$

For illustration we take the following two products:

$$\begin{pmatrix} a_{11} & a_{12} \\ a_{21} & a_{22} \end{pmatrix}\begin{pmatrix} b_{11} & b_{12} \\ b_{21} & b_{22} \end{pmatrix} = \begin{pmatrix} a_{11}b_{11}+a_{12}b_{21} & a_{11}b_{12}+a_{12}b_{22} \\ a_{21}b_{11}+a_{22}b_{21} & a_{21}b_{12}+a_{22}b_{22} \end{pmatrix}$$

$$\begin{pmatrix} b_{11} & b_{12} \\ b_{21} & b_{22} \end{pmatrix}\begin{pmatrix} a_{11} & a_{12} \\ a_{21} & a_{22} \end{pmatrix} = \begin{pmatrix} a_{11}b_{11}+a_{21}b_{12} & a_{12}b_{11}+a_{22}b_{12} \\ a_{11}b_{21}+a_{21}b_{22} & a_{12}b_{21}+a_{22}b_{22} \end{pmatrix} .$$

Clearly, the two matrices are different.

We often refer to eq. (3.11) by saying that <u>two linear operators do not necessarily commute</u>. This is a significant property distinct from that of the product of two numbers a and b since any two numbers commute with each other. It is observed from the second equation of (3.4) that any number commutes with a linear operator.

Consider now a row vector $\langle C|$. If we multiply it from the right by matrix α, we obtain a row vector $\langle C|\alpha$. The scalar product of this row vector $\langle C|\alpha$ and a column vector $|B\rangle$ is by definition a number, denoted by $(\langle C|\alpha)|B\rangle$. It is easy to show that the same number can be obtained if we first operate α on $|B\rangle$ and then make a scalar product from $\langle C|$ and $\alpha|B\rangle$:

$$(\langle C|\alpha)|B\rangle = \langle C|(\alpha|B\rangle) \equiv \langle C|\alpha|B\rangle \,. \tag{3.12}$$

For n=2, this equality may be exhibited as follows:

$$\langle C| \equiv (c_1^* \; c_2^*), \quad \alpha \equiv \begin{pmatrix} a_{11} & a_{12} \\ a_{21} & a_{22} \end{pmatrix}, \quad |B\rangle \equiv \begin{pmatrix} b_1 \\ b_2 \end{pmatrix}$$

$$\begin{aligned}
(\langle C|\alpha)|B\rangle &= \left[(c_1^* \; c_2^*) \begin{pmatrix} a_{11} & a_{12} \\ a_{21} & a_{22} \end{pmatrix} \right] \begin{pmatrix} b_1 \\ b_2 \end{pmatrix} \\
&= \begin{pmatrix} c_1^* a_{11} + c_2^* a_{21} & c_1^* a_{12} + c_2^* a_{22} \end{pmatrix} \begin{pmatrix} b_1 \\ b_2 \end{pmatrix} \\
&= (c_1^* a_{11} + c_2^* a_{21}) b_1 + (c_1^* a_{12} + c_2^* a_{22}) b_2
\end{aligned} \tag{3.13a}$$

$$\begin{aligned}
\langle C|(\alpha|B\rangle) &= (c_1^* \; c_2^*) \left[\begin{pmatrix} a_{11} & a_{12} \\ a_{21} & a_{22} \end{pmatrix} \begin{pmatrix} b_1 \\ b_2 \end{pmatrix} \right] \\
&= (c_1^* \; c_2^*) \begin{pmatrix} a_{11}b_1 + a_{12}b_2 \\ a_{21}b_1 + a_{22}b_2 \end{pmatrix} \\
&= c_1^*(a_{11}b_1 + a_{12}b_2) + c_2^*(a_{21}b_1 + a_{22}b_2) \,.
\end{aligned} \tag{3.13b}$$

We see that the last lines of (3.13a) and (3.13b) are equal.

The Hermitean conjugate or adjoint of a linear operator α, denoted by $\alpha^\dagger$, is defined by

$$<B|\alpha^\dagger|C> \;\equiv\; (<C|\alpha|B>)^* \tag{3.14}$$

for every $|B>$ and $<C|$.

If α is represented by a matrix, the matrix representation of $\alpha^\dagger$ is the Hermitean conjugate of that matrix. This may be seen as follows.

Using the same example used earlier, we have

$$\begin{aligned}[<C|\alpha|B>]^* &= [c_1^*(a_{11}b_1 + a_{12}b_2) + c_2^*(a_{21}b_1 + a_{22}b_2)]^* \\ &= c_1(a_{11}^* b_1^* + a_{12}^* b_2^*) + c_2(a_{21}^* b_1^* + a_{22}^* b_2^*) \;. \end{aligned} \tag{3.15a}$$

Let $\alpha^\dagger$ be represented by the matrix $\{x_{ij}\}$. Then,

$$\begin{aligned}<B|\alpha^\dagger|C> &= (b_1^* \; b_2^*)\begin{pmatrix} x_{11} & x_{12} \\ x_{21} & x_{22}\end{pmatrix}\begin{pmatrix} c_1 \\ c_2 \end{pmatrix} \\ &= (b_1^* x_{11} + b_2^* x_{21})c_1 + (b_1^* x_{12} + b_2^* x_{22})c_2 \;. \end{aligned} \tag{3.15b}$$

Comparing the last lines of (3.15a) and (3.15b) and noting that c_1, c_2, b_1^* and b_2^* are chosen arbitrarily, we obtain

$$\begin{pmatrix} x_{11} & x_{12} \\ x_{21} & x_{22}\end{pmatrix} = \begin{pmatrix} a_{11}^* & a_{21}^* \\ a_{12}^* & a_{22}^*\end{pmatrix} \;. \tag{3.16}$$

This means that the matrix (x_{ij}) representing $\alpha^\dagger$ is the Hermitean conjugate of the matrix (a_{ij}) representing α:

$$x_{ij} = a_{ji}^* \;. \tag{3.16a}$$

A linear operator which equals its adjoint

$$\alpha^{\dagger} = \alpha . \tag{3.17}$$

is called a self-adjoint operator or a Hermitean operator. Hermitean operators play very important roles in quantum mechanics and other branches of physics.

Problem 3.1 Pauli's spin matrices $(\sigma_x, \sigma_y, \sigma_z)$ are defined by

$$\sigma_x \equiv \begin{pmatrix} 0 & 1 \\ 1 & 0 \end{pmatrix}, \quad \sigma_y \equiv \begin{pmatrix} 0 & -i \\ i & 0 \end{pmatrix}, \quad \sigma_z \equiv \begin{pmatrix} 1 & 0 \\ 0 & -1 \end{pmatrix}.$$

Show that

(a) $\sigma_x^2 = \sigma_y^2 = \sigma_z^2 = 1$

(b) $[\sigma_y, \sigma_z] = 2i\sigma_x$, $[\sigma_z, \sigma_x] = 2i\sigma_y$, $[\sigma_x, \sigma_y] = 2i\sigma_z$

(c) $\sigma_y\sigma_z + \sigma_z\sigma_y = 0$

$\sigma_z\sigma_x + \sigma_x\sigma_z = 0$

$\sigma_x\sigma_y + \sigma_y\sigma_x = 0 .$

8.4° The Eigenvalue Problem

A square matrix (a_{ij}) whose elements satisfy

$$a_{kj} = a_{jk} \tag{4.1}$$

is called a symmetric matrix. A real, symmetric matrix is Hermitean since

$$(a_{kj})^* = a_{kj} = a_{jk} \,. \tag{4.2}$$

Let us take a general Hermitean matrix α, which satisfies the relation

$$\alpha^\dagger = \alpha \quad \text{or} \quad a_{ij}^* = a_{ji} \,. \tag{4.3}$$

The non-zero column vector $|B>[<B|B> \neq 0]$, which satisfies

$$\alpha|B> = \lambda|B> \,, \tag{4.4}$$

is called the eigenvector of α belonging to the eigenvalue λ. Equivalently, this equation may be written in the form

$$(\alpha - \lambda I)|B> = 0 \,. \tag{4.5}$$

For n=3, this means that

$$\begin{aligned} (a_{11}-\lambda)b_1 + a_{12}b_2 + a_{13}b_3 &= 0 \\ a_{21}b_1 + (a_{22}-\lambda)b_2 + a_{23}b_3 &= 0 \\ a_{31}b_1 + a_{32}b_2 + (a_{33}-\lambda)b_3 &= 0 \,. \end{aligned} \tag{4.6}$$

The necessary and sufficient condition for the existence of a non-zero column vector $|B>$ is that

$$\det(\alpha - \lambda I) = 0 \,,$$

that is,

$$\begin{vmatrix} a_{11}-\lambda & a_{12} & a_{13} \\ a_{21} & a_{22}-\lambda & a_{23} \\ a_{31} & a_{32} & a_{33}-\lambda \end{vmatrix} = 0 \,. \tag{4.7}$$

When expanded, this determinant is a cubic equation in λ. There are therefore three roots for λ. We will now show that all roots λ are real (if α is Hermitean).

Multiplying eq. (4.4) from the left by $\langle B|$, we obtain

$$\langle B|\alpha|B\rangle = \lambda\langle B|B\rangle \ . \tag{4.8}$$

The complex conjugate of the left-hand term can be calculated as follows:

$$[\langle B|\alpha|B\rangle]^*$$

$$= \langle B|\alpha^\dagger|B\rangle \qquad [\text{use of } (3.14)]$$

$$= \langle B|\alpha|B\rangle \qquad [\text{use of } (4.3)]$$

$$= \lambda\langle B|B\rangle \ . \qquad [\text{use of } (4.4)]$$

Complex conjugation of the right-hand term yields

$$[\lambda\langle B|B\rangle]^*$$

$$= \lambda^*\langle B|B\rangle^*$$

$$= \lambda^*\langle B|B\rangle \ . \qquad [\text{use of } (2.29)]$$

Therefore, we obtain

$$\lambda\langle B|B\rangle = \lambda^*\langle B|B\rangle \quad \text{or} \quad (\lambda-\lambda^*)\langle B|B\rangle = 0 \ .$$

Since $\langle B|B\rangle > 0$ (by assumption), we must have

$$\lambda = \lambda^* \ . \tag{4.9}$$

Thus, as we wished to show, the eigenvalue λ is real. It should be noted that the above proof does not depend on the degree of the matrix. The theorem therefore is valid for any arbitrary degree.

Corresponding to different eigenvalues, there will be different eigenvectors. Let $|B_1\rangle$ and $|B_2\rangle$ be eigenvectors corresponding to two different eigenvalues λ_1 and λ_2,

$$\alpha|B_1\rangle = \lambda_1|B_1\rangle$$

$$\alpha|B_2\rangle = \lambda_2|B_2\rangle$$

$$\lambda_1 \neq \lambda_2 . \qquad (4.10)$$

We now show that <u>the eigenvectors corresponding to two different eigenvalues are othogonal</u>, i.e., <u>their scalar product vanishes</u>:

$$\boxed{\langle B_2|B_1\rangle = 0 .} \qquad (\lambda_1 \neq \lambda_2) \qquad (4.11)$$

Multiplying the first equation of (4.10) by $\langle B_2|$ from the left, we obtain

$$\langle B_2|(\alpha|B_1\rangle) = \langle B_2|\lambda_1|B_1\rangle \qquad \text{[use of (4.10)]}$$

$$= \lambda_1\langle B_2|B_1\rangle .$$

The first member of these equations can also be calculated as follows:

$$\langle B_2|\alpha|B_1\rangle$$

$$= (\langle B_1|\alpha^\dagger|B_2\rangle)^* \qquad \text{[use of (3.14)]}$$

$$= (\langle B_1|\alpha|B_2\rangle)^* \qquad \text{[use of (4.3)]}$$

$$= (\lambda_2\langle B_1|B_2\rangle)^* \qquad \text{[use of (4.10)]}$$

$$= \lambda_2\langle B_2|B_1\rangle . \qquad \text{[use of (4.9) and (2.29)]}$$

Comparing the two results, we obtain

$$(\lambda_1-\lambda_2)\langle B_2|B_1\rangle = 0 .$$

Dividing this by $\lambda_1-\lambda_2$ ($\neq$ 0), we obtain

$$\langle B_2|B_1\rangle = 0 . \qquad \text{Q.E.D.}$$

It is clear from eq.(4.4) that if $|B\rangle$ is an eigenvector associated with a certain eigenvalue λ, then a complex number c times the vector $|B\rangle$, that is, $c|B\rangle$ is also an eigenvector with the same eigenvalue. The two vectors, $|B\rangle$

and $c|B\rangle$, have the same algebraic properties except for their magnitudes (and phases, see below). The set of such similar eigenvectors can all be generated from the eigenvector

$$\frac{|B\rangle}{\sqrt{\langle B|B\rangle}} \equiv |\lambda\rangle , \tag{4.12}$$

which is normalized to unity,

$$\langle\lambda|\lambda\rangle = 1 . \tag{4.13}$$

It should be noted that the normalization condition (4.13) alone does not determine the eigenvector $|\lambda\rangle$ in a unique manner. In fact, if $|\lambda\rangle$ is the vector defined by (4.12), then $e^{i\phi}|\lambda\rangle$, where ϕ is a real number called a <u>phase angle</u>, also satisfies the same normalization condition since

$$\langle\lambda|e^{-i\phi} \cdot e^{i\phi}|\lambda\rangle = \langle\lambda|\lambda\rangle = 1 . \tag{4.14}$$

If α is a n x n matrix, the order of the determinant equation (4.7) is n. There will therefore be n real roots. If the n eigenvalues are all distinct, then we can construct a set of n <u>orthogonal and normalized</u>, or <u>orthonormal</u>, eigenvectors, by means of (4.12).

Even if the eigenvalues are not distinct, we can always find n orthonormal eigenvectors. This may be understood by studying the following example. If the eigenvalue equation (4.7) has a repeated root, we can find two (or more) independent eigenvectors corresponding to this root. (By "independent" we mean that if $|B\rangle$ and $|C\rangle$ are two vectors, then $|B\rangle$ cannot be obtained from $|C\rangle$ by merely multiplying $|C\rangle$ by a number, and vice versa.) Let $|\lambda_a\rangle$ and $|\lambda_b\rangle$ be such eigenvectors belonging to the same eigenvalue λ:

$$\alpha|\lambda_a\rangle = \lambda|\lambda_a\rangle , \qquad \alpha|\lambda_b\rangle = \lambda|\lambda_b\rangle .$$

Then, any linear combination of these vectors,

$$a|\lambda_a> + b|\lambda_b> \tag{4.15}$$

is also an eigenvector. We can always find a pair of orthonormal vectors among the linear combinations; for example, $|\lambda_a>$ and

$$|\lambda_b'> \equiv |\lambda_b> - |\lambda_a><\lambda_a|\lambda_b> \tag{4.16}$$

are orthonormal, as we can verify easily (Problem 4.3).

In summary, for any n x n Hermitean matrix, we can in principle find n eigenvalues $\lambda_1, \lambda_2, \ldots, \lambda_n$ and n orthonormal eigenvectors $|\lambda_1>$, $|\lambda_2>$, ... , $|\lambda_n>$. The values λ_i may be coincidentally equal. The correspondence between the eigenvalues and eigenvectors then is not unique.

The eigenvalue problem is very important in quantum mechanics. In fact as we will see in later sections, §§8.6-8.12, it is inseparably connected with the fundamental postulates of quantum mechanics. The eigenvalue problem is also quite important in some branches of classical mechanics. For example, earlier in § 3.9* we discussed the vibration of particles on a stretched string in terms of normal modes. Finding the characteristic frequencies and the associated amplitude relations can be regarded as the solution of the eigenvalue problem of a particular type. Another significant example of a classical eigenvalue problem is the principal - axis transformation associated with the inertia tensor of a rigid body, which may be found in the standard textbook of intermediate mechanics. [2].

8.5 Orthogonal Representation

In § 8.2 we introduced generalized vectors by means of matrix algebra. A vector in the ordinary space of three dimensions, called an ordinary vector, can be represented by a 3 x 1 real matrix [or by a 1 x 3 real matrix]. The principal algebraic properties satisfied by 3 x 1 real matrices are also satisfied by n x 1 column matrices with complex elements. The latter may then be thought to represent quantities which we call generalized vectors. The generalization is twofold. First, n can be any positive integer. Secondly, the elements of column matrices can be complex numbers. These two generalizations are closely related as we will see in a moment.

Let us take the case n = 1, i.e., the 1 x 1 matrix with a complex element, which is just a complex number. A complex number z can always be written as

$$z = x + iy, \tag{5.1}$$

where x and y are real numbers. The length of z is, by definition,

$$|z| \equiv \sqrt{z^*z} = \sqrt{x^2+y^2} \quad . \tag{5.2}$$

It is well-known that any complex number z = x + iy can be represented by a point on the Gaussian plane as shown in Fig. 8.2. It is

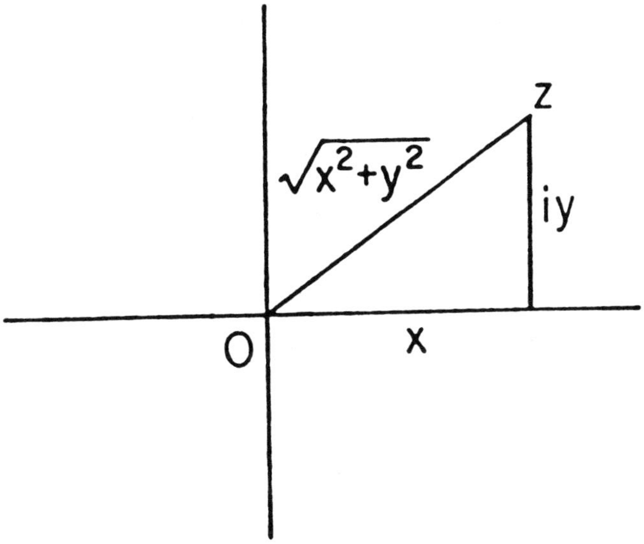

Fig. 8.2 The Gaussian representation of a complex number.

customary to choose the real axis along the horizontal direction and the imaginary axis along the vertical direction.

The complex number z now can be regarded as a vector in two dimensions; the two independent real variables (x,y) which represent the real and imaginary parts of z, correspond to the two Cartesian variables.

On the other hand, by analogy with ordinary vectors, n x 1 real matrices can be considered as vectors in an n-dimensional space. We may, therefore, say that a column vector with n complex elements can be looked at as a vector in the 2n-dimensional space, called the Hilbert space. We note that by allowing the matrix elements to be complex, we could increase the number of independent variables by the factor two.

Any ordinary vector $\vec{A}$ can be expressed as the sum of three orthogonal vectors

$$\vec{A} = \vec{i}\, A_x + \vec{j}\, A_y + \vec{k}\, A_z \,, \tag{5.3}$$

where $\vec{i}$, $\vec{j}$, $\vec{k}$ are a set of three vectors satisfying the <u>orthonormality relations</u>,

$$\begin{aligned} \vec{i}\cdot\vec{i} &= \vec{j}\cdot\vec{j} = \vec{k}\cdot\vec{k} = 1 \\ \vec{i}\cdot\vec{j} &= \vec{j}\cdot\vec{k} = \vec{k}\cdot\vec{i} = 0; \end{aligned} \tag{5.4}$$

the components A_x, A_y, A_z are projections of $\vec{A}$ along $\vec{i}, \vec{j}, \vec{k}$, respectively, and can be written as

$$A_x = \vec{i}\cdot\vec{A}, \quad A_y = \vec{j}\cdot\vec{A}\,, \quad A_z = \vec{k}\cdot\vec{A}. \tag{5.5}$$

Using these expressions, we can rewrite eq. (5.3) as

$$\vec{A} = \vec{i}\,(\vec{i}\cdot\vec{A}) + \vec{j}\,(\vec{j}\cdot\vec{A}) + \vec{k}\,(\vec{k}\cdot\vec{A}). \tag{5.6}$$

Expression (5.6) is valid for any set of basis vectors $(\vec{i}, \vec{j}, \vec{k})$ satisfying (5.4). The choice of such a set is often very important in solving a particular problem. For example, in treating the motion of a falling body we may choose the vertical direction to coincide with the direction of $\vec{k}$.

[In general, specification of an abstract quantity, such as a vector, by a suitable set of numbers is called a <u>representation</u> of that quantity. In the above example, we used the orthonormal bases $(\vec{i}, \vec{j}, \vec{k})$ to obtain the representation.]

For later convenience, let us introduce new notations for the orthonormal vectors:

$$(\vec{i}, \vec{j}, \vec{k}) \equiv (\vec{e}_1, \vec{e}_2, \vec{e}_3)\,. \tag{5.7}$$

We can then express eq. (5.6) as

$$\begin{aligned} \vec{A} &= \vec{e}_1\,(\vec{e}_1\cdot\vec{A}) + \vec{e}_2\,(\vec{e}_2\cdot\vec{A}) + \vec{e}_3\,(\vec{e}_3\cdot\vec{A}) \\ &= \sum_{j=1}^{3} \vec{e}_j\,(\vec{e}_j\cdot\vec{A})\,. \end{aligned} \tag{5.8}$$

Let us now represent the set of the orthonormal vectors $(\vec{e}_1, \vec{e}_2, \vec{e}_3)$ by the normalized kets:

$$|e_1> = \begin{pmatrix} 1 \\ 0 \\ 0 \end{pmatrix}, \quad |e_2> = \begin{pmatrix} 0 \\ 1 \\ 0 \end{pmatrix}, \quad |e_3> = \begin{pmatrix} 0 \\ 0 \\ 1 \end{pmatrix}. \tag{5.9}$$

The orthonormality relations (5.4) can then be represented by

$$<e_i|e_j> = \delta_{ij} . \tag{5.10}$$

If we represent an ordinary vector $\vec{A}$ by the column matrix $|A>$, we can express the vector decomposition property (5.8) in the form:

$$|A> = |e_1> <e_1|A> + |e_2> <e_2|A> + |e_3> <e_3|A>$$

or

$$\boxed{|A> = \sum_{j=1}^{3} |e_j> <e_j|A>.} \tag{5.11}$$

This relation means that we <u>can expand</u> an arbitrary ket vector $|A>$ in terms of the orthonormal ket vectors $|e_1>$, $|e_2>$ and $|e_3>$. By re-expressing eq. (5.11) in the form

$$|A> = (\sum_j |e_j> <e_j|) |A> , \tag{5.12}$$

we may formally represent the "expandability" by

$$1 = \sum_j |e_j> <e_j| . \tag{5.13}$$

This relation is often called the <u>completeness relation</u>. We will also say that $|e_1>$, $|e_2>$ and $|e_3>$ form a <u>complete</u> set of orthonormal vectors.

We now wish to discuss an orthogonal representation of the generalized vector in Hilbert space. We can in principle use any set of n orthonormal vectors for this purpose. For a particular physical problem, however, we will do better if we choose a special set such that the problem can be handled in a simplest possible manner.

Let us take a certain Hermitean operator ξ. Assume that n orthonormal eigenvectors of ξ,

$$|\xi_1>, |\xi_2>, \ldots, |\xi_n> \tag{5.14}$$

are found. We will in general use the same letter for both operator and its eigenvalues with the distinction being given by the subscripts or primes. The eigenvectors are denoted by the corresponding eigenvalues with ket or bra symbols according to whether they are column or row vectors.

Using the n vectors in (5.14), we can expand an arbitrary vector $|\psi>$ as the sum of orthogonal vectors in the form

$$|\psi> = |\xi_1><\xi_1|\psi> + |\xi_2><\xi_2|\psi> + \ldots + |\xi_n><\xi_n|\psi>$$

or

$$\boxed{|\psi> \equiv \sum_{j=1}^{n} |\xi_j><\xi_j|\psi>.} \tag{5.15}$$

This is a straightforward generalization of (5.11). The orthonormal eigenvectors can be represented by column vectors having only one non-vanishing element whose value is unity. If we choose

$$|\xi_1> \equiv \begin{pmatrix} 1 \\ 0 \\ 0 \\ \cdot \\ \cdot \\ 0 \end{pmatrix}, \quad |\xi_2> \equiv \begin{pmatrix} 0 \\ 1 \\ 0 \\ \cdot \\ \cdot \\ 0 \end{pmatrix}, \ldots, |\xi_n> \equiv \begin{pmatrix} 0 \\ 0 \\ 0 \\ \cdot \\ \cdot \\ 1 \end{pmatrix}, \tag{5.16}$$

then we can express the ket-vector $|\psi>$ as follows:

$$|\psi\rangle = \sum_j |\xi_j\rangle\langle\xi_j|\psi\rangle$$

$$= \begin{pmatrix} \langle\xi_1|\psi\rangle \\ \langle\xi_2|\psi\rangle \\ \cdot \\ \cdot \\ \langle\xi_n|\psi\rangle \end{pmatrix} . \tag{5.17}$$

Here we see that the vector $|\psi\rangle$ is represented by the set of n complex numbers $\langle\xi_1|\psi\rangle$, $\langle\xi_2|\psi\rangle$, ... , $\langle\xi_n|\psi\rangle$ just as an ordinary vector $\vec{A}$ can be represented by the set of 3 real numbers

$$A_x \equiv \vec{i} \cdot \vec{A} \equiv \langle e_1|A\rangle \;, \; A_y \equiv \langle e_2|A\rangle \;, \; A_z \equiv \langle e_3|A\rangle \;. \tag{5.18}$$

From (5.15) we can write down the <u>completeness relation</u> in the form:

$$\boxed{\sum_{j=1}^{n} |\xi_j\rangle\langle\xi_j| = 1 .} \tag{5.19}$$

We now wish to represent an arbitrary linear operator α in terms of orthonormal bases, $\{\xi_j\}$.

Let us apply the linear operator α to a ket-vector $|\psi\rangle$ and denote the resulting ket-vector by $|\phi\rangle$:

$$\alpha|\psi\rangle = |\phi\rangle. \tag{5.20}$$

We expand the r.h.s. in the form of eq. (5.15):

$$|\phi\rangle = \sum_{j=1}^{n} |\xi_j\rangle\langle\xi_j|\phi\rangle. \tag{5.21}$$

We note that this expansion can be obtained by simply multiplying $|\phi\rangle$ by the completeness relation (5.19) from the left. In fact, this simple operation is the principal convenience of having the completeness relation.

We expand the l.h.s. of eq. (5.20) by using the completeness relation twice:

$$\alpha|\psi\rangle = \left(\sum_{j=1}^{n} |\xi_j\rangle\langle\xi_j|\right) \alpha \left(\sum_{k=1}^{n} |\xi_k\rangle\langle\xi_k|\right) |\psi\rangle$$

$$= \sum_{j=1}^{n} |\xi_j\rangle \sum_{k=1}^{n} \langle\xi_j|\alpha|\xi_k\rangle\langle\xi_k|\psi\rangle . \tag{5.22}$$

Comparing the coefficients of the vector $|\xi_j\rangle$ in the two expansions (5.21) and (5.22), we obtain

$$\langle\xi_j|\phi\rangle = \sum_{k=1}^{n} \langle\xi_j|\alpha|\xi_k\rangle\langle\xi_k|\psi\rangle , \quad j = 1,2,\ldots,n. \tag{5.23}$$

This means that if we represent the operator α by the n x n matrix

$$\alpha \sim \begin{pmatrix} \langle\xi_1|\alpha|\xi_1\rangle & \langle\xi_1|\alpha|\xi_2\rangle & \cdots & \langle\xi_1|\alpha|\xi_n\rangle \\ \langle\xi_2|\alpha|\xi_1\rangle & \langle\xi_2|\alpha|\xi_2\rangle & \cdots & \langle\xi_2|\alpha|\xi_n\rangle \\ \cdot & \cdot & & \cdot \\ \cdot & \cdot & & \cdot \\ \langle\xi_n|\alpha|\xi_1\rangle & \langle\xi_n|\alpha|\xi_2\rangle & \cdots & \langle\xi_n|\alpha|\xi_n\rangle \end{pmatrix}, \tag{5.24}$$

then the elements of $|\phi\rangle$ and $|\psi\rangle$ are connected by the known matrix multiplication rule as exhibited in eq. (5.23).

In summary, an arbitrary vector in the Hilbert space can be represented by a column matrix of the form (5.17). An arbitrary linear operator α can be represented by a square matrix as indicated in (5.24). The orthogonal bases in which vectors and linear operators are represented, can be chosen so that the particular problem in question may be treated in the simplest manner.

8.6 Quantum Mechanical Description of Linear Motion

The quantum description of the dynamics of a system is quite different from, but also much related to, the classical description. We may develop a quantum theory in close analogy with the classical Hamiltonian mechanics. This was done by Dirac in his classic book [1], Principles of Quantum Mechanics . Following this development, we will present a quantum theory in one dimension in the present section.

Let us consider a particle moving along a straight line of length L in the range (0,L). In classical mechanics, the system is characterized by the Hamiltonian

$$H(x,p) = \frac{1}{2m} p^2 + V(x), \tag{6.1}$$

In § 3.2 we saw that the dynamical state of this particle can be represented by the set of canonical variables x and p, which change with time in accordance with Hamilton's equations of motion:

$$\dot{x} = \frac{\partial H(x,p)}{\partial p} = \frac{p}{m}$$

$$\dot{p} = - \frac{\partial H(x,p)}{\partial x} = - \frac{dV}{dx} . \tag{6.2}$$

In quantum mechanics, observable dynamical quantities such as position, momentum, and energy, by assumption, are represented by Hermitean operators . In particular, let x be the Hermitean operator representing the position. The eigenvalue equation for the position x is

$$x|x'\rangle = x'|x'\rangle, \tag{6.3}$$

where x' is eigenvalue and may take any value between 0 and L.

If we take any two eigenstates associated with two different eigenvalues x' and x'' and form their scalar product, the product vanishes according to the orthogonality theorem (4.11),

$$\langle x''|x'\rangle = 0 \quad \text{if } x' \neq x''. \tag{6.4}$$

When the eigenvalues form a continuum as in the present case, we cannot normalize the ket vectors $|x'\rangle$ simply. This is so because the scalar product $\langle x'|x'\rangle$ becomes infinitely large as we will see presently. It is then found convenient to express the orthonormality relation by

$$\boxed{\langle x''|x'\rangle = \delta(x'-x''),} \tag{6.5}$$

where $\delta(y)$ is the Dirac delta-function, which was defined earlier in § 4.2. The principal definition properties of this function are [see (4.2.5)]

$$\delta(y) = 0 \quad \text{if } y \neq 0$$

$$\int_{-\infty}^{\infty} f(y)\, \delta(y)\, dy = f(0), \tag{6.6}$$

where $f(y)$ is an arbitrary function which is continuous at $y=0$.

The convenience of (6.5) can be demonstrated as follows. Let us assume that the vectors $\{|x'\rangle\}$, for $0 \leqq x' \leqq L$, form a complete set such that an arbitrary ket $|\psi\rangle$ can be expanded as follows:

$$\boxed{|\psi\rangle = \int_0^L dx'\, |x'\rangle\langle x'|\psi\rangle.} \tag{6.7}$$

This is the generalization of (5.15) for the continuous variable x'. The absolute square of $|\psi\rangle$ is

$$\langle\psi|\psi\rangle = \left(\int_0^L dx''\, \langle\psi|x''\rangle\langle x''|\right) \cdot \left(\int_0^L dx'\, |x'\rangle\langle x'|\psi\rangle\right)$$

$$= \int_0^L dx''\, \langle\psi|x''\rangle \left[\int_0^L dx'\, \langle x''|x'\rangle\langle x'|\psi\rangle\right].$$

Consider now the quantity in the square brackets:

$$\int_0^L dx' \quad \langle x''|x'\rangle \langle x'|\psi\rangle .$$

According to (6.4) the factor $\langle x''|x'\rangle$ in the integrand vanishes unless $x' = x''$. The integral however must have the value $\langle x''|\psi\rangle$ to be consistent with the relation (6.7). This is possible only if $\langle x''|x'\rangle$ is infinitely large at $x' = x''$. Furthermore, the desired result can be obtained if we assume (6.5) and apply (6.6), that is,

$$\int_0^L dx' \langle x''|x'\rangle \langle x'|\psi\rangle = \int_0^L dx' \quad \delta(x''-x')\langle x'|\psi\rangle = \langle x''|\psi\rangle .$$

We now consider the matrix representation in terms of the eigenstates of the position operator x. This representation is called the position representation. Using (6.3) and (6.5), we obtain

$$\langle x''|x|x'\rangle = \langle x''|x'|x'\rangle = x'\langle x''|x'\rangle = x' \ \delta(x''-x'). \tag{6.8}$$

We can easily see that

$$\langle x''|x^n|x'\rangle = \langle x''|(x^{n-1})x|x'\rangle = x'\langle x'' \ x^{n-1}|x'\rangle = x'^n\langle x''|x'\rangle$$
$$= x'^n \ \delta(x''-x')$$
$$n = 1,2, \ldots \tag{6.9}$$

If f(x) represents a polynomial or series in x:

$$f(x) \equiv a_o + a_1 x + a_2 \ x^2 + \ldots . \tag{6.10}$$

then we have

$$\langle x''|f(x)|x'\rangle = f(x') \ \delta(x''-x'). \tag{6.11}$$

In particular, the potential energy operator V(x), which depend on x only, has non-vanishing values only along the diagonal ($x' = x''$) in the position representation.

In a sharp contrast to classical mechanics, the quantum operators x and p representing the position and momentum of the particle, do not commute but, by postulate, satisfy the following commutation relation:

$$\boxed{xp - px = i\hbar,} \tag{6.12}$$

where $\hbar$ is Planck's constant divided by 2π: $\hbar \equiv h/2\pi$. It can be shown from (6.12) [Problem 6.1] that the operator p is equivalent to the following operator except for an unimportant phase factor:

$$\boxed{p = -i\hbar \frac{d}{dx},} \tag{6.13}$$

where the operator $\frac{d}{dx}$ is defined by

$$\langle x'| \frac{d}{dx} |\psi\rangle \equiv \frac{d}{dx'} \langle x'|\psi\rangle \tag{6.14}$$

for any $|\psi\rangle$.

After a series of calculations,

$$\begin{aligned}
\langle\psi| \frac{d}{dx} |x'\rangle &= \int \langle\psi|x''\rangle\, dx''\, \langle x''| \frac{d}{dx} |x'\rangle && \text{[use of (6.7)]} \\
&= \int \langle\psi|x''\rangle\, dx''\, \frac{d}{dx''} \langle x''|x'\rangle && \text{[use of (6.14)]} \\
&= \int \langle\psi|x''\rangle\, dx''\, \frac{d}{dx''} \delta(x''-x') && \text{[use of (6.5)]} \\
&= \int \langle\psi|x''\rangle\, dx''\left[-\frac{d}{dx'}\delta(x''-x')\right] && \text{[use of (4.2.11)]} \\
&= -\frac{d}{dx'} \langle\psi|x'\rangle && \text{[use of (4.3.7)]}
\end{aligned}$$

we obtain

$$\langle\psi|\frac{d}{dx} |x'\rangle = -\frac{d}{dx'} \langle\psi|x'\rangle. \tag{6.14a}$$

Using the last two equations, we obtain

$$[\langle x'|\frac{d}{dx}|\psi\rangle]^* = (\frac{d}{dx'}\langle x'|\psi\rangle)^* \qquad \text{[use of (6.14)]}$$

$$= \frac{d}{dx'}\langle\psi|x'\rangle \qquad \text{[use of (2.21)]}$$

$$= \langle\psi|(-\frac{d}{dx})|x'\rangle . \qquad \text{[use of (6.14a)]}$$

This means that the operator d/dx is not Hermitean but anti-Hermitean:

$$(\frac{d}{dx})^\dagger = -\frac{d}{dx} . \tag{6.15}$$

Since

$$(-i\hbar\frac{d}{dx})^\dagger = (-i\hbar)^*(\frac{d}{dx})^\dagger = (i\hbar)(-\frac{d}{dx}) = -i\hbar\frac{d}{dx} ,$$

the operator $-i\hbar d/dx$, which is equivalent to p, is Hermitean.

Using (6.13), we obtain

$$\langle x'|f(x,p)|\psi\rangle = f(x', -i\hbar\frac{d}{dx'})\langle x'|\psi\rangle . \tag{6.16}$$

For example, by choosing $f = p^2/2m + V(x)$, we have

$$\langle x'|[\frac{p^2}{2m} + V(x)]|\psi\rangle = [\frac{1}{2m}(-i\hbar\frac{d}{dx'})^2 + V(x')]\langle x'|\psi\rangle . \tag{6.17}$$

In classical mechanics the dynamical state is represented by a point in the phase space. In quantum mechanics, the dynamical state, by postulate, will be represented by a ket vector, say $|\psi\rangle$, which may be viewed as a multi-dimensional column vector.

The dynamical state in general changes with time. In classical mechanics, this change is governed by Hamilton's equations of motion (6.2). In quantum mechanics, the change is ruled, by postulate, by Schrödinger's equation of motion:

$$\boxed{i\hbar\frac{d}{dt}|\psi,t\rangle = H(x,p)|\psi,t\rangle ,} \tag{6.18}$$

where $H(x,p)$ is the Hamiltonian operator, the Hermitean operator which has the same functional form as the classical Hamiltonian. Multiplying eq. (6.18) from the left by $\langle x'|$, we obtain

$$i\hbar \langle x'|\frac{d}{dt}|\psi, t\rangle = \langle x'|H(x,p)|\psi, t\rangle . \tag{6.19}$$

We may write the term on the l.h.s. as

$$i\hbar \frac{\partial}{\partial t} \psi(x',t), \tag{6.20}$$

where

$$\psi(x',t) \equiv \langle x'|\psi,t\rangle \tag{6.21}$$

is a function of both x' and t, and because of this the partial time derivative $\partial/\partial t$ was indicated. Using (6.17) we can write the r.h.s. of (6.19) as

$$[-\frac{\hbar^2}{2m} \frac{\partial^2}{\partial x'^2} + V(x')] \psi(x',t).$$

Therefore, eq. (6.19) can be re-expressed as follows :

$$i\hbar \frac{\partial}{\partial t} \psi(x,t) = [-\frac{\hbar^2}{2m} \frac{\partial^2}{\partial x^2} + V(x)] \psi(x,t), \tag{6.22}$$

where we dropped the primes indicating the position eigenvalues.

The last equation (6.22) is called Schrödinger's wave equation. The function ψ, defined in (6.21), is called the wave function. By postulate its absolute square

$$|\psi(x,t)|^2 = |\langle x|\psi,t\rangle|^2 \equiv P(x,t) \tag{6.23}$$

represents the probability distribution function in position. That is,

$$\boxed{|\psi(x,t)|^2 \, dx = \text{the probability of finding the particle in } (x,x+dx),} \tag{6.24}$$

normalized such that

$$\int_0^L dx \ \psi^*(x,t) \ \psi(x,t) = \int_0^L dx \ P(x,t) = 1. \tag{6.25}$$

The average position of the particle is then given by

$$\begin{aligned} \langle x \rangle &\equiv \int_0^L dx \ x \ P(x,t) \\ &= \int_0^L dx \ x \ \psi^*(x,t) \ \psi(x,t). \end{aligned} \tag{6.26}$$

More generally if a dynamical quantity ξ is given as a function of x and p, then its average is given by

$$\boxed{\langle \xi \rangle_t = \int_0^L dx \ \psi^*(x,t) \ \xi(x, -i\hbar \frac{\partial}{\partial x}) \ \psi(x,t).} \tag{6.27}$$

For example, if we take the Hamiltonian H for ξ, we obtain

$$\langle H \rangle_t = \int_0^L dx \ \psi^*(x,t) \ [-\frac{\hbar^2}{2m} \frac{\partial^2}{\partial x^2} + V(x)] \ \psi(x,t). \tag{6.28}$$

Since

$$\psi^*(x,t) = \langle \psi, t | x \rangle, \quad \langle x | \xi | \psi, t \rangle = \xi(x, -i\hbar \frac{d}{dx}) \ \langle x | \psi, t \rangle,$$

we can rewrite eq. (6.27) as follows:

$$\begin{aligned} \langle \xi \rangle_t &= \int dx \ \langle \psi, t | x \rangle \ \langle x | \xi | \psi, t \rangle \\ &= \langle \psi, t | \xi | \psi, t \rangle \qquad \text{[use of (6.7)]} \end{aligned}$$

or

$$\boxed{\langle \xi \rangle_t = \langle \psi, t | \xi | \psi, t \rangle .} \tag{6.29}$$

The average of the dynamical function ξ, $\langle\xi\rangle_t$, as given by (6.27) or by (6.29), is called the expectation value of the dynamical function ξ. It gives the average value of the dynamical function ξ after repeated experiments when the particle is in the state ψ.

Problem 6.1 Show that

$$\langle x'|[x(-i\hbar\, d/dx) - (-i\hbar\, d/dx)x]|\psi\rangle = i\hbar \langle x'|\psi\rangle$$

for any ket $|\psi\rangle$.

Problem 6.2 The operators, x and p, for the position and momentum, are, by definition, Hermitean. (a) Is the product xp Hermitean? (b) Find the Hermitean conjugates of xp - px and xp + px. Are they Hermitean?

Problem 6.3 Show that the normalization for a wave function $\psi(x,t)$

$$f(t) \equiv \int_{-\infty}^{+\infty} dx\, \psi^*(x,t)\, \psi(x,t) = 1$$

is maintained for all time.

Hint: By using Schrödinger's wave equation and its Hermitean conjugate, show that

$$\frac{df}{dt} = 0 \quad .$$

8.7 The Momentum Eigenvalue Problem

The momentum p is a basic dynamical variable. In quantum mechanics its role is just as important as that of the position x. This can be expected from the fact that they both appear in the commutation rule (a fundamental postulate)

$$xp - px = i\hbar. \tag{7.1}$$

The eigenvalue equation for the momentum (operator) p is

$$p\,|\,p'\rangle = p'\,|\,p'\rangle\ . \tag{7.2}$$

Multiplying this equation from the left by $\langle x'|$, we obtain

$$\langle x'|\ p\,|\,p'\rangle = p'\ \langle x'\,|\,p'\rangle.$$

Noting the equivalence relation : $p = -i\hbar\ d\,/dx$ [(6.13)] and using (6.14), we obtain

$$-\,i\hbar\,\frac{\partial}{\partial x'}\,\langle x'\,|\,p'\rangle = p'\ \langle x'\,|\,p'\rangle. \tag{7.3}$$

This is nothing but a differential equation for the ordinary, c-number (classical number) function $\langle x'\,|\,p'\rangle$. We can write its solution in the form :

$$\langle x\,|\,p'\rangle = c\ e^{ip'x/\hbar}\ . \tag{7.4}$$

Here we dropped the prime from the position eigenvalue ; the factor c does not depend on x but may possibly depend on p'. This possibility, however, can be excluded since the fundamental commutation rule (7.1) is symmetric in x and p. Thus, c must be a numerical constant.

In many branches of physics including statistical physics, we deal with properties which do not depend on specific boundary conditions. In such cases it is advantageous to choose a boundary condition which makes the development of a theory as simple as possible. The periodic boundary condition

$$\langle x + L \mid p' \rangle = \langle x \mid p' \rangle, \quad -\infty < x < \infty, \tag{7.5}$$

is such a boundary condition. Substitution of (7.4) in (7.5) yields

$$e^{ip'(x+L)/\hbar} = e^{ip'x/\hbar} \quad \text{for all } x,$$

from which we obtain $e^{ip'L/\hbar} = 1$ or

$$p' = \frac{2\pi\hbar}{L} n \equiv p_n , \quad n = 0, \pm 1, \pm 2 \ \dots \quad . \tag{7.6}$$

The eigenvalues p_n for the momentum p are discrete, and are given by integral multiples of $2\pi\hbar/L$.

The constant c in (7.4) can be determined from the orthonormality relation :

$$\boxed{\langle p_n \mid p_m \rangle = \delta_{n,m}} . \tag{7.7}$$

Since

$$\begin{aligned} &\langle p_n \mid p_n \rangle \\ &= \int \langle p_n \mid x \rangle \, dx \, \langle x \mid p_n \rangle && [\text{ use of (6.7) }] \\ &= \int_0^L (c^* e^{-ip_n x/\hbar}) \, dx \, (c\, e^{ip_n x/\hbar}) && [\text{ use of (7.4) }] \\ &= |c|^2 L = 1, && [\text{ use of (7.7) }] \end{aligned}$$

we may choose the normalization constant

$$c = L^{-\frac{1}{2}} , \tag{7.8}$$

and obtain

$$\boxed{\begin{aligned} \langle x \mid p_n \rangle &= L^{-\frac{1}{2}} \, e^{ip_n x/\hbar} \\ \langle p_n \mid x \rangle &= L^{-\frac{1}{2}} \, e^{-ip_n x/\hbar} . \end{aligned}} \tag{7.9}$$

These are called transformation functions, which are useful when we wish to change from the position representation to the momentum representation or vice-versa.

The eigenvlaues of the momentum, $\{p_n\}$, form a complete set. That is, an arbitrary ket $|\psi\rangle$ can be expanded in the form:

$$|\psi\rangle = \sum_n |p_n\rangle\langle p_n|\psi\rangle. \tag{7.10}$$

or equivalently we may represent this property by the completeness relation:

$$\sum_n |p_n\rangle\langle p_n| = 1. \tag{7.11}$$

We note that the completeness of the momentum eigenstates $\{p_n\}$ is related to that of the position eigenstates characterized by the continuous eigenvlaue x, $0 \leq x \leq L$, by the Fourier transformation [3]. In fact, we have

$$\langle p_n|\psi\rangle = \int \langle p_n|x\rangle dx\langle x|\psi\rangle \qquad \text{[use of (6.7)]}$$

$$= L^{-\frac{1}{2}} \int_0^L dx\, e^{-ip_n x/\hbar} \langle x|\psi\rangle \qquad \text{[use of (7.9)]} \tag{7.12}$$

$$\langle x|\psi\rangle = \sum_n \langle x|p_n\rangle \langle p_n|\psi\rangle \qquad \text{[use of (7.10)]}$$

$$= \sum_n L^{-\frac{1}{2}} e^{ip_n x/\hbar} \langle p_n|\psi\rangle. \qquad \text{[use of (7.9)]} \tag{7.13}$$

When the length L is made greater, the spacing in momentum, $2\pi\hbar/L$, becomes smaller. In the limit, the momentum eigenvalues form a continuum. There, the situation becomes similar to the case of the continuous position eigenvalue. We must then reformulate the orthonormality condition since the length of the momentum ket becomes infinitely large.

In analogy with (7.11), let us assume the completeness relation of the form

$$\int_{-\infty}^{\infty} dp'\, |p'\rangle\langle p'| = 1. \tag{7.14}$$

Using this relation and (6.7), we have

$$\langle p'' | \psi \rangle = \int_{-\infty}^{\infty} dx \int_{-\infty}^{\infty} dp' \quad \langle p'' | x \rangle \langle x | p' \rangle \langle p' | \psi \rangle$$

[use of (6.7) and (7.14)]

$$= \int_{-\infty}^{\infty} dx \int_{-\infty}^{\infty} dp' \, (c^* \, e^{-ip''x/\hbar}) \, (c \, e^{ip'x/\hbar}) \, \langle p' | \psi \rangle \quad \text{[use of (7.4)]}$$

$$= |c|^2 \int_{-\infty}^{\infty} dx \int_{-\infty}^{\infty} dp \; e^{-i(p''-p')x/\hbar} \, \langle p' | \psi \rangle. \tag{7.15}$$

Let us compare this expression with the identity (Fourier's integral theorem)

$$f(k'') = \frac{1}{2\pi} \int_{-\infty}^{\infty} dx \int_{-\infty}^{\infty} dk' \; e^{-i(k''-k')x} \; f(k'), \tag{7.16}$$

where f represents a continuous function of k. Assuming that

$$p' = \hbar k', \quad p'' = \hbar k'', \quad \langle p'' | \psi \rangle \equiv \langle \hbar k'' | \psi \rangle = f(k''),$$

$$\langle p' | \psi \rangle \equiv \langle \hbar k' | \psi \rangle = f(k'), \quad dp' = \hbar \, dk',$$

we obtain $|c|^2 = (2\pi \hbar)^{-1}$. We may therefore choose

$$c = (2\pi\hbar)^{-\frac{1}{2}}, \tag{7.17}$$

and obtain

$$\langle x | p \rangle = (2\pi\hbar)^{-\frac{1}{2}} \, e^{ipx/\hbar}$$

$$\langle p | x \rangle = (2\pi\hbar)^{-\frac{1}{2}} \, e^{-ipx/\hbar}. \qquad \text{(continuous p)} \tag{7.18}$$

Here we dropped the primes on p.

Let us now go back and look at (7.16), that is, the Fourier integral theorem. A way of obtaining the l.h.s. in a formal manner from the r.h.s. is (a) to allow the change of the order of integrations and (b) introduce the delta-function $\delta(k'' - k')$ by

$$\delta(k'' - k') = \frac{1}{2\pi} \int_{-\infty}^{\infty} dx\, e^{-i(k''-k')x}, \tag{7.19}$$

which makes (7.16) appear as

$$f(k'') = \int_{-\infty}^{\infty} dk' \;\delta(k'' - k') \; f(k'). \tag{7.20}$$

This formal manipulation is often useful. It is stressed here that the validity of (7.19) depends on the precondition of the Fourier integral theorem (7.16) and not on the formal identity between the left and right hand sides. In fact, the formal identity (7.19) should be regarded merely as a device.

It is convenient to express the orthonormality in the form

$$\boxed{\langle p' | p'' \rangle = \delta(p' - p'')} \quad \text{[continuous } p' \text{ and } p''\text{]}, \tag{7.21}$$

In fact, this can be verified formally as follows :

$$\langle p' | p'' \rangle$$

$$= \int_{-\infty}^{\infty} dx \quad \langle p' | x \rangle \langle x | p'' \rangle \qquad \text{[use of (6.7)]}$$

$$= \int_{-\infty}^{\infty} dx \; [\, (2\pi\hbar)^{-\frac{1}{2}} e^{-ip'x/\hbar} \,][\,(2\pi\hbar)^{-\frac{1}{2}} e^{ip''x/\hbar}\,] \; \text{[use of (7.18)]}$$

$$= \frac{1}{2\pi\hbar} \int_{-\infty}^{\infty} dx\, e^{-i(p'-p'')x/\hbar}$$

$$= \delta(p' - p''). \qquad \text{[use of (7.19)]}$$

Note that the orthonormality relation (7.21) for the continuous momentum eigenvalues is quite analogous to the corresponding relation (6.5) for the position.

8.8 The Energy Eigenvalue Problem

The Hamiltonian H plays a central role in the description of the dynamics. Most often, the Hamiltonian H represents the total energy, that is, the sum of the kinetic and potential energies. In the present section, we will discuss the eigenvalue problem for the Hamiltonian H.

Let us start with the energy eigenvalue equation :

$$H|E\rangle = E|E\rangle, \tag{8.1}$$

where H is the Hamiltonian and E its eigenvalue. Multiplying this from the left by $\langle x|$ we obtain

$$\langle x|\, H(x,p)\, |E\rangle = H(x, -i\hbar \frac{d}{dx}) \langle x|\, E\rangle = E \langle x|\, E\rangle. \tag{8.2}$$

If we regard the c - number $\langle x|E\rangle$ as a function of x and write

$$\varphi_E(x) \equiv \langle x|\, E\rangle, \tag{8.3}$$

we can rewrite eq.(8.2) in the form

$$\boxed{H(x,-i\hbar \frac{d}{dx})\, \varphi_E(x) = E\, \varphi_E(x).} \tag{8.4}$$

If $H = p^2 / 2m + V(x)$, we then obtain

$$\boxed{\left[-\frac{\hbar^2}{2m} \frac{d^2}{dx^2} + V(x)\right] \varphi_E(x) = E\, \varphi_E(x).} \tag{8.5}$$

This equation is called Schrödinger's equation for the energy-eigenvalue problem. The function φ_E defined in (8.4) is referred to as the probability amplitude function. By postulate, its absolute square $|\varphi_E(x)|^2$ gives the (relative) position-space distribution function when the system is in the eigenstate E.

In general, there are many eigenvalues. Let us denote them by $\{E_j\}$ and their associated eigenfunctions by $\{\varphi_j\}$. We then have

$$H \varphi_j (x) = E_j \varphi_j(x). \tag{8.6}$$

If we assume a wave function of separable form : $\psi_j(x,t) = a(t)\, \varphi_j(x)$, and substitute it into the Schrödinger wave equation (6.22), we obtain

$$i\hbar \frac{\partial}{\partial t} [a(t)\, \varphi_j(x)] = H [a(t)\, \varphi_j(x)]$$

$$i\hbar \frac{da}{dt} \varphi_j(x) = E_j\, a\, \varphi_j(x) \qquad \text{[use of (8.6)]}$$

or

$$i\hbar \frac{da}{dt} = E_j\, a,$$

whose solution is given by

$$a(t) = \text{constant} \times e^{-i E_j t/\hbar}.$$

Therefore, we obtain the wave function of the form :

$$\psi_j (x,t) = e^{-i E_j t/\hbar} \varphi_j(x). \tag{8.7}$$

Note that this wave function has a simple time-dependence. It changes only in phase : $\psi_j \propto e^{-i E_j t/\hbar}$. In fact, the probability distribution function $P(x,t)$ corresponding to this ψ_j :

$$P(x,t) = |\psi_j (x,t)|^2$$

$$= [e^{i E_j t/\hbar} \varphi_j^* (x)]\, [e^{-i E_j t/\hbar} \varphi_j(x)]$$

$$= |\varphi_j(x)|^2 , \tag{8.8}$$

does not depend on the time t. This means that if the particle is in any one of the energy-eigenstates the probability distribution is stationary.

As we will see in the examples in this and other sections, the eigenvalues E_j can often take discrete values. Combination of these two statements implies existence of stationary states of discrete energies for a quantum particle as postulated in Bohr's model of an atomic structure [see §8.1].

For illustration, let us take a free particle whose Hamiltonian is given by

$$H = p^2/2m . \tag{8.9}$$

This H is a function of p alone, and therefore it is diagonal in the p-representation. Since the momentum eigenvalues are given by $p_n = 2\pi\hbar\, n/L$, (n, integer) the energy eigenvalues are

$$\varepsilon_n \equiv \frac{p_n^2}{2m} = \frac{1}{2m} \left(\frac{2\pi\hbar}{L} \right)^2 n^2. \tag{8.10}$$

We note that to each eigenvalue ε_n there correspond two eigenstates characterized by p_n and p_{-n} except for n = 0. When two or more eigenstates correspond to the same energy eigenvalue, these eigenstates are said to be degenerate. Thus, the states of energy ε_n are degenerate except for n = 0. For each n, the momentum wave function is given by (7.9) :

$$\varphi_n(x) = \frac{1}{L^{\frac{1}{2}}}\, e^{i\, p_n x/\hbar} . \tag{8.11}$$

From (8.7), we can write the wave function $\psi_n(x,t)$ in the form

$$\psi(x,t) = \frac{1}{L^{\frac{1}{2}}} e^{-i\varepsilon_n t/\hbar + i p_n x/\hbar} . \tag{8.12}$$

By comparing it with the standard form of a running wave [see §3.12*]

$$A\ e^{-i\omega t + ikx}, \tag{8.13}$$

where ω, k and A are respectively angular frequency (2π times frequency), k - number (2π times wave number per unit length) and amplitude, we obtain

$$k \text{ (wave number)} = \frac{2\pi}{\lambda} = \frac{p_n}{\hbar} \tag{8.14}$$

$$\omega \text{ (angular frequency)} = \frac{\varepsilon_n}{\hbar} . \tag{8.15}$$

These two relations are exactly those which appeared explicitly in the discussions of the quantum mechanical particle-wave duality presented in §8.1.

8.9 Simple Harmonic Oscillator

In the present section, we will look at the quantum mechanical description of a simple harmonic oscillator. This system plays a particularly interesting and important role in many applications of quantum statistical mechanics.

In classical mechanics a simple harmonic oscillator obeys the equation of motion :

$$m \frac{d^2x}{dt^2} = - kx. \tag{9.1}$$

The general solution of this equation is given by

$$x = A \sin (\omega_o t + \delta), \tag{9.2}$$

where

$$\omega_o \equiv (k / m)^{\frac{1}{2}} \tag{9.3}$$

is the angular frequency; A and δ are respectively the amplitude and phase angle which may be determined from the given initial condition.

The harmonic force -kx, can be derived from the potential function

$$V = \frac{1}{2} k x^2$$

through the standard formula $F = - dV / dx = - kx$. The potential V(x) versus the position x is shown in Figure 8.3. The Hamiltonian H for the system is given by

$$H (x,p) = \frac{1}{2m} p^2 + \frac{1}{2} k x^2$$

$$= \frac{1}{2m} p^2 + \frac{1}{2} m \omega_o^2 x^2 . \text{ [use of (9.3)]} \tag{9.4}$$

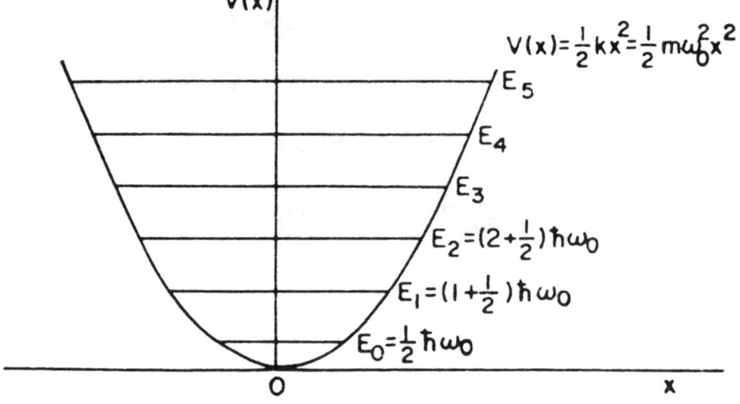

Fig. 8.3 The harmonic potential $V = \frac{1}{2} k x^2 = \frac{1}{2} m \omega_0^2 x^2$ and the energy eigenvalues $E_n = (n + \frac{1}{2}) \hbar \omega_0$ for the harmonic oscillator.

Passing now to quantum mechanics, Schrödinger's equation (8.5) for the energy eigenvalue can be written down from (9.4) as follows :

$$[- \frac{\hbar^2}{2m} \frac{d^2}{dx^2} + \frac{1}{2} m \omega_0^2 x^2] \quad \varphi(x) = E \varphi(x), \qquad (9.5)$$

where E is the energy eigenvalue.

In order to solve eq.(9.5), let us try the function

$$\varphi_0(x) = e^{-\alpha x^2}. \tag{9.6}$$

After differentiation, we obtain

$$\frac{d\varphi_0}{dx} = -2\alpha\, x\, e^{-\alpha x^2}$$

$$\frac{d^2 \varphi_0}{dx^2} = -\, 2\alpha\, e^{-\alpha x^2} + 4\alpha^2\, x^2\, e^{-\alpha x^2}. \tag{9.7}$$

By choosing

$$\alpha = m\omega_0 /2\hbar, \tag{9.8}$$

eq.(9.7) can be rewritten in the form :

$$- \frac{\hbar^2}{2m} \frac{d^2 \varphi_0}{dx^2} = \left(\frac{1}{2} \hbar\omega_0 - \frac{1}{2} m\omega_0^2 x^2\right) \varphi_0$$

or

$$- \frac{\hbar^2}{2m} \frac{d^2 \varphi_0}{dx^2} + \frac{1}{2} m\, \omega_0^2\, x^2\, \varphi_0 = \frac{1}{2} \hbar\, \omega_0\, \varphi_0. \tag{9.9}$$

This equation is the same as eq.(9.5) if we take

$$E_0 = \frac{1}{2} \hbar\, \omega_0. \tag{9.10}$$

That is, our trial function φ_0 is an eigenfunction with the eigenvalue $\frac{1}{2} \hbar\, \omega_0$.

Now consider another function

$$\varphi_1(x) = x\, e^{-\alpha x^2} . \tag{9.11}$$

The reader may verify that this function φ_1 is also an eigenfunction with the eigenvalue [Problem 9.1]

$$E_1 = \frac{3}{2}\hbar\omega_0 . \tag{9.12}$$

The eigenvalues and eigenfunctions can, of course, be found by a more systematic procedure. This is somewhat lengthy, however, and may be referred to in a standard textbook of quantum mechanics [4]. Systematic calculations yield the set of energy eigenvalues given by

$$\boxed{E_n = [\, n + \frac{1}{2}\,]\,\hbar\omega_0 , \qquad n = 0,1,2, \ldots} \tag{9.13}$$

Here we see that the energy eigenvalues are discrete. The eigenvalues are equally spaced with the distance-in-energy $\hbar\omega_0$, where ω_0 is the classical angular frequency given by (9.3).

The eigenfunctions φ_0 and φ_1 in (9.6) and (9.11) are orthogonal to each other. More generally, they satisfy the orthogonality relation:

$$\int_{-\infty}^{\infty} dx \;\; \varphi_n^*(x)\, \varphi_m(x) = 0, \qquad n \neq m . \tag{9.14}$$

However, they are not normalized. Using the normalization condition :

$$\int_{\infty}^{\infty} dx\, \varphi_n^*(x)\, \varphi_n(x) = 1, \tag{9.15}$$

we can obtain the normalized wave-amplitude (or wave) functions as follows : [Problem 9.2]

$$\varphi_0(x) = \left(\frac{2\alpha}{\pi}\right)^{1/4} e^{-\alpha x^2} \tag{9.16}$$

$$\varphi_1(x) = 2\left(\frac{2\alpha^3}{\pi}\right)^{1/4} x\, e^{-\alpha x^2} . \tag{9.17}$$

These wave functions are real. The wave functions of the lowest three energies are shown in Figure 8.4.

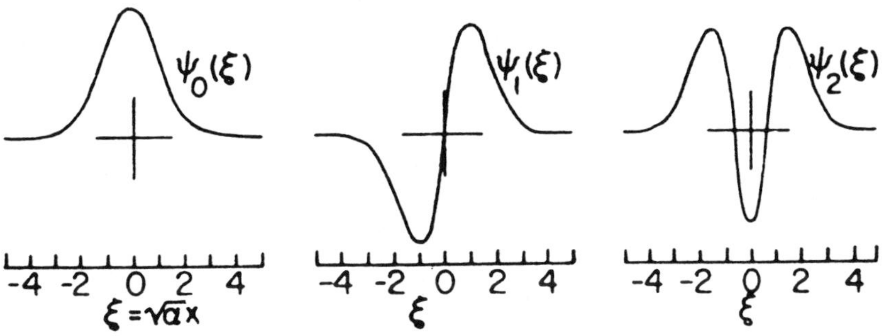

Fig. 8.4 The wave functions of the lowest three energies for the harmonic oscillator.

It is interesting to observe that the ground-state wave function $\varphi_0(x)$ with the lowest energy $\frac{1}{2}\hbar\omega_0$ is positive throughout and therefore has a single sign while the first excited-state wave function $\varphi_1(x)$ changes sign once (at the origin). In general, the wave functions with higher energies have more zeros and change sign more times. This feature is connected with the orthogonality relation for the wave functions.

The probability distribution functions

$$P_n(x) \equiv |\varphi_n(x)|^2 \qquad (9.18)$$

corresponding to these wave functions are shown in Figure 8.5. It is interesting to observe that the quantum probability distributions extend

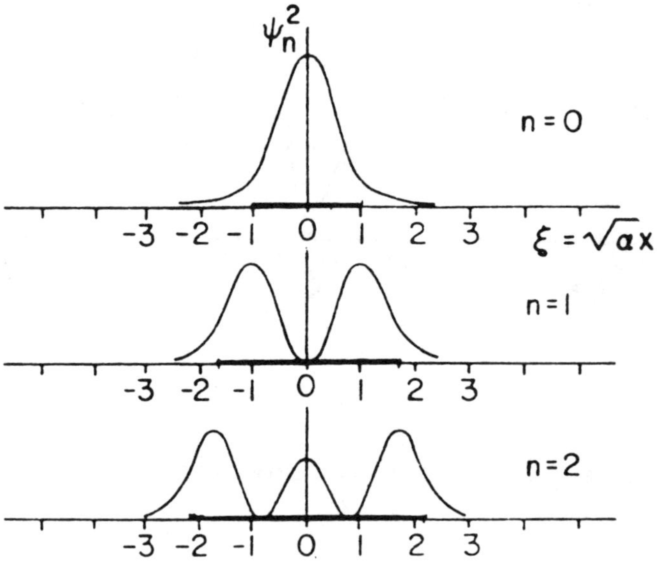

Fig. 8.5 Probability distribution functions $P_n = \psi_n^2$ against the dimensionless displacement $\xi = \sqrt{\alpha}x$. The horizontal bars indicate the classically permissible ranges of motion.

further than the permissible ranges of the classical motion of the same total energies; these ranges are indicated by horizontal line segments in the figure.

Another point worth noting is the following : according to classical mechanics the oscillator could have zero energy in which case the particle would be at rest at the origin. According to quantum mechanics, however this is impossible. The lowest energy that the oscillator can have is $(1/2)\hbar\omega_0$, which is called the zero-point energy of the oscillator. This fact has important practical consequences. For example, the atoms (or ions) in a crystal lattice vibrate like a collection of harmonic oscillators [see §11.6]. At the absolute zero of temperature, when classical theory would have predicted a state in which all the atoms are at rest, they must still move about with their zero-point energies. Such zero-point motion leads to the striking existence of quantum liquids, liquid He^3 and He^4, which do not freeze even at 0 K (at atmospheric pressure). We will briefly discuss such quantum fluids in chapter 10.

Problem 9.1 Show that

$$\varphi_1(x) = x\, e^{-\alpha x^2}, \qquad \alpha \equiv \frac{1}{2\hbar} m\omega_0 ,$$

satisfies the eigenvalue equation (9.5) with the energy $E = \frac{3}{2}\hbar\omega_0$.

Problem 9.2 Verify the normalization factors for φ_0 and φ_1 in (9.16) and (9.17).

8.10* Heisenberg's Uncertainty Principle

The most significant distinction of quantum mechanics from classical mechanics lies in the commutation rule for the position operator x and the momentum operator p :

$$[\, x, p \,] = i\hbar \,. \tag{10.1}$$

This leads to fundamental uncertainty with regard to the simultaneous observation of the position and momentum. This is known as Heisenberg's uncertainty principle, and will be discussed in the present section.

For demonstration, let us take a wave function $\psi(x)$ of the form

$$\psi(x) \equiv \langle x | \psi \rangle$$

$$= \left(\frac{1}{\pi a^2} \right)^{\frac{1}{4}} \exp \left[- \frac{(x - x_0)^2}{2\, a^2} \right], \tag{10.2}$$

where x_0 and a are both real constants. [Comparison with eq.(9.6) shows that, this function $\psi(x)$ can represent a ground-state wave function for harmonic oscillator with the center at x_0.] The position distribution function P(x) corresponding to this wave function $\psi(x)$ is given by

$$P(x) = |\, \psi(x) \,|^2 = \left[\frac{1}{\pi a^2} \right]^{\frac{1}{2}} \exp \left[- \frac{(x - x_0)^2}{a^2} \right]. \tag{10.3}$$

After simple calculations, we find that [Problem 10.1]

$$\langle x \rangle \equiv \int_{-\infty}^{\infty} dx \;\; x \, P(x) = x_0 \tag{10.4}$$

$$\langle \Delta x^2 \rangle \equiv \langle (x - x_0)^2 \rangle = \frac{1}{2}\, a^2. \tag{10.5}$$

Eqs.(10.3) - (10.5) mean that the wave function $\psi(x)$ represents a quantum state ψ of a particle in which the particle is localized around the position x_o with the root-mean-square deviation :

$$(\Delta x)_{rms} \equiv \sqrt{\langle \Delta x^2 \rangle} = a / \sqrt{2} . \qquad (10.6)$$

We now wish to find the momentum distribution function for the quantum state ψ . To do this we first find the "wave-function in the momentum space", $\langle p | \psi \rangle$, which is related to the wave function $\psi(x) \equiv \langle x | \psi \rangle$ by

$$\langle p | \psi \rangle = \int_{-\infty}^{\infty} dx \quad \langle p | x \rangle \langle x | \psi \rangle. \text{ [use of (7.11)]} \qquad (10.7)$$

By assuming the infinite-length normalization, we can calculate this quantity as follows :

$$\langle p | \psi \rangle$$

$$= \int_{-\infty}^{\infty} dx \frac{e^{-ipx/\hbar}}{(2\pi\hbar)^{\frac{1}{2}}} \left(\frac{1}{\pi a^2} \right)^{\frac{1}{4}} \exp \left[- \frac{(x - x_o)^2}{2a^2} \right]$$

[use of (7.14) and (10.2)]

$$= \frac{1}{(2\pi\hbar)^{\frac{1}{2}}} \left(\frac{1}{\pi a^2} \right)^{\frac{1}{4}} \exp \left[- \frac{a^2 p^2}{2 \hbar^2} \right] \int_{-\infty}^{\infty} dx \exp \left[- \frac{1}{2a^2} (x - x_o + \right.$$

$$\left. + ia^2 \hbar^{-1} p)^2 \right]$$

$$= \left(\frac{a}{\sqrt{\pi}\hbar} \right)^{\frac{1}{2}} \exp \left(- \frac{a^2 p^2}{2 \hbar^2} \right) \quad \text{[use of (A.7)]} \qquad (10.8)$$

Taking the absolute square of this expression, we obtain the momentum distribution function P(p) as

$$P(p) \equiv |\langle p|\psi\rangle|^2$$

$$= \frac{a}{\sqrt{\pi}\,\hbar} \exp\left[- \frac{a^2 p^2}{\hbar^2} \right], \qquad (10.9)$$

which is normalized such that

$$\int_{-\infty}^{\infty} dp \;\; P(p) = 1. \qquad (10.10)$$

Using (10.9), we obtain

$$\langle p \rangle \equiv \int_{-\infty}^{\infty} dp \;\; p\,P(p) = 0 \;\; [\text{integrand : odd}] \qquad (10.11)$$

and [Problem 9.1]

$$\langle \Delta p^2 \rangle \equiv \langle (p - \bar{p})^2 \rangle = \langle p^2 \rangle$$

$$= \frac{1}{2}\,\hbar^2/a^2 \,. \qquad (10.12)$$

Eq.(10.11) means that the average momentum for the quantum state vanishes. The root-mean-square deviation of the momentum, $(\Delta p)_{rms}$, calculated from (10.12), is given by

$$(\Delta p)_{rms} \equiv \sqrt{\langle \Delta p^2 \rangle} = \frac{1}{\sqrt{2}} \frac{\hbar}{a} \,. \qquad (10.13)$$

Multiplying this expression by (10.6), we obtain

$$(\Delta x)_{rms} \ (\Delta p)_{rms} = \frac{1}{2}\hbar \ . \tag{10.14}$$

This equation implies that the uncertainties(or standard deviations) in position and momentum are correlated, and their product is finite. In fact if we reduce the uncertainty in the position $(\Delta x)_{rms} = a/\sqrt{2}$ by halving the parameter a, the uncertainty in momentum, $(\Delta p)_{rms} = \frac{\hbar}{\sqrt{2}a}$, doubles. This most remarkable feature of a quantum state is known as Heisenberg's uncertainty principle.

The uncertainty relation (10.14) is not restricted to the wave function (10.2) which we chose in this particular example. In fact, for any non-commuting observables q and p satisfying a commutation relation of the form (10.1), the following inequality can be established

$$\Delta q \ \ \Delta p \ \geqq \ (\text{ order } \hbar) > 0, \tag{10.15}$$

where Δq and Δp are the uncertainties associated with the measurement of q and p.

Problem 10.1 Verify eqs.(10.4), (10.5) and (10.12).

Problem 10.2 Assume that $\langle x | \psi \rangle \equiv \psi(x) = c \dfrac{a}{x^2+a^2}$.

(a) Find the constant c from the normalization

$$\int_{-\infty}^{\infty} dx |\ \psi(x)\ |^2 = 1.$$

(b) Evaluate

$$\langle x\rangle \equiv x_0 \text{ and } \langle(\Delta x)^2\rangle \equiv \langle(x-x_0)^2\rangle.$$

(c) Find the corresponding momentum distribution function

$$|\langle p|\psi\rangle|^2, \text{ where } \langle p|\psi\rangle = \int_{-\infty}^{\infty} dx\, \langle p|x\rangle \langle x|\psi\rangle.$$

(d) Evaluate

$$\langle p\rangle = p_0 \text{ and } \langle\Delta p^2\rangle \equiv \langle(p-p_0)^2\rangle.$$

(e) Evaluate

$$(\Delta x)_{rms}\,(\Delta p)_{rms} \equiv \sqrt{\langle\Delta x^2\rangle}\,\sqrt{\langle\Delta p^2\rangle}\,.$$

8.11 Particle Moving in Three-Dimensional Space.

So far we have considered a particle in linear motion. In the present section, we extend the theory to the three-dimensional motion.

By postulate, the fundamental commutation rules for the Cartesian components of position and momentum (operators): (x, y, z, p_x, p_y, p_z) are given by

$$
\boxed{
\begin{aligned}
&[x,p_x] \equiv xp_x - p_x x = [y,p_y] = [z,p_z] = i\hbar \\
&[x,y] = [y,z] = [z,x] = [p_x,p_y] = [p_y,p_z] = [p_z,p_x] = 0 \\
&[x,p_y] = [x,p_z] = [y,p_z] = [y,p_x] = [z,p_x] = [z,p_y] = 0.
\end{aligned}
}
\tag{10.1}
$$

Note that only the same Cartesian components of position and momentum, e.g. x and p_x do not commute.

In Hamiltonian mechanics, the basic dynamical variables can be any set of canonical variables $(q_1, q_2, q_3, p_1, p_2, p_3)$ [, a special choice of which may be convenient for the solution of the motion]. In quantum theory, the set of Cartesian variables (x, y, z, p_x, p_y, p_z) plays a most fundamental role; only they have the simple commutation relations given here. The commutation rules for other quantum variables, if needed, must be derived from the commutation rules (10.1) [or must be postulated as in the case of the spin angular momentum components (see § 11.3)].

We can see from (10.1) that the Cartesian components of the position, (x, y, z) commute among them. In analogy with the one-dimensional case, we may assume that the eigenvalues for x, y, and z are continuous, and that the set of the eigenvalues (x', y', z') covers the three dimensional space.

The quantum position vector $\vec{r}$ may be decomposed just as the clasical position vector :

$$\vec{r} = x\,\vec{i} + y\,\vec{j} + z\,\vec{k}, \tag{10.2}$$

where $\vec{i}$, $\vec{j}$ and $\vec{k}$ are Cartesian unit vectors. The eigenvalue equation for the quantum position $\vec{r}$ can be written as follows :

$$(x\vec{i} + y\vec{j} + z\vec{k})\,|\,x',y',z'\,\rangle = (x'\vec{i} + y'\vec{j} + z'\vec{k})\,|\,x',y',z'\rangle. \tag{10.3}$$

By introducing the short-hand notation

$$|\,x',\,y',\,z'\,\rangle \equiv |\,\vec{r}\,'\,\rangle\,, \tag{10.4}$$

we can re-express eq.(11.3) as

$$\boxed{\vec{r}\,|\,\vec{r}\,'\,\rangle = \vec{r}\,'\,|\,\vec{r}\,'\,\rangle.} \tag{10.5}$$

If we take any two eigenstates characterized by two different sets of eigenvalues (x', y', z') and (x", y", z"), and form their scalar product, the product vanishes according to the orthogonality theorem (4.11). Since the eigenvalues form a continuum, we cannot normalize the ket-vector $|\,x',y',z'\,\rangle \equiv |\,\vec{r}\,'\,\rangle$ in the usual manner. We know, however, how to treat such a case for the one-dimensional system. In analogy with this case, we can express the orthonormality relation in terms of a three-dimensional Dirac delta function as follows :

$$\boxed{\langle\,\vec{r}\,'\,|\,\vec{r}\,''\,\rangle = \delta^{(3)}\,(\vec{r}\,' - \vec{r}\,'').} \tag{10.6}$$

where $\delta^{(3)}(\vec{r}) \equiv \delta(x)\,\delta(y)\,\delta(z)$, [see (4.2.25)].

We now look at the position representation, using the position eigenstates as the bases. The position vector operator $\vec{r} \equiv (x, y, z)$ is, by definition, diagonal, that is,

$$
\begin{aligned}
\langle \vec{r}''|\vec{r}|\vec{r}'\rangle &\equiv \langle x'',y'',z''|(x\vec{i} + y\vec{j} + z\vec{k})|x',y',z'\rangle && \text{[use of (11.4)]} \\
&\equiv (x'\vec{i} + y'\vec{j} + z'\vec{k})\,\langle x'',y'',z''|x',y',z'\rangle && \text{[use of (11.3)]} \\
&\equiv \vec{r}'\ \delta(x''-x')\,\delta(y''-y')\,\delta(z''-z') && \text{[use of (11.6)]} \\
&\equiv \vec{r}'\ \delta^{(3)}(\vec{r}'' - \vec{r}'). && (11.7)
\end{aligned}
$$

In general any function f of the position $\vec{r}$ alone, is diagonal :

$$
\begin{aligned}
\langle \vec{r}''|f(\vec{r})|\vec{r}'\rangle &= f(\vec{r}')\,\langle \vec{r}''|\vec{r}'\rangle \\
&= f(\vec{r}')\,\delta^{(3)}(\vec{r}'' - \vec{r}'). \qquad (11.8)
\end{aligned}
$$

In particular this means that a potential energy $V(\vec{r})$, which depends on the position $\vec{r}$ only, is diagonal in the position representation.

A general dynamical function such as the Hamiltonian depends not only on the position $\vec{r}$ but also on the momentum $\vec{p}$. Then such a function is no longer diagonal in the position representation. Dynamical functions of interest however are simple functions of the momentum components p_x, p_y, p_z. In fact, the kinetic energy, $\varepsilon_p \equiv (p_x^2 + p_y^2 + p_z^2)/\,2m$ is a quadratic function of p_x, p_y, p_z. On the other hand, the x-, y-, z- dependence of a general potential energy function can be much more complicated. Taking advantage of these characters of dynamical functions, we can develop a quantum theory in a special representation in which dynamical functions are represented by differential operators and dynamical states by wave functions. This formulation of quantum mechanics, often called the wave mechanics, will now be discussed.

We begin with the equivalence relations :

$$p_x = -i\hbar \frac{\partial}{\partial x}, \quad p_y = -i\hbar \frac{\partial}{\partial y}, \quad p_z = -i\hbar \frac{\partial}{\partial z}. \qquad (11.9)$$

From here, we obtain for example

$$\langle x',y',z'| \, p_x|\psi\rangle = \langle x',y',z'| \, (-i\hbar)\frac{\partial}{\partial x} \, |\psi\rangle = (-i\hbar)\frac{\partial}{\partial x'}\langle x',y',z' \, |\psi\rangle. \qquad (11.10)$$

If the Hamiltonian H is given in the form

$$H(x,y,z,p_x,p_y,p_z) = \frac{1}{2m}(p_x^2 + p_y^2 + p_z^2) + V(x,y,z), \qquad (11.11)$$

then we obtain

$$\langle x,y,z \mid H|\psi\rangle = \left[-\frac{\hbar^2}{2m}\left(\frac{\partial^2}{\partial x^2} + \frac{\partial^2}{\partial y^2} + \frac{\partial^2}{\partial z^2} \right) + V(x,y,z) \right]\langle x,y,z \, |\psi\rangle.$$

$$\equiv \left[-\frac{\hbar^2}{2m}\nabla^2 + V \right] \langle \vec{r}|\psi\rangle. \qquad (11.12)$$

In Dirac's formulation, the quantum state of a particle is represented by a ket vector $|\psi,t\rangle$, and the motion of the latter is ruled by the equation of motion [see(6.18)]

$$i\hbar \frac{\partial}{\partial t}|\psi, t\rangle = H|\psi,t\rangle. \qquad (11.13)$$

In the wave-mechanical formulation, the quantum state is represented by the wave function

$$\psi(x,y,z,t) \equiv \psi(\vec{r},t) \equiv \langle\vec{r} \, |\psi,t\rangle, \qquad (11.14)$$

whose change is governed by Schrödinger's equation of motion : [see(6.22)]

$$i\hbar \frac{\partial}{\partial t} \psi(\vec{r},t) = [-\frac{\hbar^2}{2m} \nabla^2 + V] \psi(\vec{r},t). \qquad (11.15)$$

By postulate, the absolute square of the normalized wave function $\psi(\vec{r},t)$ gives the probability distribution function [see(6.24)] :

$$|\psi(\vec{r},t)|^2 d^3r = |\langle \vec{r} | \psi,t\rangle |^2 d^3r \equiv P(\vec{r},t) d^3r$$

$$= \text{probability of finding the particle in } d^3r. \qquad (11.16)$$

The distribution function $P(\vec{r},t)$ in general changes with time, and its rate of change follows the equation :

$$\frac{\partial P}{\partial t} + \nabla\cdot \vec{j}(\vec{r},t) = 0, \qquad (11.17)$$

where

$$\vec{j}(\vec{r},t) \equiv \frac{\hbar}{2mi} [\psi^*(\vec{r},t) \nabla \psi(\vec{r},t) - \psi(\vec{r},t) \nabla \psi^*(\vec{r},t)] \qquad (11.18)$$

is called the probability current density. Note that eq.(11.17) has the same form as the equation of continuity (3.7.8) in fluid mechanics, and it is called by the same name. This equation can be proved by using the Schrödinger equation of motion (11.15) and its complex conjugate [Problem 10.1]. Physically it represents the conservation of the probability. It is interesting to note that the current density $\vec{j}(\vec{r},t)$ vanishes if the wave function $\psi(\vec{r},t)$ is real, that is, if $\psi^*(\vec{r},t) = \psi(\vec{r},t)$.

In wave mechanics, the basic point of departure is Schrödinger's equation of motion, (11.15). Here we deal with a partial differential equation of the second order in the space-variables (x,y,z) and of the first order in the time t. Similar equations appear in classical physics. The diffusion equation [see eq.(13.13.3)] is precisely of this type. However because of the imaginary unit i appearing in the Schrödinger wave equation, the solution of this equation is closer to that of the wave equations of classical physics [see eqs.(3.11.10) and (6.11.12), which are partial differential equations of the second order in both space and time variables.] Various mathematical techniques for solving partial differential equations may therefore be applied for quantum problems.

It is known that a quantum problem does not always have a classical analogue. In particular, every elementary particle is known to have an intrinsic angular momentum called the spin angular momentum, which will be discussed in §12.3. The Dirac formulation in which dynamical functions are represented by linear operators and dynamical states by bra-and ket-vectors, can handle this new situation in a unified manner as we will see later in §§12.3 and 12.4. This is one of the significant advantages of the Dirac formulation.

Problem 11.1 Verify the equation of continuity (10.19).

Hint : use the Schrödinger's equation of motion and its Hermitean conjugate.

Problem 11.2 Calculate the current densities corresponding to

(a) $\psi(\vec{r},t) = (2\pi\hbar)^{-3/2} e^{-i\omega t + i\vec{k}\cdot\vec{r}}$

(b) $\psi(\vec{r},t) = q\, e^{-i\omega t} \dfrac{e^{ik_o r}}{r}$,

where ω, $\vec{k}$, k_o and a are constants.

Use $\nabla\psi(\vec{r},\vartheta,\phi) = \vec{n}\dfrac{\partial\psi}{\partial r} + \vec{\ell}\,\dfrac{1}{r}\,\dfrac{\partial\psi}{\partial\vartheta} + \vec{m}\,\dfrac{1}{r\sin\vartheta}\dfrac{\partial\psi}{\partial\phi}$ for both problems.

8.12 Free Particle in Space

In the present section, we will look at a quantum particle moving in free space. The Hamiltonian H for the system is

$$H \equiv \frac{1}{2m} p^2 \equiv \frac{1}{2m} (p_x^2 + p_y^2 + p_z^2). \qquad (12.1)$$

Since this Hamiltonian depends on the momentum $\vec{p} \equiv (p_x, p_y, p_z)$ only, we can find the energy eigenvalues simply by working with the momentum representation.

Earlier in § 8.7, we solved the momentum eigenvalue problem for linear motion. The eigenvalues depend on the boundary condition.

Let us introduce a cube-shaped periodic boundary condition as follows. We take a cube of length L on a side, and choose a Cartesian frame of reference along the cubic edges. Next, we assume that the quantum state for a particle is repeated in the x-, y- and z- directions with the linear periodicity L in each direction such that

$$\langle x + L, y,z|\psi \rangle = \langle x,y + L,z|\psi \rangle = \langle x,y,z + L|\psi \rangle$$

$$= \langle x,y,z|\psi \rangle , \qquad -\infty < x,y,z < \infty . \qquad (12.2)$$

In Figure 8. 6, we indicate the one-dimensional analogue of such a periodic boundary condition.

We can now extend our earlier theory to each momentum component p_a, $a = x,y,z$, and obtain the eigenvalues as follows

$$p'_x = \frac{2\pi\hbar}{L} j , \qquad j = \ldots, -1, 0, 1, 2, \ldots$$

$$p'_y = \frac{2\pi\hbar}{L} k , \quad k = \ldots, -1, 0, 1, 2, \ldots$$

$$p'_z = \frac{2\pi\hbar}{L} \ell , \quad \ell = \ldots, -1, 0, 1, 2, \ldots \qquad (12.3)$$

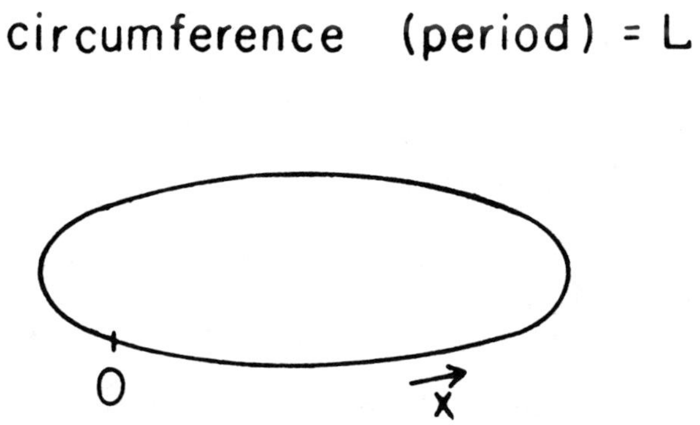

Fig. 8.6 A periodic boundary condition : $\langle x + L|\psi\rangle = \langle x|\psi\rangle$ in one dimension can be represented by the loop shown.

We note that the momentum eigenvalues are discrete and the set of eigenvalues ($2\pi\hbar j/L$, $2\pi\hbar k/L$, $2\pi\hbar\ell/L$) form a simple cubic lattice of lattice constant $2\pi\hbar/L$ in the momentum space, which is represented in Figure 8.7.

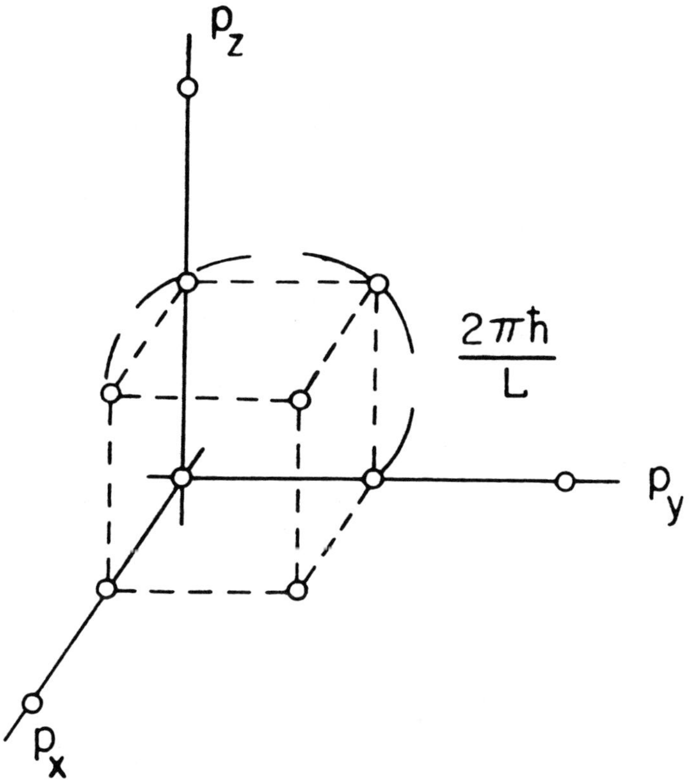

Fig. 8.7 The momentum eigenvalues as given by (12.3) form a simple cubic lattice with the lattice constant $2\pi\hbar/L$.

Let us denote the simultaneous eigenket of (p_x, p_y, p_z) corresponding to the eigenvalue (p'_x, p'_y, p'_z) by

$$| p'_x, p'_y, p'_z > \equiv | \vec{p}' > . \tag{12.4}$$

The eigenvalue equation for the momentum vector $\vec{p} \equiv \vec{i}\, p_x + \vec{j}\, p_y + \vec{k}\, p_z$ is given by

$$(\vec{i}p_x + \vec{j}p_y + \vec{k}p_z) \mid p'_x,p'_y,p'_z > = (\vec{i}p'_x + \vec{j}p'_y + \vec{k}p'_z) \mid p'_x,p'_y,p'_z >$$

or

$$\boxed{\vec{p} \mid \vec{p}' > = \vec{p}' \mid \vec{p}' >.} \qquad (12.5)$$

The momentum eigenvectors will be normalized in the standard form :

$$\boxed{<\vec{p}'' \mid \vec{p}'> = \delta^{(3)}_{\vec{p}'',\vec{p}'},} \qquad (12.6)$$

where $\delta^{(3)}$ denotes the three-dimensional Kronecker's delta :

$$\delta^{(3)}_{\vec{p}'',\vec{p}'} = \begin{cases} 1 & \text{if } \vec{p}'' = \vec{p}' \\ 0 & \text{otherwise.} \end{cases}$$

Using the last few equations, we can obtain [Problem 12.1]

$$\boxed{< \vec{r} \mid \vec{p} > = \frac{1}{L^{3/2}} e^{i\vec{r}\cdot\vec{p}/\hbar}, \quad < \vec{p} \mid \vec{r} > = \frac{1}{L^{3/2}} e^{-i\vec{r}\cdot\vec{p}/\hbar} .} \qquad (12.7)$$

<u>Transformation functions</u>, $< \vec{r} \mid \vec{p} >$ and $< \vec{p} \mid \vec{r} >$, obtained here are generalizations of (7.9).

It is important to note that the momentum eigenvalues depend on the size of the box. The eigenvalue spectrum (in the momentum space) becomes more dense for a larger box or greater L as can be seen from Figure 8.7. This is in accordance with the Heisenberg uncertainty principle. If the particle is confined to a linear dimension L, the uncertainty in momentum of the order $\hbar/L$ must occur and this uncertainty is reflected by the spacing in momentum : $2\pi\hbar/L$.

When the box of the volume L^3 is made greater, the spacing between the momentum eigenvalues becomes smaller. In the limit : $L \to \infty$, the spectrum of the momentum eigenvalues becomes continuous and covers the entire momentum space. In this continuum limit, it is convenient to choose the orthonormality condition in the following form :

$$\langle \vec{p}'' \mid \vec{p}' \rangle = \delta^{(3)} (\vec{p}'' - \vec{p}'). \qquad (\text{continuous } \vec{p}' \text{ and } \vec{p}'') \qquad (12.8)$$

In this case, the transformation functions $\langle \vec{r} \mid \vec{p} \rangle$ and $\langle \vec{p} \mid \vec{r} \rangle$ must be re-examined. They are now given by [Problem 12.1]

$$\langle \vec{r} \mid \vec{p} \rangle = (2\pi\hbar)^{-3/2} e^{i\vec{r}\cdot\vec{p}/\hbar}, \quad \langle \vec{p} \mid \vec{r} \rangle = (2\pi\hbar)^{-3/2} e^{-i\vec{r}\cdot\vec{p}/\hbar}. \qquad (\text{continuous } \vec{p}) \qquad (12.9)$$

The theoretical developments so far concerning the momentum eigenvalue problem are quite general, and in fact, can be applied to a quantum particle in an arbitrary potential field. For a free particle characterized by the Hamiltonian $H = (2m)^{-1} (p_x{}^2 + p_y{}^2 + p_z{}^2)$, we can obtain the energy eigenvalues as follows :

$$H \mid \vec{p}' \rangle = \varepsilon_{p'} \mid \vec{p}' \rangle, \qquad (12.10)$$

where

$$\varepsilon_{p'} \equiv \frac{1}{2m} (p_x'^2 + p_y'^2 + p_z'^2)$$

$$= \frac{1}{2m} \left\{ \left[\frac{2\pi\hbar}{L}\right]^2 j^2 + \left[\frac{2\pi\hbar}{L}\right]^2 k^2 + \left[\frac{2\pi\hbar}{L}\right]^2 \ell^2 \right\} \qquad [\text{use of } (12.3)]$$

$$\equiv \varepsilon_{(j,k,\ell)} \cdot \qquad (12.11)$$

Note that the energies are characterized by the three quantum numbers (j,k,ℓ) or equivalently $\vec{p}' \equiv (p_x', p_y', p_z')$.

Problem 12.1 (a) Verify that the expression (12.7) for $\langle\vec{r}|\vec{p}\rangle$ is consistent with the orthonormality relation (12.6).

(b) Verify the consistency between (12.8) and (12.9).

Hint : Similar problems were worked out explicitly for the one-dimensional case in § 12.7.

Problem 12.2 Find the momentum eigenvalues for a particle moving in a rectangular box with unequal side-lengths (L_1, L_2, L_3). Assume the periodic boundary condition.

8.13* Five Fundamental Postulates in Quantum Mechanics

In the last several sections, we have dealt with quantum mechanics of a particle moving in one and three dimensions. In this and the following two sections we will look at the fundamental structure of quantum mechanics from a general point of view.

Quantum mechanics is quite distinct from, but at the same time closely related to , classical mechanics. In fact, the latter can be regarded as a limiting case of the former. [See further discussions in §8.15.] The opposite however is not true. This means in particular that we cannot logically deduce quantum mechanics from classical mechanics. In the present section, we will enumerate fundamental postulates in the quantum mechanics of a single particle. [To treat a system of more than one particle, we will need an additional postulate, the quantum statistical postulate which will be discussed in § 9.4].

1. Quantum mechanical states

The dynamical state of a quantum particle is represented by a ket vector, denoted by $|\psi\rangle$, which may be viewed as a multidimensional column vector. The two kets $|\psi\rangle$ and $c|\psi\rangle$, where c is a non-zero complex number, represent the same state. The adjoint (or Hermitean conjugate) of the ket vector $|\psi\rangle$ is denoted by the bra vector $\langle\psi|$ which can be regarded as a multidimensional row vector. The bra and ket vectors so-defined, $\langle\psi|$ and $|\psi\rangle$, by assumption, represent the same state ψ . The ket with vanishing magnitude is called the null ket. The null ket does not correspond to any quantum state.

2. Quantum dynamical operators

Observable dynamical functions such as position, momentum, energy, are represented by Hermitean operators.

If such an operator is denoted by ξ, we can formulate the eigenvalue problem by

$$\xi \mid \xi'\rangle = \xi' \mid \xi'\rangle, \tag{13.1}$$

where ξ' and $|\xi'\rangle$ are respectively eigenvalue and eigenket vector. In general, there will be a number of eigenvalues, which are all real [see (4.9)]. The eigenvectors will be normalized such that

$$\langle \xi' | \xi'' \rangle = \begin{cases} \delta_{\xi',\, \xi''} & \text{for discrete } (\xi',\, \xi'') \\ \delta(\xi' - \xi'') & \text{for continuous } (\xi',\, \xi''). \end{cases} \tag{13.2}$$

We may expand an arbitrary ket $|\psi\rangle$ as follows :

$$|\psi\rangle = \sum_{\xi'} |\xi'\rangle \langle\xi'|\psi\rangle \qquad \text{(discrete } \xi'\text{)}$$

$$|\psi\rangle = \int d\xi' \; |\xi'\rangle \langle\xi'|\psi\rangle \qquad \text{(continuous } \xi'\text{)}. \tag{13.3}$$

We can also represent this "expandability" by the completeness relation :

$$\sum_{\xi'} |\xi'\rangle \langle\xi'| = 1 \quad \text{(discrete } \xi'\text{)}$$

$$\int d\xi' \; |\xi'\rangle \langle\xi'| = 1 \quad \text{(continuous } \xi'\text{)}. \tag{13.4}$$

3. Fundamental commutation rules

Linear operators do not necessarily commute with each other. To make the quantum description complete we must supply commutation relations. If the Cartesian components of the position and momentum operators are denoted by $(x,y,z,p_x,p_y,p_z) \equiv (q_1,q_2,q_3,p_1,p_2,p_3)$, they obey, by postulate, the following fundamental commutation rules :

$$[q_r, p_s] \equiv q_r p_s - p_s q_r = i\hbar \delta_{rs}$$

$$[q_r, q_s] = [p_r, p_s] = 0. \qquad (13.5)$$

These relations (13.5) have a great similarity with the fundamental Poisson bracket relations (3.6.10) in Hamiltonian mechanics. [See §3.6] In dealing with the spin angular momentum, special rules will be required. [See further discussions in § 12.3].

4. Quantum mechanical equation of motion

In the "Schrödinger picture" [see § 8.14 for the definition of this picture], dynamical functions ξ do not change with time but the dynamical state ψ changes. If the state is represented by the ket-vector $|\psi, t\rangle$, its change, by postulate, follows the quantum mechanical equation of motion :

$$i\hbar \frac{d}{dt} |\psi, t\rangle = H |\psi, t\rangle, \qquad (13.6)$$

where H is the Hamiltonian operator which has the same functional form as the classical Hamiltonian.

5. Probability postulate

If the particle is in the eigenstate ξ' of a certain observable [such as the momentum p_x and the energy H], then the measurement of the observable ξ is certain to yield the value ξ' . If, at the moment when the measurement takes place, the dynamical state of the system (particle) is denoted by $|\psi, t\rangle$, then the probability of finding the system in the state ξ', by postulate, is given by

$$\frac{|\langle \xi' | \psi, t \rangle|^2}{\langle \psi, t | \psi, t \rangle} \quad \left(\geqq 0 \right) . \tag{13.7}$$

If the state vectors are normalized : $\langle \psi | \psi \rangle = 1$, this expresion can be reduced to

$$|\langle \xi | \psi, t \rangle|^2 \qquad (\text{normalized } \psi). \tag{13.7a}$$

Multiplying ξ' by this probability and summing over all the states $\{\xi'\}$, we obtain

$$\begin{aligned} &\sum_{\xi'} \xi' \, |\langle \xi' | \psi, t \rangle|^2 \\ &= \sum_{\xi'} \langle \psi, t | \xi | \xi' \rangle \langle \xi' | \psi, t \rangle && [\text{use of (13.1)}] \\ &= \langle \psi, t | \xi | \psi, t \rangle \equiv \langle \xi \rangle_t . && [\text{use of (13.4)}] \end{aligned} \tag{13.8}$$

This quantity represents the average of the measured values of ξ after repeated experiments, and is called the expectation value of the observable ξ.

Dirac's formulation of quantum mechanics constructed on the basis of the above five fundamental postulates is equivalent to the more familiar wave-mechanical formulation. We may summarize the interconnections in the following manner.

The set of operators, (x, y, z), representing the Cartesian components of the position, mutually commute according to the fundamental commutation rules

(13.5). The continuous eigenvalues (x',y',z') of (x,y,z) cover the allowed position space and describe the possible states of the particle position. The eigenstates of the position are represented by the ket $| x', y', z'>$ or the bra $< x',y',z'|$ which satisfy the orthonormality relation :

$$< x',y',z' \mid x'',y'',z'' > \; = \; \delta(x' - x'')\, \delta(y' - y'')\, \delta(z' - z''). \tag{13.9}$$

Assume that the dynamical state at the time t is represented by the ket $|\psi,t >$. Multipling it from the left by $< x',y',z'|$, we obtain

$$< x',y',z'| \psi,t > \; \equiv \; \psi(x',y',z',t) . \tag{13.10}$$

This quantity can be regarded as a c-number continuous function of the space and time variables (x',y',z',t), and is called the <u>wave function</u>. Applying the probability postulate (13.8), we obtain

$$|< x,y,z \mid \psi,t >|^2 \; dx\, dy\, dz$$

$$\equiv \; | \psi(x,y,z,t) |^2 \, dx\, dy\, dz$$

= the (relative) probability of finding the particle in the volume element dx dy dz around the point (x,y,z) at the time t. (13.11)

From the fundamental commutation relations (13.5), we can derive the equivalence relations :

$$p_x = - i\hbar \frac{\partial}{\partial x} , \quad p_y = - i\hbar \frac{\partial}{\partial y} , \quad p_z = - i\hbar \frac{\partial}{\partial z} . \tag{13.12}$$

Using these relations, we construct a differential operator representing a dynamical function ξ by the rule :

$$\xi(x,y,z,p_x,p_y,p_z) \rightarrow \xi(x,y,z,\ -i\hbar\frac{\partial}{\partial x},\ -i\hbar\frac{\partial}{\partial y},\ -i\hbar\frac{\partial}{\partial z}) \tag{13.13}$$

In particular if the Hamiltonian $H = \frac{p^2}{2m} + V(r)$ is given, the corresponding Hamiltonian operator is

$$H^{(qu)} = -\frac{\hbar^2}{2m}\left(\frac{\partial^2}{\partial x^2} + \frac{\partial^2}{\partial y^2} + \frac{\partial^2}{\partial z^2}\right) + V(x,y,z)$$

$$= -\frac{\hbar^2}{2m}\nabla^2 + V. \tag{13.14}$$

Using this quantum differential operator, we can write the equation of motion for the wave function $\psi(\vec{r},t)$ as

$$\boxed{i\hbar\frac{\partial}{\partial t}\psi(\vec{r},t) = H^{(qu)}\psi(\vec{r},t) \equiv \left[-\frac{\hbar^2}{2m}\nabla^2 + V\right]\psi(\vec{r},t).} \tag{13.15}$$

This is <u>Schrödinger's wave equation</u>, which is the basic starting point for all the wave-mechanical calculations.

8. 14* The Heisenberg Picture

In the quantum mechanical postulate 4, it is stated that the dynamical functions are fixed in time and the dynamical state $|\psi ,t\rangle$ changes with time, following Schrödinger's equation of motion :

$$i\hbar \frac{d}{dt} \; |\psi ,t\rangle = H\, |\psi ,t\rangle. \tag{14.1}$$

This way of describing the time-evolution of the system (particle) is called the quantum description in the <u>Schrödinger picture</u>. It is possible to describe the time evolution of the system from different angles. The expectation value of a dynamical function ξ at the time t is, by postulate, given by $\langle\psi ,t|\xi\, |\psi,t\rangle$ according to eq.(13.8). In quantum mechanics, this is the only type of quantity that can be tested with experiments. In general, there are a number of descriptions in which the same expectation value can be calculated in a systematic manner. One of the most important equivalent descriptions is the one in the Heisenberg picture, which will be discussed in the present section.

Let us denote the Schrödinger ket at a fixed initial time t_o by $|\psi,t_o\rangle$ and the ket at a later time $t \geqq t_o$ by $|\psi,t\rangle$. We write the connection between the two in the form :

$$|\psi, t\rangle = U(t,t_o)\, |\psi, t_o\rangle. \tag{14.2}$$

Substituting this in eq.(14.1), we have

$$i\hbar \frac{dU}{dt} \; |\psi, t_o\rangle = H\; U\,|\psi, t_o\rangle.$$

Dropping the common vector $|\psi, t_o\rangle$, we obtain

$$i\hbar \frac{d}{dt} U(t,t_o) = H\, U(t,t_o). \tag{14.3}$$

For $t = t_o$, we must have from (14.2)

$$U(t_o, t_o) = 1. \tag{14.4}$$

The last two equations determine the evolution operator $U(t,t_o)$ in a unique manner.

The evolution operator U has a few significant properties :

Theorem The operator U is unitary, that is, U satisfies

$$U(t,t_o)\, U^\dagger(t,t_o) = 1 \tag{14.5a}$$

$$U^\dagger(t,t_o)\, U(t,t_o) = 1. \tag{14.5b}$$

Theorem The operator U has the reduction property such that

$$U(t_2, t_1)\, U(t_1,t_o) = U(t_2, t_o), \quad t_2 \geqq t_1 \geqq t_o. \tag{14.6}$$

Let us prove the unitarity property (14.5) by mathematical induction for a continuous variable. [A more common case of mathematical induction for discrete variable n (usually positive integers) can be found in Appendix E.] The proof proceeds in two steps :

(step 1) At the initial time t_o, eq.(14.5a) is true since

$$U(t_o, t_o)\, U^\dagger(t_o, t_o) = (1)\,(1) = 1. \quad \text{[use of (14.4)]} \tag{14.7}$$

(step 2) Let us look at the Taylor expansion :

$$U(t + dt, t_o)\, U^\dagger(t + dt, t_o)$$

$$= U(t,t_o)\ U^\dagger(t,t_o) + \frac{d}{dt}[U(t,t_o)\ U^\dagger(t,t_o)]\,dt + O(dt^2). \qquad (14.8)$$

If

$$\frac{d}{dt}[U(t,t_o)\ U^\dagger(t,t_o)] = 0, \qquad (14.9)$$

eq.(14.8) indicates that the desired property at the time t : $U(t,t_o)\ U^\dagger(t,t_o) = 1$ is maintained a small time dt later :

$$U(t + dt,\ t_o)\ U^\dagger(t + dt, t_o) = 1. \qquad (14.10)$$

Eq.(14.9) can be shown as follows. Dropping the fixed variable t_o, we obtain

$$\frac{d}{dt}[U(t)\ U^\dagger(t)]$$

$$= \left\{\frac{d}{dt}U(t)\right\} U^\dagger(t) + U(t)\left\{\frac{d}{dt}U^\dagger(t)\right\}. \qquad (14.11)$$

Using eq.(14.3) and its Hermitean conjugate :

$$-i\hbar\frac{d}{dt}U^\dagger(t) = U^\dagger(t)H, \qquad (14.12)$$

we obtain from (14.11)

$$\frac{d}{dt}[U(t)\ U^\dagger(t)] = \left[\frac{1}{i\hbar}HU\right]U^\dagger + U\left[\frac{1}{-i\hbar}U^\dagger H\right]$$

$$= \frac{1}{i\hbar}(H - H) = 0. \qquad \text{[use of (14.5a)]}$$

Combination of the two properties, (14.7) and (14.10), then extends the validity of the desired property (14.5a) to any time $t \geq t_o$.

[The advantage of the proof by mathematical induction lies in the fact that we can use, if convenient as in the present case, the disired property (14.5a) in the demonstration of eq.(14.9)].

By a similar method, we can show eqs.(14.5b) and (14.6), which will be left as excercises for the reader [Problems 14.1 and 14.2.]

In general for a given operator U, there may exist an operator which satisfies

$$U\,U^{-1} = U^{-1}\,U = 1. \tag{14.13}$$

Such operator U^{-1}(if exists) is called the inverse of U. Comparing eqs.(14.5): $U\,U^{\dagger} = U^{\dagger}U = 1$ with eqs.(14.13), we see that the Hermitean conjugate $U^{\dagger}$ equals the inverse U^{-1}:

$$U^{\dagger} = U^{-1}. \tag{14.14}$$

This equation can also be regarded as the definition of the unitarity.

A most significant property of a unitary operator is that such an operator U, upon acting on a ket-vector $|\psi\rangle$, does not change the magnitude of that ket-vector since

$$(\langle\psi|\,U^{\dagger})\,(U\,|\psi\rangle) = \langle\psi|\,U^{-1}\,U\,|\psi\rangle = \langle\psi\,|\psi\rangle. \tag{14.15}$$

In fact , using this property, we can assure the conservation of the normalization condition :

$$\langle \psi, t | \psi, t \rangle = [\langle \psi, t_o | U(t,t_o)] [U(t,t_o) | \psi, t_o \rangle]$$

$$= \langle \psi, t_o | \psi, t_o \rangle. \tag{14.16}$$

Let us now introduce a new ket-vector $|\psi\rangle_H$ defined by

$$|\psi\rangle_H \equiv U^\dagger(t,t_o) | \psi, t \rangle$$

$$= U^\dagger(t,t_o) \; U(t,t_o) \; | \psi, t_o\rangle \quad \text{[use of (14.2)]}$$

$$= |\psi, t_o\rangle. \qquad \text{[use of (14.5b)]} \tag{14.17}$$

This ket $|\psi\rangle_H$, called the <u>Heisenberg ket</u>, therefore does not change with time. We further introduce a new operator $\xi_H(t)$:

$$\xi_H(t) \equiv U^\dagger(t,t_o)\, \xi \, U(t,t_o), \tag{14.18}$$

which is time-dependent.

We can then establish the identity :

$${}_H\langle \xi | \xi_H(t) | \psi\rangle_H$$

$$= \langle \xi, t_o | [U^\dagger(t,t_o)\, \xi \, U(t,t_o)] | \psi, t_o\rangle \quad \text{[use of (14.17) and (14.18)]}$$

$$= \langle \psi, t | \xi | \psi, t\rangle \quad \text{[use of (14.2)]}$$

or

$$\boxed{{}_H\langle \psi | \xi_H(t) | \psi\rangle_H = \langle \psi, t | \xi | \psi, t\rangle} \tag{14.19}$$

Differentiating eq.(14.18) with respect to t and using (14.3) and (14.12), we can obtain the following differential equation :

$$\boxed{\frac{d\xi_H(t)}{dt} = \frac{i}{\hbar}[\,H_H\,,\ \xi_H(t)\,].} \tag{14.20}$$

We may now interpret the last four equations in the following manner. Instead of having fixed observable dynamical operators ξ and time-dependent state vector $|\psi,t\rangle$, we have, in the new scheme, time-dependent observables $\xi_H(t)$ and fixed state vector $|\psi\rangle_H$. The temporal evolution of $\xi_H(t)$ is governed by the Heisenberg equation of motion (14.20). The expectation value of the observable ξ is given by (14.19). This new mode of description is called the Heisenberg picture.

The significance of the theory presented in this section will be expounded in the following section.

Problem 14.1 Prove (13.5b) by mathematical induction.

Problem 14.2 Prove the reduction property (14.6)

8.15* Correspondence between Quantum and Classical Mechanics

The postulates and concepts of quantum mechanics are quite distinct from those of classical mechanics. In classical mechanics all dynamical variables (q,p) commute whereas in quantum mechanics dynamical variables do not necessarily commute ; in particular, Cartesian coordinates and momentum components obey the following fundamental commutation relations : [see eqs.(3.5)]

$$[q_r, p_s] \equiv q_r p_s - p_s q_r = i\hbar\, \delta_{rs}$$

$$[q_r, q_s] = [p_r, p_s] = 0 \quad \text{(quantum mechanics).} \qquad (15.1)$$

These relations have a striking formal resemblance to the fundamental Poisson bracket relations (3.6.10) for a set of canonical variables :

$$\{ q_r, p_s \} = \delta_{rs}$$

$$\{ q_r, q_s \} = \{ p_r, p_s \} = 0 \quad \text{(classical mechanics).} \qquad (15.2)$$

The correspondence may symbolically be expressed by

$$\boxed{\frac{1}{i\hbar}\, [A,B] \leftrightarrow \{ A,B \},} \qquad (15.3)$$

where A and B represent any two of the Cartesian variables. By straightforward calculations we can verify

$$[u,v] = - [v,u], \qquad [u,c] = 0 \qquad (15.4)$$

$$[u,c_1 v + c_2 w] = c_1 [u,v] + c_2 [u,w] \qquad (15.5)$$

$$[uv,w] = u[v,w] + [u,w]v$$

$$[u,vw] = v[u,w] + [u,v]w \tag{15.6}$$

$$[u,[v,w]] + [v,[w,u]] + [w,[u,v]] = 0, \tag{15.7}$$

where u,v,w are linear operators, and c,c_1,c_2 are numbers. For example, the first identity of eqs.(15.6) can be shown as follows :

$$\begin{aligned}[u\ v,w] &\equiv uvw - wuv \\ &= uvw - uwv + uwv - wuv \\ &= u[v,w] + [u,w]v.\end{aligned}$$

Other identities can be checked in a similar manner. [Problems 15.1 and 15.2]. Note that the algebraic properties (15.4-7) for the commutator brackets are identical in form with those for the Poisson brackets. [See §3.6]

This means that the analogy between quantum and classical mechanics is nearly complete except for the fact that the fundamental commutation relations (15.1) hold only for Cartesian coordinates and momentum components.

In general, postulates in a physical theory must ultimately be justified by comparing the theoretical predictions based on them with experimental oservations. We have seen in §8.10 that the non-commutativity of conjugate variables such as x, p_x, leads to the fundamental uncertainty with regard to the simultaneous observation of the corresponding physical quantities. All the experimental observation has directly or indirectly supported Heisenberg's uncertainty principle, and today, we look at the fundamental commutation rules as the backbone of quantum mechanics.

The analogy between quantum and classical mechanics can be extended further. As we saw in the preceding section, the equations of motion for a Heisenberg variable ξ_H and the classical equations of motion for the corresponding variable ξ_{cl} have analogous forms :

$$\frac{d\xi_H(t)}{dt} = \frac{i}{\hbar} [H_H (t), \ \xi_H(t)] \quad \text{(quantum mechanics)} \qquad (15.8)$$

$$\frac{d\xi_{cl}(t)}{dt} = - \{ H, \xi_{cl} \}. \quad \text{(classical mechanics)} \qquad (15.9)$$

Historically, this analogy was used by Dirac to logically support the quantum mechanical equation of motion (15.8). Since Heisenberg and Schrödinger descriptions of quantum mechanics are equivalent, this analogy also supports Schrödinger's equation of motion , a fundamental quantum mechanical postulate.

Unfortunately, the formal correspondence between quantum and classical mechanics does not help solve quantum mechanical problems. The detailed mathematical structures associated with eqs.(15.8) and (15.9) are quite different; eq.(15.8) deals with linear operators while eq.(15.9) relates ordinary c-number functions .

From the logical and more importantly practical point of view, the relation between the two mechanics can be summarized as follows : <u>Classical mechanics can be regarded as a limiting case of quantum mechanics when Planck's constant h is artificially reduced to zero</u>. Indeed, in this limit, all the commutators appearing in (15.1) vanish. And it then follows that all dynamical functions commute with each other, and therefore can be represented by ordinary numbers.

Finally, we point out that some quantum mechanical concepts do not have classical analogues. This is the case, for example, for the spin angular momentum associated with each elementary particle.

Problem 15.1 Prove the identities (15.4) - (15.6).

Problem 15.2 Prove the identity (15.7).

8.16 The Gibbs Ensemble in Quantum Mechanics

Earlier in §4.2, we discussed the Gibbs ensemble of classical systems. In the present section we will discuss the Gibbs ensemble for quantum-mechanical single particles.

If a particle at time t is in the state ψ represented by the ket-vector $|\psi,t>$ which is normalized such that

$$<\psi,t|\psi,t> = 1, \tag{16.1}$$

then the expectation value of a dynamical function ξ is given by [see (13.8)]

$$<\xi>_t = <\psi,t|\xi|\psi,t>. \tag{16.2}$$

Using the completeness relation, we can re-express this as follows :

$$<\xi>_t = \int d^3r \; <\psi,t|\vec{r}><\vec{r}|\xi|\psi,t> \quad \text{[use of (13.4)]}$$

$$= \int d^3r \; <\vec{r}|\xi|\psi,t><\psi,t|\vec{r}>. \quad \text{[exchange of the order of two numbers]} \tag{16.3}$$

We now look at the quantity

$$\rho(t) \equiv |\psi,t><\psi,t|. \tag{16.4}$$

This $\rho(t)$ is not a number like $<\psi,t|\psi,t>$ but a useful time-dependent operator. We note that ρ is Hermitean :

$$\rho^\dagger(t) = \rho(t). \tag{16.5}$$

We will call it the quantum density operator (or statistical operator).

Let us first take the diagonal element of $\rho(t)$ in the position representation :

$$< \vec{r} \mid \rho(t) \mid \vec{r} >$$

$$= < \vec{r} \mid \psi,t > < \psi,t \mid \vec{r} > = \mid < \vec{r} \mid \psi,t > \mid^2$$

$$= \mid \psi(\vec{r},t) \mid^2 . \qquad \text{[use of (11.14)]} \qquad (16.6)$$

This quantity, by postulate, represents the probability distribution function in the position space. Second, we consider the diagonal element of $\rho(t)$ in the momentum representation :

$$< \vec{p} \mid \rho(t) \mid \vec{p} > = < \vec{p} \mid \psi,t > < \psi,t \mid \vec{p} >$$

$$= \mid < \vec{p} \mid \psi,t > \mid^2 , \qquad (16.7)$$

which gives the probability of finding the system in the momentum state $\vec{p}$. In fact, we can generalize these results, (16.6) and (16.7), by taking an arbitrary representation, and obtain

$< \alpha' \mid \rho \mid \alpha' >$

= the probability of finding the system in the quantum state α',

(16.8)

where $< \alpha' \mid$ and $\mid \alpha' >$ are normalized bra- and ket-vectors representing the state α'.

Summing (16.8) over all states α' , we obtain

$$\sum_{\alpha'} \langle\alpha' | \rho|\alpha' \rangle$$

$$= \sum_{\alpha'} \langle \alpha' | \psi,t\rangle \langle \psi,t |\alpha' \rangle \quad \text{[use of (16.4)]}$$

$$= \sum_{\alpha'} \langle \psi,t | \alpha'\rangle \langle \alpha'| \psi,t \rangle \quad \text{[exchange of two numbers]}$$

$$= \langle \psi,t| \psi,t \rangle \quad \text{[completeness relation]}$$

$$= 1. \quad \text{[use of (16.1)]} \qquad (16.9)$$

The l. h. s. is the diagonal sum or trace of the operator ρ in the α-representation. We can show that the trace is the same for any representation. In fact, let us take the diagonal sum of an arbitrary matrix A in the α-representation, $\Sigma_{\alpha'} \langle \alpha'| A |\alpha' \rangle$. By expanding the ket $|\alpha'\rangle$ in terms of other bases :

$$|\alpha'\rangle = \sum_{\gamma'} |\gamma'\rangle \langle \gamma' |\alpha' \rangle,$$

we obtain

$$\sum_{\alpha'} \langle \alpha' |A |\alpha' \rangle = \sum_{\alpha'} \sum_{\gamma'} \langle \alpha'| A |\gamma' \rangle \langle \gamma' |\alpha' \rangle$$

$$= \sum_{\gamma'} \sum_{\alpha'} \langle \gamma' |\alpha'\rangle \langle \alpha' |A |\gamma' \rangle$$

$$= \sum_{\gamma'} \langle \gamma' | A |\gamma' \rangle \equiv \text{Tr}\,\{A\}. \qquad (16.10)$$

The last member is the diagonal sum in the γ-representation. Q.E.D.

To stress this useful property we will denote the diagonal sum of matrix A by $\mathrm{Tr}\{A\}$ read "trace of A". Using this notation we can rewrite (16.9) as

$$\boxed{\mathrm{Tr}\{\rho\} = 1.} \tag{16.11}$$

This represents the normalization condition of the density operator ρ.

We now go back to eq.(16.3). Using (16.4), we can re-express it as

$$\langle \xi \rangle_t = \int d^3r \, \langle \vec{r} | \xi \, \rho(t) | \vec{r} \rangle . \tag{16.12}$$

The integral on the r.h.s. can be regarded as the diagonal "sum" (integral) in the position space. We can therefore re-express eq.(16.12) in the form:

$$\boxed{\langle \xi \rangle_t = \mathrm{Tr}\{ \xi\rho(t) \}.} \tag{16.13}$$

This means that to find the expectation value of the observable ξ we merely calculate the trace of the product of ξ and the density operator $\rho(t)$ in any representation.

For a product of two arbitrary operators A and B, the following useful property holds :

$$\mathrm{Tr}\{ AB \} = \mathrm{Tr}\{ BA \}, \tag{16.14}$$

which can be proved by writing out the sums explicitly

$$\sum_j (AB)_{jj} = \sum_j \sum_k A_{jk} B_{kj} = \sum_k [\sum_j B_{kj} A_{jk}] = \sum_k (BA)_{kk} .$$

The kets and bras, $|\psi,t>$ and $<\psi,t|$, change with time, following Schrödinger's equation of motion :

$$i\hbar \frac{d}{dt}|\psi , t> = H|\psi,t> \tag{16.15}$$

and its adjoint :

$$-i\hbar \frac{d}{dt}<\psi,t| = <\psi,t| H. \tag{16.15a}$$

Using these equations, we can obtain the evolution equation for $\rho(t)$ as follows :

$$i\hbar \frac{d\rho}{dt} = i\hbar\left[\left(\frac{d}{dt}|\psi,t>\right)<\psi,t| + |\psi,t>\frac{d}{dt}<\psi,t| \right]$$

$$= H|\psi,t><\psi,t| - |\psi,t><\psi,t| H \quad \text{[use of (16.15) and (16.15a)]}$$

$$= H\rho - \rho H \quad \text{[use of (16.4)]}$$

or

$$\boxed{i\hbar \frac{d\rho}{dt} = [H, \rho].} \tag{16.16}$$

This is called the <u>quantum Liouville equation</u>. It is the quantum analogue of the classical Liouville equation (4.5.3). It is also referred to as <u>Landau equation</u> or <u>Von Neumann equation</u> in some literature. Apart from the sign, this equation is of the same form as the equation of motion (14.11) for a Heisenberg dynamical operator.

Let us now introduce a new density operator ρ' which is different from ρ by a large positive integer N_G :

$$\rho' \equiv N_G\, \rho. \tag{16.17}$$

We may rewrite eqs.(16.11) and (16.13) in the form :

$$\mathrm{Tr}\{\rho'\} = N_G \tag{16.18}$$

$$\langle \xi \rangle = \frac{\mathrm{Tr}\{\xi \rho'\}}{\mathrm{Tr}\{\rho'\}}. \tag{16.19}$$

With the new density operator ρ' we may reinterpret the statistics in the following manner : We have N_G similar quantum mechanical systems, whose dynamical states change independently, following Schrödinger's equation of motion; the diagonal element in the α-representation $\langle \alpha' | \rho' | \alpha' \rangle$, yields the number of systems occupying the quantum state α'; the ratio $\mathrm{Tr}\{\xi\rho'\}$ / $\mathrm{Tr}\{\rho'\}$ represents the average of the observable ξ over all the systems. The set of such dynamical systems is called the quantum mechanical Gibbs ensemble.

The operators ρ and ρ' differ only in normalization. In practice the interpretation associated with ρ' will be extended to the case of ρ when convenient.

In order to find the density operator $\rho(t)$, we must solve the quantum Liouville equation (16.16) subject to a given initial condition.

We now discuss some frequently used initial condition.

(a) Let us consider a quantum particle in linear motion .
Let p_n be the momentum eigenvalues. We assume that at the initial time the particle is in one or other of the momentum states p_n with given probabilities Q_n. Note that this is a reasonable assumption for a particle in an arbitrary potential field. The sum of Q_n is normalized to unity :

$$\sum_n Q_n = 1. \tag{16.20}$$

According to (16.8), the diagonal element of the density matrix, $\langle p_n | \rho(0) | p_n \rangle$, must be equal to the occupation probability Q_n :

$$\langle p_n | \rho(0) | p_n \rangle = Q_n. \qquad (16.21)$$

We may then choose

$$\rho(0) = \sum_n | p_n \rangle Q_n \langle p_n | . \qquad (16.22)$$

(b) Let us assume that the particle is distributed over the energy eigen-states ε_k with given probabilities P_k . From the same sort of reasoning as in (a), we may define the initial density operator $\rho(0)$ as

$$\rho(0) = \sum_{k(\text{states})} | \varepsilon_k \rangle P_k \langle \varepsilon_k |. \qquad (16.23)$$

The density operator constructed here is sationary : $d\rho/dt = 0$. In fact, we can see this property in the following manner :

$$i\hbar \frac{d\rho}{dt} = [H, \rho] \qquad [\text{use of } (16.16)]$$

$$= \sum_k \{ H | \varepsilon_k \rangle P_k \langle \varepsilon_k | - | \varepsilon_k \rangle P_k \langle \varepsilon_k | H \} \ [\text{use of } (16.23)]$$

$$= \sum_k \{ \varepsilon_k | \varepsilon_k \rangle P_k \langle \varepsilon_k | - | \varepsilon_k \rangle P_k \langle \varepsilon_k | \varepsilon_k \}$$

$$= 0. \qquad (16.23)$$

In this section, we discussed the Gibbs ensemble of single particles . For a system containg more than one particle, we must take account of the quantum statistical postulate [see §9.4]. This will introduce a small change in the definition of the density operator. But most formulas including (16.8), (16.11) and (16.13) will retain the same mathematical structure.

Problem 16.1 Show that

(a) $\langle \alpha | \rho | \alpha \rangle \geqq 0$ for any state α .

(b) $\langle \xi \rangle = \mathrm{Tr}\{ \xi \rho \}$ = a real number for any observable ξ .

References

[1] Dirac, P.A.M., Principles of Quantum Mechanics, Oxford University Press, London, 4th Edition (1958)

[2] e.g. Symon, K.R., Mechanics, Addison-Wesley, Reading, Mass. 3rd Edition (1971)

[3] e.g. Arfken, G., Mathematical Methods for Physicists, Academic Press, New York (1970) ; see General Bibliography for other comparable books.

[4] e.g. Pauling, L. and Wilson, E.B., Introduction to Quantum Mechanics, McGraw-Hill, New York (1935)

Review Questions

1. Write down mathematical symbols, expressions and / or figures, and explain briefly:

(a) linear operators. Cite at least two examples

(b) Hermitean conjugates of a ket vector with two elements, $|A\rangle = \begin{pmatrix} 1 \\ i \end{pmatrix}$.

(c) eigenvalue problem for $\sigma_y = \begin{pmatrix} 0 & -i \\ i & 0 \end{pmatrix}$. Find the eigenvalues.

(d) momentum eigenvalue problem

(e) five quantum mechanical postulates for the description of a single particle in motion

(f) wave function $\psi(\vec{r},t)$ and the quantum state vector $|\psi,t\rangle$

(g)* unitary operator. Definition and its principal properties

(h)* equation of motion for a Heisenberg variable

(i) classical limit

(j)* quantum Liouville equation.

General Problems

1. Pauli's spin matrices $(\sigma_x, \sigma_y, \sigma_z) \equiv (\sigma_1, \sigma_2, \sigma_3)$ are given explicitly in p.8.20. The Dirac matrices also called the gamma matrices (4 x 4 matrices) are defined by

$$\gamma_k \equiv \begin{bmatrix} 0 & -i\sigma_k \\ i\sigma_k & 0 \end{bmatrix}, \qquad k = 1,2,3$$

$$\gamma_4 \equiv \begin{bmatrix} I & 0 \\ 0 & -I \end{bmatrix},$$

which really mean

$$\gamma_3 = \begin{bmatrix} 0 & 0 & -i & 0 \\ 0 & 0 & 0 & i \\ i & 0 & 0 & 0 \\ 0 & -i & 0 & 0 \end{bmatrix}, \qquad \gamma_4 = \begin{bmatrix} 1 & 0 & 0 & 0 \\ 0 & 1 & 0 & 0 \\ 0 & 0 & -1 & 0 \\ 0 & 0 & 0 & -1 \end{bmatrix}, \text{ etc.}$$

Show that

(a) $\gamma_\mu^2 = 1$, $\mu = 1,2,3,4.$

(b) $\gamma_\mu \gamma_\nu + \gamma_\nu \gamma_\mu = 0$ if $\mu \neq \nu$.

2. Let us consider a thin uniform triangle of sides (3 cm, 4 cm, 5 cm). Take a convenient set of Cartesian axes, and find

(a) Center of mass

(b) Inertia tensor $\overleftrightarrow{I} \equiv \Sigma \; m\,(r^2\overleftrightarrow{E} - \vec{r}\,\vec{r})$ about the center of mass

(c) Principal moments of inertia

(d) Principal axes of inertia

3. Let us take a simple harmonic oscillator.

(a) Write down the Hamiltonian.

(b) Write down Schrödinger's wave equation.

(c) Write down Schrödinger's equation for the energy eigenvalue.

(d) Try $\psi_0(x) = c\, e^{-\alpha x^2}$ and find the corresponding eigenvalue.

(e) Normalize the wave function ψ_0.

(f) Evaluate the average of x, x^2, p, p^2 and xp, using the wave function $\psi_0(x)$.

4. Show that the simple harmonic oscillator characterized by the Hamiltonian $H = p^2(2m)^{-1} + \frac{1}{2}k\,(x-x_0)^2$ has the same energy eigenvalues irrespective of the position of the center, x_0.

5.* The momentum eigenvalues for a quantum particle in linear motion is given by $p_n \equiv 2\pi\hbar n/L$, where L is the periodicity length. We now look at the operator $\xi(p_n) \equiv |p_n\rangle\langle p_n|$. Show that

(a) $\xi^\dagger = \xi$ (b) $\xi^2 = \xi$

(c) $\xi(p_n)\,|p_m\rangle = \delta_{p_n,\,p_m}\,|p_m\rangle$ (d) $\sum_n \xi(p_n) = 1.$

These properties indicate that the operator $\xi(p_n)$ may be identified as $\delta_{p,\,p_n}$, where p represents the momentum operator. Let us now take the infinite-length limit ($L \to \infty$). By similar calculations, show that $|p'\rangle\langle p'| = \delta(p - p')$, where p' and $|p'\rangle$ are continuous momentum eigenvalue and eigenket, respectively.

Chapter 9. QUANTUM STATISTICAL MECHANICS. BASIC PRINCIPLES

In this chapter we will discuss basic principles of quantum statistical mechanics. We begin in the two preparatory sections, §§9.1 and 9.2, with introductory discussions of the permutation group, and odd and even permutations. In the following sections, §§9.3 - 9.6, the basic ground rules for the treatment of a system of identical particles are laid down. The indistinguishability of particles for both quantum and classical mechanical systems can be formulated in terms of the symmetry properties of the Hamiltonian and the many-particle distribution function. By the quantum statistical postulate, every quantum particle is either a boson or fermion. Bosons and fermions are defined with the stipulation that the quantum state for a system of identical bosons (fermions) be symmetric (antisymmetric). This definition immediately leads to Pauli's exclusion principle for fermions.

The quantum state for a many-particle system can best be represented by the set of the numbers of particles occupying the single-particle quantum states. This so-called occupation number representation is discussed in §9.7. The number of fermions occupying any quantum state is restricted to 0 or 1 while the occupation number of bosons can be 0 or any positive integer. In §§9.8 - 9.10, we extend various concepts and techniques introduced for classical statistical mechanics to quantum statistical systems. By considering a grand canonical ensemble of free bosons (fermions), we calculate the average occupation number for a momentum state and express it in terms of the Bose (Fermi) distribution function. The Bose and Fermi distribution functions are distinct from each other, and they are also different from the (classical) Boltzmann distribution function.

All results of quantum statistical mechanics approach those of classical statistical mechanics in the classical limit: $h \to 0$. General arguments for this are given in §9.12. Practical guidelines as to when classical statistical mechanics can give essentially correct results are discussed in §9.13.

9.1 Permutation Group

Let us consider the set of ordered numerals $[1,2,3]$. By interchanging 1 and 2, we obtain $[2,1,3]$. Let us denote this operation by

$$(2,1)[1,2,3] \equiv (1,2)[1,2,3] = [2,1,3] ; \tag{1.1}$$

the operator $(2,1) \equiv (1,2)$ is called the <u>interchange</u> or <u>transposition</u> between 1 and 2. We can express the same operation another way:

$$\begin{pmatrix} 1 & 2 & 3 \\ 2 & 1 & 3 \end{pmatrix} [1,2,3] = [2,1,3] , \tag{1.2}$$

where $\begin{pmatrix} 1 & 2 & 3 \\ 2 & 1 & 3 \end{pmatrix}$ indicates the change in order from $[1,2,3]$ to $[2,1,3]$ and will be called a <u>permutation</u>. The order of writing the numerals in columns is immaterial, so that

$$\begin{pmatrix}1&2&3\\2&1&3\end{pmatrix} \equiv \begin{pmatrix}1&3&2\\2&3&1\end{pmatrix} \equiv \begin{pmatrix}2&1&3\\1&2&3\end{pmatrix} \equiv \begin{pmatrix}2&3&1\\1&3&2\end{pmatrix} \equiv \begin{pmatrix}3&1&2\\3&2&1\end{pmatrix} \equiv \begin{pmatrix}3&2&1\\3&1&2\end{pmatrix} . \tag{1.3}$$

We can define other permutations in a similar manner. Distinct permutations of three numerals are given by

$$\begin{pmatrix}1&2&3\\1&2&3\end{pmatrix}, \begin{pmatrix}1&2&3\\2&3&1\end{pmatrix}, \begin{pmatrix}1&2&3\\3&1&2\end{pmatrix}, \begin{pmatrix}1&2&3\\2&1&3\end{pmatrix}, \begin{pmatrix}1&2&3\\1&3&2\end{pmatrix}, \begin{pmatrix}1&2&3\\3&2&1\end{pmatrix} \tag{1.4}$$

which will be symbolically denoted by $\{P_1,P_2,\ldots,P_6\}$.

The product of any P_j and P_k is defined by

$$(P_jP_k)[1,2,3] \equiv P_j(P_k[1,2,3]) . \tag{1.5}$$

The set of six elements $\{P_1,P_2,\ldots,P_6\}$ in (1.4) has the group properties (see below) and is called a <u>permutation group</u> of degree 3. The said <u>group properties</u> are as follows:

(i) Composition

Any two members of the set, P_j and P_k, have the <u>composition</u> property that the product P_jP_k is also a member of the set, say, P_i:

$$P_jP_k = P_i . \tag{1.6}$$

We demonstrate this property by examples. Let us choose

$$P_j = \begin{pmatrix}1&2&3\\2&1&3\end{pmatrix} \quad \text{and} \quad P_k = \begin{pmatrix}1&2&3\\1&3&2\end{pmatrix} .$$

By direct calculations, we obtain

$$\begin{aligned}\left\{\begin{pmatrix}1&2&3\\2&1&3\end{pmatrix}\begin{pmatrix}1&2&3\\1&3&2\end{pmatrix}\right\}[1,2,3] &\equiv \begin{pmatrix}1&2&3\\2&1&3\end{pmatrix}\left\{\begin{pmatrix}1&2&3\\1&3&2\end{pmatrix}[1,2,3]\right\}\\ &= \begin{pmatrix}1&2&3\\2&1&3\end{pmatrix}[1,3,2]\\ &= [2,3,1] .\end{aligned} \tag{1.7}$$

Since

$$\begin{pmatrix}1&2&3\\2&3&1\end{pmatrix}[1,2,3] = [2,3,1] , \tag{1.8}$$

we can express the product $\begin{pmatrix}1&2&3\\2&1&3\end{pmatrix}\begin{pmatrix}1&2&3\\1&3&2\end{pmatrix}$ by the single permutation $\begin{pmatrix}1&2&3\\2&3&1\end{pmatrix}$, which is a member of the set.

We may obtain the same result by examining the permutations directly as follows.

Start with the permutation $P_k = \begin{pmatrix}1&2&3\\1&3&2\end{pmatrix}$. This moves 1 to 1. Then, the permutation $P_j = \begin{pmatrix}1&2&3\\2&1&3\end{pmatrix}$ moves 1 to 2. The net move is therefore $1\to1\to2$. We repeat the same reasoning to observe that $2\to3\to3$ and $3\to2\to1$. We can then represent the net move by the permutation $\begin{pmatrix}1&2&3\\2&3&1\end{pmatrix}$.

(ii) Association

If P_i, P_j and P_k are any three elements of the set, then (See Problem 3.1):

$$(P_iP_j)P_k = P_i(P_jP_k) \equiv P_iP_jP_k . \tag{1.9}$$

For example, by applying the second method of calculation, we obtain

$$\left\{\begin{pmatrix}1&2&3\\2&1&3\end{pmatrix}\begin{pmatrix}1&2&3\\1&3&2\end{pmatrix}\right\}\begin{pmatrix}1&2&3\\2&1&3\end{pmatrix} = \begin{pmatrix}1&2&3\\2&3&1\end{pmatrix}\begin{pmatrix}1&2&3\\2&1&3\end{pmatrix} = \begin{pmatrix}1&2&3\\3&2&1\end{pmatrix}$$

$$\begin{pmatrix}1&2&3\\2&1&3\end{pmatrix}\left\{\begin{pmatrix}1&2&3\\1&3&2\end{pmatrix}\begin{pmatrix}1&2&3\\2&1&3\end{pmatrix}\right\} = \begin{pmatrix}1&2&3\\2&1&3\end{pmatrix}\begin{pmatrix}1&2&3\\3&1&2\end{pmatrix} = \begin{pmatrix}1&2&3\\3&2&1\end{pmatrix} .$$

Thus, we obtain

$$\left\{\begin{pmatrix}1&2&3\\2&1&3\end{pmatrix}\begin{pmatrix}1&2&3\\1&3&2\end{pmatrix}\right\}\begin{pmatrix}1&2&3\\2&1&3\end{pmatrix} = \begin{pmatrix}1&2&3\\2&1&3\end{pmatrix}\left\{\begin{pmatrix}1&2&3\\1&3&2\end{pmatrix}\begin{pmatrix}1&2&3\\2&1&3\end{pmatrix}\right\} .$$

(iii) Identity

There exists a unique element, called the identity, and denoted by E, which has the property

$$P_jE \equiv P_j = EP_j \quad \text{for any } P_j \; . \tag{1.10}$$

In the present case,

$$E = \begin{pmatrix} 1 & 2 & 3 \\ 1 & 2 & 3 \end{pmatrix} . \tag{1.11}$$

(iv) Inverse

For every element P_j, there exists an element P_j^{-1}, called the inverse of P_j, in the set, which satisfies

$$P_jP_j^{-1} = E = P_j^{-1}P_j \; . \tag{1.12}$$

For example, if $P = \begin{pmatrix} 1 & 2 & 3 \\ 2 & 3 & 1 \end{pmatrix}$, $P^{-1} = \begin{pmatrix} 1 & 2 & 3 \\ 3 & 1 & 2 \end{pmatrix}$.

It is stressed that the group properties (i) - (iv) all refer to the principal operation defined by (1.5).

Let us consider all integers 0, ±1, ±2, ... With respect to the summation j + k, the four group properties are satisfied; the identity is represented by 0 and the inverse of j is -j. Therefore, the set of integers form a group with respect to the summation. The same set however, does not form a group with respect to multiplication since the inverse of j, say 2, cannot be found within the set.

It is important to note that the product P_jP_k is not in general equal to the product of the reversed order:

$$P_jP_k \neq P_kP_j \; . \qquad \text{(in general)} \tag{1.13}$$

For example,

$$\begin{pmatrix}1&2&3\\1&3&2\end{pmatrix}\begin{pmatrix}1&2&3\\2&1&3\end{pmatrix} = \begin{pmatrix}1&2&3\\3&1&2\end{pmatrix}$$

$$\begin{pmatrix}1&2&3\\2&1&3\end{pmatrix}\begin{pmatrix}1&2&3\\1&3&2\end{pmatrix} = \begin{pmatrix}1&2&3\\2&3&1\end{pmatrix}.$$

Therefore,

$$\begin{pmatrix}1&2&3\\1&3&2\end{pmatrix}\begin{pmatrix}1&2&3\\2&1&3\end{pmatrix} \neq \begin{pmatrix}1&2&3\\2&1&3\end{pmatrix}\begin{pmatrix}1&2&3\\1&3&2\end{pmatrix}.$$

The number of elements in a group is called the <u>order</u> of the group. The order of the permutation group of degree N is N!. In fact, let us write a permutation in the form $\begin{pmatrix}1&2&\dots&N\\j_1&j_2&\dots&j_N\end{pmatrix}$, where $\{j_k\}$ are different numerals taken from (1, 2, ..., N). Clearly there are N! ways of ordering these j's. Such ordering generates N! distinct permutations.

Problem 1.1 Reduce each of the following into a single permutation.

(a) $\begin{pmatrix}1&2&3\\2&3&1\end{pmatrix}\begin{pmatrix}1&2&3\\2&1&3\end{pmatrix}$ (b) $\begin{pmatrix}1&2&3\\1&3&2\end{pmatrix}\begin{pmatrix}1&2&3\\3&2&1\end{pmatrix}$ (c) $\begin{pmatrix}1&2&3\\2&1&3\end{pmatrix}\begin{pmatrix}1&2&3\\2&1&3\end{pmatrix}$

Problem 1.2 Refer to Problem 1.1. Construct the product of the reversed order for each case and reduce it to a single permutation. For each case state if the two permutations are commutative or not.

Problem 1.3 Find the inverses of the following permutations:

(a) $\begin{pmatrix}1&2&3\\3&1&2\end{pmatrix}$ (b) $\begin{pmatrix}1&2&3&4\\2&3&4&1\end{pmatrix}$ (c) $\begin{pmatrix}A&B&C&D\\B&C&D&A\end{pmatrix}$.

Problem 1.4 The set of permutations of degree 3 is given in (1.4)

(a) Multiply each permutation by (1,2) from the left, and reduce it to a single permutation. Confirm that the set of all permutations so obtained is the same as the original set.

(b) The property (a) can be verified no matter what permutation is used to construct the new set. Explain the reason for this.

9.2 Odd and Even Permutations

As we can see from the two alternative expressions (1.1) and (1.2) for the same operation, permutations and interchanges are related. The relation may be summarized by the following two theorems.

Theorem 1. Any permutation can be equivalently expressed as a product of interchanges.

For example,

$$\begin{pmatrix} 1 & 2 & 3 \\ 2 & 1 & 3 \end{pmatrix} = (2,1) \ ,$$

$$\begin{pmatrix} 1 & 2 & 3 \\ 2 & 3 & 1 \end{pmatrix} = (2,3)(3,1) = (1,3)(1,2) = (1,2)(2,3) \ .$$

This theorem is intuitively obvious. As we see in the second example, there generally exist a number of equivalent expressions for a given permutation. However, these expressions are not arbitrary, but are subject to the following important restriction:

If a permutation P is expressed in terms of an odd (even) number of interchanges, all of the equivalent decompositions of P have odd (even) numbers of interchanges.

We may rephrase this property as

Theorem 2. <u>Any permutation may be classified as an odd or even permutation according to whether it can be built up from odd or even numbers of interchanges.</u>

We will demonstrate this as follows.

Let us call a permutation of the form $\begin{pmatrix} i_1 & i_2 & \cdots & i_n \\ i_2 & i_3 & \cdots & i_1 \end{pmatrix}$ which transforms $i_1 \to i_2$, $i_2 \to i_3$, ..., $i_n \to i_1$ a <u>cycle</u>, and denote it by $(i_1, i_2, \ldots, i_n)$, that is,

$$(i_1, i_2, \ldots, i_n) \equiv \begin{pmatrix} i_1 & i_2 & \cdots & i_n \\ i_2 & i_3 & \cdots & i_1 \end{pmatrix} . \qquad (2.1)$$

The number n is called the <u>length</u> of the cycle. Such a cycle can be decomposed into a product of interchanges (cycles of length 2). For example,

$$(i_1, i_2, \ldots, i_n) = (i_1, i_2)(i_2, i_3) \cdots (i_{n-1}, i_n) , \qquad (2.2)$$

which shows that a cycle of length n can be decomposed into a product of n-1 interchanges. Therefore, a cycle is even or odd according to whether its length n is odd or even. Any permutation can be expressed as a product of cycles whose arguments are mutually exclusive. For example,

$$\begin{pmatrix} 1 & 2 & 3 & 4 & 5 \\ 5 & 3 & 2 & 1 & 4 \end{pmatrix} = \begin{pmatrix} 1 & 5 & 4 \\ 5 & 4 & 1 \end{pmatrix} \begin{pmatrix} 2 & 3 \\ 3 & 2 \end{pmatrix} = (1,5,4)(2,3) .$$

The parity of the permutation can now be determined by finding the parity of each cycle and applying the rules:

$$\begin{aligned} (\text{even})(\text{even}) &= (\text{odd})(\text{odd}) = \text{even} \\ (\text{odd})(\text{even}) &= \text{odd} , \end{aligned} \qquad (2.3)$$

which follows simply from the definition of parity. In the above example, the parity of (1,5,4) is even and the parity of (2,3) is odd. Therefore the parity of $\begin{pmatrix} 1 & 2 & 3 & 4 & 5 \\ 5 & 3 & 2 & 1 & 4 \end{pmatrix}$ is odd according to the rule: (even)(odd) = odd. Clearly, the parity of any permutation determined in this manner is unique. Q.E.D.

[It is noted that the rules (2.3) are of the same form as those referring to the parities of the single-variable functions as described in §2.5. See (2.5.6).]

The sign δ_P of the parity of a permutation P is defined by

$$\delta_P = \begin{cases} 1 & \text{if } P \text{ is even} \\ -1 & \text{if } P \text{ is odd}. \end{cases} \tag{2.4}$$

From this definition, we obtain

$$\delta_P^2 = (\pm 1)^2 = 1 . \tag{2.5}$$

We can further establish that

$$\delta_{P^{-1}} = \delta_P \tag{2.6}$$

$$\delta_{PQ} = \delta_P \delta_Q . \tag{2.7}$$

The last equation may be proved as follows. If P and Q are both odd, then $\delta_P \delta_Q = (-1)(-1) = 1$. By definition, P and Q can be expressed in terms of odd numbers of interchanges. Using these expressions for P and Q, we can express the product PQ in terms of interchanges. The number of interchanges here is even since the sum of two odd numbers is even. Hence $\delta_{PQ} = 1$. We can repeat similar arguments for each of the other possibilities.

Let us now consider a function of several variables, $f(x_1,x_2,\ldots,x_n)$. If this function satisfies

$$P\, f(x_1,x_2,\ldots,x_n) = f(x_1,x_2,\ldots,x_n) \quad \text{for all } P, \tag{2.8}$$

where P's are permutations of n indices, then it is called a symmetric function. For example, $x_1 + x_2 + x_3$, $x_1^2 + x_2^2 + x_3^2$, and $x_1x_2x_3$ are all symmetric functions. If a function $f(x_1,x_2,\ldots,x_n)$ satisfies

$$P\, f(x_1,x_2,\ldots,x_n) \;=\; \delta_P\, f(x_1,x_2,\ldots,x_n) \quad \text{for all } P, \tag{2.9}$$

then f is called an <u>antisymmetric</u> function. For example, $(x_1-x_2)(x_1-x_3)(x_2-x_3)$ is antisymmetric. Note that some functions may satisfy neither (2.8) nor (2.9). (See Problem 2.4.)

Consider now two functions $f(x_1,x_2)$ and $g(x_1,x_2)$. Applying the permutation $\begin{pmatrix}1 & 2\\ 2 & 1\end{pmatrix}$ on the product $f(x_1,x_2)\; g(x_1,x_2)$, we obtain

$$\begin{pmatrix}1 & 2\\ 2 & 1\end{pmatrix}\{f(x_1,x_2)\; g(x_1,x_2)\} \;=\; f(x_2,x_1)\; g(x_2,x_1)\;. \tag{2.10}$$

We may write the last member as

$$\left[\begin{pmatrix}1 & 2\\ 2 & 1\end{pmatrix} f(x_1,x_2)\right]\left[\begin{pmatrix}1 & 2\\ 2 & 1\end{pmatrix} g(x_1,x_2)\right] ,$$

where the angular brackets mean that the permutation should be completed within the brackets. We then obtain from (2.10)

$$\left[\begin{pmatrix}1 & 2\\ 2 & 1\end{pmatrix} f(x_1,x_2)\; g(x_1,x_2)\right] \;=\; \left[\begin{pmatrix}1 & 2\\ 2 & 1\end{pmatrix} f(x_1,x_2)\right]\left[\begin{pmatrix}1 & 2\\ 2 & 1\end{pmatrix} g(x_1,x_2)\right] .$$

This example indicates that, for a general P,

$$[P(fg)] \;=\; [P\, f][P\, g] \tag{2.11}$$

holds. The restriction implied by the angular brackets may be removed by the following device.

With the understanding that permutation operations always act toward the right, we may write (2.11) as

$$P(fg) \;=\; [P\, f]\; P\, g \tag{2.12}$$

But

$$fg \;=\; f\, P^{-1}(P\, g) \;\equiv\; f\, P^{-1} P g\;. \qquad \text{[use of (1.12)]} \tag{2.13}$$

Multiplying this equation by P from the left , we obtain

$$P\ f\ P^{-1} P\ g \;=\; P(fg) \qquad \text{[use of (2.13)]}$$

$$= [P\ f]\ P\ g\ . \qquad \text{[use of (2.12)]}$$

Comparison between the first and last lines shows that

$$\boxed{P\ f\ P^{-1} \;=\; [P\ f]\ .} \tag{2.14}$$

Using this result, we may re-express eq.(2.11) as

$$\boxed{P(fg)P^{-1} \;=\; (P\ f\ P^{-1})(P\ g\ P^{-1})\ .} \tag{2.15}$$

We note that the term on the r.h.s. can be obtained from $P(fg)P^{-1}$ by inserting the identity $P^{-1}P$ between f and g, and re-associating the resulting factors:

$$P(fg)P^{-1} \;=\; P\ f(P^{-1}P)g\ P^{-1} \;=\; (P\ f\ P^{-1})(P\ g\ P^{-1})\ .$$

Eq.(2.15) suggests that the symmetry property of the product fg can be determined by studying each factor separately, that is, by calculating $P\ f\ P^{-1}$ and $P\ g\ P^{-1}$. Thus, if a quantity to be studied appears as a factor, it is convenient to define the permutation symmetry in the following form:

$$P\ f(x_1,x_2,\ldots,x_n)P^{-1} \;=\; f(x_1,x_2,\ldots,x_n) \quad \text{(symmetric function)} \tag{2.16}$$

$$P\ f(x_1,x_2,\ldots,x_n)P^{-1} \;=\; \delta_P\ f(x_1,x_2,\ldots,x_n) \quad \text{(antisymmetric function).} \tag{2.17}$$

Multiplying eq.(2.16) by P from the right, we obtain

$$P\ f \;=\; f\ P \quad \text{or} \quad [P,f] \;\equiv\; P\ f - f\ P \;=\; 0\ . \quad \text{(symmetric function)} \tag{2.18}$$

This form can also be used for the definition of a symmetric function.

Problem 2.1 Decompose the following permutations into products of interchanges:

(a) $\begin{pmatrix} 1 & 2 & 3 & 4 \\ 2 & 3 & 1 & 4 \end{pmatrix}$ (b) $\begin{pmatrix} 1 & 2 & 3 & 4 & 5 \\ 3 & 1 & 2 & 5 & 4 \end{pmatrix}$.

Problem 2.2 Find the parities of the following permutations:

(a) $\begin{pmatrix} 1 & 2 & 3 & 4 \\ 2 & 3 & 1 & 4 \end{pmatrix}$ (b) $\begin{pmatrix} 1 & 2 & 3 & 4 & 5 \\ 3 & 1 & 2 & 5 & 4 \end{pmatrix}$

(c) $\begin{pmatrix} 1 & 2 & 3 & 4 & 5 \\ 5 & 4 & 3 & 2 & 1 \end{pmatrix}$ (d) $\begin{pmatrix} a & b & c & d & e & f \\ b & c & a & d & f & e \end{pmatrix}$.

Problem 2.3 For any permutation group, the number of even permutations is equal to the number of odd permutations.

(a) Confirm this statement for $N = 2$ and $N = 3$.

(b) Prove it for a general N.

Problem 2.4 Let $f(x_1,x_2)$ be an arbitrary function of x_1 and x_2.

(a) Show that $f(x_1,x_2) + f(x_2,x_1)$ and $f(x_1,x_2) - f(x_2,x_1)$ are, respectively, symmetric and antisymmetric.

(b) Using the results from (a), express f as the sum of symmetric and antisymmetric functions.

9.3 Indistinguishable Classical Particles

When a system is composed of a number of particles of the same kind, these particles cannot be distinguished from each other. If numbered particle-coordinates are used in a theoretical formulation, a prediction from the theory must be independent of the numbering of the particles. For example, the N-body distribution function for the system must be invariant under permutations of the particle indices. Such a concept of indistinguishability can be applied to both classical and quantum mechanical systems. In this section we will consider classical systems only.

9.14

Let us take a system of three identical particles moving in one dimension. The dynamical state of the system can be represented by three points in the μ-space as indicated in Figure 9.1. Let us denote the location of these points by $(x_1, p_1, x_2, p_2, x_3, p_3)$.

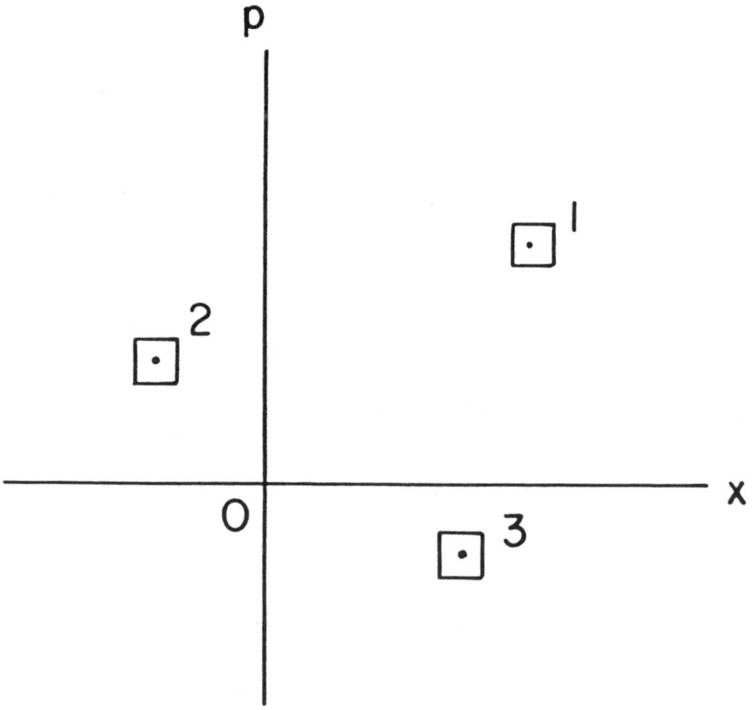

Fig.9.1. The dynamical state for three identical classical particles can be represented by three particle-points in the μ-space.

Consider an infinitesimal phase area $dx_j\ dp_j$ surrounding each point (x_j, p_j). The probability of finding the system in a state in which three particle-points are within the element $dx_1 dp_1\ dx_2 dp_2\ dx_3 dp_3$ can be written in the form

$$\rho(x_1,p_1,x_2,p_2,x_3,p_3)dx_1dp_1\ dx_2dp_2\ dx_3dp_3\,(2\pi\hbar)^{-3}(3!)^{-1}$$

$$\equiv\ \rho(1,2,3)dx_1dp_1\ dx_2dp_2\ dx_3dp_3\ (2\pi\hbar)^{-3}\ (3!)^{-1}, \tag{3.1}$$

where ρ represents the distribution function for the system. For identical particles, we require that the distribution function $\rho(1,2,3)$ be symmetric, that is,

$$\boxed{P\ \rho(1,2,3)\ =\ \rho(1,2,3)\quad \text{for all permutations } P\ .} \tag{3.2}$$

Explicitly, this means that

$$\rho(1,2,3) = \rho(1,3,2) = \rho(3,2,1) = \rho(2,1,3) = \rho(3,1,2) = \rho(2,3,1)\ . \tag{3.2a}$$

In general, the distribution function ρ changes with time. However, <u>the symmetry of the distribution function is maintained if the Hamiltonian H of the system is symmetric</u>:

$$\boxed{PH\ =\ HP\quad \text{for all } P\ .} \tag{3.3}$$

This may be shown as follows. The function ρ obeys the Liouville equation

$$\frac{\partial\rho}{\partial t}\ =\ \{H,\rho\}\ \equiv\ \sum_{k=1}^{3}\left(\frac{\partial H}{\partial x_k}\frac{\partial\rho}{\partial p_k}-\frac{\partial\rho}{\partial x_k}\frac{\partial H}{\partial p_k}\right)\ . \tag{3.4}$$

Let us write an arbitrary permutation P in the form

$$P\ \equiv\ \begin{pmatrix}1 & 2 & 3\\ i_1 & i_2 & i_3\end{pmatrix}, \tag{3.5}$$

where (i_1,i_2,i_3) are numerals taken from (1,2,3). By definition,

$$\frac{\partial H}{\partial x_1}\ \equiv\ \lim_{\Delta x\to 0}\frac{H(x_1+\Delta x,p_1,x_2,p_2,x_3,p_3)\ -\ H(x_1,p_1,x_2,p_2,x_3,p_3)}{\Delta x}\ .$$

Applying P from the left, we obtain

$$P \frac{\partial H}{\partial x_1} = \lim_{\Delta x \to 0} \frac{1}{\Delta x} \Big[H(x_{i_1}+\Delta x, p_{i_1}, x_{i_2}, p_{i_2}, x_{i_3}, p_{i_3})$$

$$- H(x_{i_1}, p_{i_1}, x_{i_2}, p_{i_2}, x_{i_3}, p_{i_3}) \Big] P \ .$$

Since H is symmetric, we can re-order the arguments (i_1, i_2, i_3) to the natural order (1,2,3). Subsequently taking the limit $\Delta x \to 0$, we obtain

$$P \frac{\partial H}{\partial x_1} = \frac{\partial}{\partial x_{i_1}} H(x_1, p_1, x_2, p_2, x_3, p_3) \equiv \frac{\partial H}{\partial x_{i_1}} P \tag{3.6}$$

or

$$P \frac{\partial H}{\partial x_1} P^{-1} = \frac{\partial H}{\partial x_{i_1}} \ .$$

By similar calculations, we obtain

$$P \frac{\partial H}{\partial x_k} P^{-1} = \frac{\partial H}{\partial x_{i_k}} \ , \quad P \frac{\partial H}{\partial p_k} P^{-1} = \frac{\partial H}{\partial p_{i_k}} \ . \tag{3.7}$$

In the same manner we can also show that

$$P \frac{\partial \rho}{\partial x_k} P^{-1} = \frac{\partial \rho}{\partial x_{i_k}} \ , \quad P \frac{\partial \rho}{\partial p_k} P^{-1} = \frac{\partial \rho}{\partial p_{i_k}} \ . \tag{3.8}$$

Using (3.7) and (3.8), we obtain

$$P \sum_{\{k\}} \left[\frac{\partial H}{\partial x_k} \frac{\partial \rho}{\partial p_k} - \frac{\partial \rho}{\partial x_k} \frac{\partial H}{\partial p_k} \right] P^{-1} = \sum_{\{i_k\}} \left[\frac{\partial H}{\partial x_{i_k}} \frac{\partial \rho}{\partial p_{i_k}} - \frac{\partial \rho}{\partial x_{i_k}} \frac{\partial H}{\partial p_{i_k}} \right]$$

The summation over the set $\{i_k\}$ is equivalent to the summation over the set $\{k\}$. Therefore, it follows that

$$P \sum_{\{k\}} \left[\frac{\partial H}{\partial x_k} \frac{\partial \rho}{\partial p_k} - \frac{\partial \rho}{\partial x_k} \frac{\partial H}{\partial p_k} \right] P^{-1} = \sum_{\{k\}} \left[\frac{\partial H}{\partial x_k} \frac{\partial \rho}{\partial p_k} - \frac{\partial \rho}{\partial x_k} \frac{\partial H}{\partial p_k} \right]$$

or

$$\boxed{P\{H,\rho\}P^{-1} = \{H,\rho\} .} \tag{3.9}$$

Using this result, we obtain from (3.4)

$$P \frac{\partial\rho}{\partial t} P^{-1} = \frac{\partial\rho}{\partial t} . \tag{3.10}$$

By assumption, ρ is symmetric at the initial time, and we showed in eq.(3.10) that $\partial\rho/\partial t$ is symmetric. Therefore, by mathematical induction for a continuous variable [see p.82] , $\rho(t)$ is symmetric for all time.

Dynamical functions such as the total momentum depend on the particle-variables:

$$\xi = \xi(x_1,p_1,x_2,p_2,x_3,p_3) \equiv \xi(1,2,3) \tag{3.11}$$

For identical particles, we require that ξ be symmetric,

$$\boxed{P\xi(1,2,3) = \xi P \quad \text{for all } P .} \tag{3.12}$$

This equation and eq.(3.2) will insure that the average of ξ,

$$\langle\xi\rangle = \mathrm{Tr}\{\xi\rho\} , \tag{3.13}$$

is independent of the numbering of the particles.

In summary, a system of identical particles can be properly treated if (a) the distribution function is symmetric [eq.(3.2)] and (b) all dynamical functions including the Hamiltonian are symmetric [eqs.(3.12) and (3.3)]. It is clear that we can extend our discussions to any number of particles.

The significance of the symmetry requirements may be clearly understood by considering distinguishable particles. Let us take a hydrogen atom,which consists of an electron with mass m_e, and a proton with mass m_p. The Hamiltonian H for the system is

$$H(\vec{r}_1,\vec{p}_1,\vec{r}_2,\vec{p}_2) \equiv H(1,2) \equiv \frac{1}{2m_e}p_1^{\,2} + \frac{1}{2m_p}p_2^{\,2} + V(|\vec{r}_1 - \vec{r}_2|). \qquad (3.13)$$

Let us apply the interchange operator (1,2) and obtain

$$\begin{aligned}(1,2)H(1,2)\left[(1,2)\right]^{-1} &= H(2,1) \\ &= \frac{1}{2m_e}p_2^{\,2} + \frac{1}{2m_p}p_1^{\,2} + V(|\vec{r}_2-\vec{r}_1|) \\ &\neq H(1,2)\ ,\end{aligned} \qquad (3.14)$$

which shows that this Hamiltonian H is not symmetric.

This example shows that if the particles have different masses, and are therefore distinguishable, then the Hamiltonian H is not symmetric. This is no surprise. However, the important point is that the Hamiltonian H can and must contain the built-in permutation symmetry appropriate for the system.

9.4 Quantum Statistical Postulate. Symmetric States for Bosons

The symmetry requirements for a classical system of identical particles, given in the preceding section, can simply be extended to quantum mechanical many-particle systems.

Let us consider a quantum system of N identical particles. We require that the Hamiltonian of the system be a symmetric function of the particle variables, $\zeta_j \equiv (\vec{r}_j,\vec{p}_j)$:

$$P\,H(\zeta_1,\zeta_2,\ldots,\zeta_N) = H(\zeta_1,\zeta_2,\ldots,\zeta_N)\,P$$

or

$$\boxed{P\,H = H\,P \quad \text{for all } P\ .} \qquad (4.1)$$

Since the quantum Hamiltonian is a linear operator, and appears as a factor in the definition equation, the symmetry requirement must necessarily be expressed in the form (4.1) instead of $P\,H = H$.

We further require that all dynamical functions ξ be symmetric:

$$\boxed{P\,\xi = \xi\,P\,.} \tag{4.2}$$

Before discussing the symmetry requirement for the quantum distribution function, we introduce a new postulate concerning quantum mechanical many-body systems.

Every particle in quantum physics is either a boson or fermion. This is known as the quantum statistical postulate. This postulate is distinct from the five quantum mechanical postulates stated in §8.13, and should be treated as an independent postulate. As we will see later, a system of identical bosons or fermions behaves quite differently from the corresponding classical system.

In the remainder of this section we will discuss the quantum states for identical bosons. The quantum states for fermions will be discussed in the next section. For a quantum particle moving in one dimension, the eigenvalues for the momentum are

$$p_k = \frac{2\pi\hbar}{L}\,k\,, \quad k = \ldots,-1,0,1,\ldots \tag{4.3}$$

The corresponding normalized eigenkets will be denoted by $\{|p_k\rangle\}$, which satisfy the orthonormality condition: $\langle p_k|p_\ell\rangle = \delta_{k,\ell}$.

When a system contains two different particles, say A and B, we may construct a ket for the system from the direct product of the single-particle kets for A and B. Such a ket

$$|p_k^{(A)}\rangle|p_\ell^{(B)}\rangle \equiv |p_\ell^{(B)}\rangle|p_k^{(A)}\rangle \equiv |p_k^{(A)}\,p_\ell^{(B)}\rangle \tag{4.4}$$

will represent a state in which the particles A and B occupy the momentum states p_k and p_ℓ, respectively. The ket $|p_k^{(A)} p_\ell^{(B)}>$ is normalized to unity since

$$
\begin{aligned}
<p_k^{(A)} p_\ell^{(B)} | p_k^{(A)} p_\ell^{(B)}> &\equiv \left[<p_\ell^{(B)} | <p_k^{(A)} | \right] \left[| p_k^{(A)} > | p_\ell^{(B)} > \right] \\
&= <p_\ell^{(B)} | \left[<p_k^{(A)} | p_k^{(A)} > \right] | p_\ell^{(B)} > \\
&= <p_\ell^{(B)} | p_\ell^{(B)} > \quad \text{[normalization]} \\
&= 1 \, . \qquad (4.5)
\end{aligned}
$$

We now consider a system of two identical particles. Let us construct a new ket

$$
\frac{1}{\sqrt{2}} \{ |p_k^{(A)}>|p_\ell^{(B)}> + |p_k^{(B)}>|p_\ell^{(A)}> \} \equiv |p_k p_\ell>_S \, , \qquad (4.6)
$$

which is obtained by symmetrizing the ket (4.4). The new ket, by postulate, represents the state for the system in which two states p_k and p_ℓ are occupied by two particles with no further specifications. The factor $(2)^{-1/2}$ was given to facilitate the normalization. See below.

Let us now extend our theory to the case of N bosons moving in three dimensions. Let $|\alpha_a^{(j)}>|\alpha_b^{(j)}>, \ldots$ be the kets for the j-th particle occupying single-particle states $\alpha_a, \alpha_b, \ldots$ We construct a ket for N particles by taking the product of kets for each particle:

$$
|\alpha_a^{(1)}>|\alpha_b^{(2)}> \ldots |\alpha_g^{(N)}> \, . \qquad (4.7)
$$

Multiplying this by the <u>symmetrizing operator</u>

$$
S \equiv (N!)^{-\frac{1}{2}} \sum_P P \, , \qquad (4.8)
$$

where the summation is over all permutations, we obtain a ket for the system,

$$S\left[|\alpha_a{}^{(1)}>|\alpha_b{}^{(2)}>\ldots|\alpha_g{}^{(N)}>\right] \equiv |\alpha_a\alpha_b\ldots\alpha_g>_S , \qquad (4.9)$$

where the labels of particles are omitted on the r.h.s. since they are no longer relevant. The ket in (4.9) is obviously symmetric:

$$P\, S\left[|\alpha_a{}^{(1)}>|\alpha_b{}^{(2)}>\ldots|\alpha_g{}^{(N)}>\right] = S\left[|\alpha_a{}^{(1)}>|\alpha_b{}^{(2)}>\ldots|\alpha_g{}^{(N)}>\right] . \quad \text{(for any P)} \qquad (4.10)$$

The symmetric ket (4.9), by postulate, represents the state of the system of N bosons in which the single-particle states $\alpha_a, \alpha_b, \ldots, \alpha_g$ are occupied.

Problem 4.1

(a) Construct a symmetric ket for three bosons by means of (4.9).

(b) Verify explicitly that the obtained ket is symmetric under the permutation $\begin{pmatrix} 1 & 2 & 3 \\ 2 & 3 & 1 \end{pmatrix}$.

(c) Assuming that the occupied states, say $\alpha_a \equiv a,b,c,$ are all different, find the magnitude of the ket.

(d) Assuming that a = b = c, find the magnitude of the ket.

9.5 Antisymmetric States for Fermions. Pauli's Exclusion Principle

Let us define the antisymmetrizing operator A by

$$A \equiv (N!)^{-\frac{1}{2}} \sum_P \delta_P P , \qquad (5.1)$$

where δ_P is the sign of parity of the permutation P defined by (2.4):

$$\delta_P = \begin{cases} 1 & \text{if P is even} \\ -1 & \text{if P is odd .} \end{cases} \qquad (5.2)$$

Applying the operator A on the ket in (4.7), we obtain a new ket

$$A(|\alpha_a{}^{(1)}>|\alpha_b{}^{(2)}>\ldots|\alpha_g{}^{(N)}>) \equiv |\alpha_a\alpha_b\ldots\alpha_g>_A , \qquad (5.3)$$

where the subscript A denotes the antisymmetric ket.

The fact that this ket is antisymmetric can be shown in the following manner. Multiply (5.3) by an arbitrary permutation P and obtain

$$P\,A(\alpha_a{}^{(1)}>|\alpha_b{}^{(2)}>\dots|\alpha_g{}^{(N)}>)$$

$$= (N!)^{-\frac{1}{2}} \sum_Q \delta_Q\, P\,Q(|\alpha_a{}^{(1)}>|\alpha_b{}^{(2)}>\dots|\alpha_g{}^{(N)}>) \ , \qquad (5.4)$$

where we denoted the summation variable by Q. According to the composition property (1.7), the product PQ is another element, say R, of the permutation group:

$$PQ \equiv R \ . \qquad (5.5)$$

Multiplying this equation from the left by P^{-1}, we have

$$P^{-1}\,PQ = P^{-1}\,R$$

or

$$Q = P^{-1}\,R \ . \qquad \text{[use of (1.12)]} \qquad (5.6)$$

We now express δ_Q as follows:

$$\begin{aligned} \delta_Q &= \delta_{P^{-1}R} && \text{[use of (5.6)]} \\ &= \delta_{P^{-1}}\delta_R && \text{[use of (2.7)]} \\ &= \delta_P\delta_R \ . && \text{[use of (2.6)]} \end{aligned} \qquad (5.7)$$

Using this result and (5.5), we can rewrite the r.h.s. of eq.(5.4) as

$$(N!)^{-\frac{1}{2}} \sum_Q \delta_Q\, P\,Q(|\alpha_a{}^{(1)}>|\alpha_b{}^{(2)}>\dots|\alpha_g{}^{(N)}>)$$

$$= (N!)^{-\frac{1}{2}} \sum_R \delta_P\delta_R\, R(|\alpha_a{}^{(1)}>|\alpha_b{}^{(2)}>\dots|\alpha_g{}^{(N)}>)$$

$$= \delta_P\left[(N!)^{-\frac{1}{2}} \sum_R \delta_R\, R(|\alpha_a{}^{(1)}>|\alpha_b{}^{(2)}>\dots|\alpha_g{}^{(N)}>)\right]$$

$$= \delta_P \, A(|\alpha_a{}^{(1)}\rangle|\alpha_b{}^{(2)}\rangle \ldots |\alpha_g{}^{(N)}\rangle) \,, \tag{5.8}$$

which establishes the desired antisymmetric property:

$$P[A(|\alpha_a{}^{(1)}\rangle|\alpha_b{}^{(2)}\rangle \ldots |\alpha_g{}^{(N)}\rangle]$$

$$= \delta_P \, [A(|\alpha_a{}^{(1)}\rangle|\alpha_b{}^{(2)}\rangle \ldots |\alpha_g{}^{(N)}\rangle] \tag{5.9}$$

or

$$\boxed{P|\alpha_a\alpha_b \ldots \alpha_g\rangle_A = \delta_P|\alpha_a\alpha_b \ldots \alpha_g\rangle_A \,.} \tag{5.9a}$$

We postulate that <u>the antisymmetric ket in (5.3) represents the quantum state of N fermions in which particle-states a,b,...,g are occupied</u>. For illustration, let us take the case of two fermions moving in one dimension. Corresponding to the state in which the momentum states p_k and p_ℓ are occupied, we construct the ket

$$|p_k \; p_\ell\rangle_A \equiv \frac{1}{\sqrt{2}} \sum_P \delta_P \, P[|p_k{}^{(1)}\rangle|p_\ell{}^{(2)}\rangle]$$

$$= \frac{1}{\sqrt{2}} \{|p_k{}^{(1)}\rangle|p_\ell{}^{(2)}\rangle - |p_k{}^{(2)}\rangle|p_\ell{}^{(1)}\rangle\} \,. \tag{5.10}$$

This ket is clearly antisymmetric:

$$|p_k \; p_\ell\rangle_A = -|p_\ell \; p_k\rangle_A \,. \tag{5.11}$$

If $p_k = p_\ell$, we obtain

$$|p_k \; p_k\rangle_A = 0 \,. \tag{5.12}$$

By postulate, the null ket does not correspond to any quantum state. Thus, the state for the system in which two fermions occupy the same state p_k cannot qualify as a quantum state. In other words, <u>two fermions must not occupy the</u>

same state. This is known as Pauli's exclusion principle.

Let us shed more light on this point by treating the general case of N fermions.

The antisymmetric state in (5.3) can be written in the determinant form:

$$|\alpha_a\alpha_b\cdots\alpha_g\rangle_A = \frac{1}{(N!)^{\frac{1}{2}}}\begin{vmatrix} |\alpha_a{}^{(1)}\rangle & |\alpha_a{}^{(2)}\rangle \cdots & |\alpha_a{}^{(N)}\rangle \\ |\alpha_b{}^{(1)}\rangle & |\alpha_b{}^{(2)}\rangle \cdots & |\alpha_b{}^{(N)}\rangle \\ \cdots & \cdots & \cdots \\ |\alpha_g{}^{(1)}\rangle & |\alpha_g{}^{(2)}\rangle \cdots & |\alpha_g{}^{(N)}\rangle \end{vmatrix} . \quad (5.13)$$

It represents the state for the system in which single-particle states $a \equiv \alpha_a, \alpha_b, \ldots, \alpha_g$ are occupied. If two of the particle-states are identical, the ket (5.13) vanishes since two rows in the determinant become identical.

Pauli's principle applies to any single-particle state. For example, except for the spin degeneracy (see §12.4), two fermions can occupy neither the same momentum state nor the same space point. Historically, Pauli's exclusion principle was discovered from the study of atomic spectra. The original statement referred to atomic orbitals: No two electrons can occupy the same orbital with identical quantum numbers.

Problem 5.1

(a) Construct an antisymmetric ket for three fermions by means of (5.2) or (5.13).

(b) Multiply the result from the left by (1,2) and verify that the ket is indeed antisymmetric.

(c) Show that the ket is normalized to unity.

9.6 More about Bosons and Fermions. Quantum Statistics and Spin

In quantum mechanics, theory and experiment are compared only through the expectation values for observable dynamical functions. For a system of identical particles these values must then be independent of the numbering of the particles. We will study this theoretical requirement in the present section. We will also point out a close connection between quantum statistics and spin angular momentum.

Let us first consider the case of identical fermions. We saw in the last section that a quantum state for a system of fermions, by postulate, is represented by the antisymmetric ket $|\alpha_a\alpha_b\ldots\alpha_g>_A$, or equivalently by the antisymmetric bra $_A<\alpha_a\alpha_b\ldots\alpha_g|$. We also saw in §9.3 that any dynamical function ξ for a system of identical particles must be symmetric [see (3.12)]. According to the general prescription given in §8.12, the expectation value of the observable ξ is given by

$$_A<\alpha_a\alpha_b\ldots\alpha_g|\xi|\alpha_a\alpha_b\ldots\alpha_g>_A \,. \tag{6.1}$$

We now wish to examine the permutation symmetry of this quantity. By multiplying (6.1) by an arbitrary permutation P from the left and by the inverse P^{-1} from the right, we have $P[_A<\alpha_a\alpha_b\ldots\alpha_g|\xi|\alpha_a\alpha_b\ldots\alpha_g>_A]P^{-1}$, which can be calculated as follows:

$$P[_A<\alpha_a\alpha_b\ldots\alpha_g|\xi|\alpha_a\alpha_b\ldots\alpha_g>_A]P^{-1}$$

$$= (P\ _A<\alpha_a\alpha_b\ldots\alpha_g|P^{-1})(P\xi P^{-1})(P|\alpha_a\alpha_b\ldots\alpha_g>_A\ P^{-1}) \qquad \text{[use of (2.15)]}$$

$$= (\delta_P\ _A<\alpha_a\alpha_b\ldots\alpha_g|)(\xi)(\delta_P|\alpha_a\alpha_b\ldots\alpha_g>_A) \qquad \text{[use of (4.2) and (5.9a)]}$$

$$= \delta_P^{\,2}\ _A<\alpha_a\alpha_b\ldots\alpha_g|\xi|\alpha_a\alpha_b\ldots\alpha_g>_A$$

$$= {}_A<\alpha_a\alpha_b\ldots\alpha_g|\xi|\alpha_a\alpha_b\ldots\alpha_g>_A\,, \quad \text{for any P .} \qquad [\delta_P = \pm 1] \tag{6.2}$$

We thus find that expression (6.1) is <u>invariant under the permutations</u> of the particle indices.

Next, we take up the case of identical bosons. The expectation value of ξ for this system is given by

$$_S\langle\alpha_a\alpha_b\cdots\alpha_g|\xi|\alpha_a\alpha_b\cdots\alpha_g\rangle_S \,, \tag{6.3}$$

which is clearly symmetric:

$$P\left[_S\langle\alpha_a\alpha_b\cdots\alpha_g|\xi|\alpha_a\alpha_b\cdots\alpha_g\rangle_S\right]P^{-1} = {_S\langle\alpha_a\alpha_b\cdots\alpha_g|\xi|\alpha_a\alpha_b\cdots\alpha_g\rangle_S} \quad \text{for any } P \,. \tag{6.4}$$

In summary, symmetric and antisymmetric vectors, assumed for bosons and fermions respectively, satisfy the necessary invariance requirement. In principle, other vectors which satisfy the invariance requirement can be constructed, but these will be irrelevant because of the quantum statistical postulate stated earlier: all quantum particles are either bosons or fermions.

The invariance properties, (6.4) and (6.2), once assumed at an initial time, will be maintained indefinitely. This can be demonstrated as follows.

If the ket $|\psi,t\rangle_S$ is symmetric at a time t, the ket $H|\psi,t\rangle_S$, where H represents the symmetric Hamiltonian, is also symmetric. In the Schrödinger picture the ket $|\psi,t\rangle_S$ varies with time according to the Schrödinger equation of motion:

$$i\hbar(\partial/\partial t)|\psi,t\rangle_S = H|\psi,t\rangle_S \,.$$

Therefore, $(\partial/\partial t)|\psi,t\rangle$ is also symmetric. By mathematical induction for a continuous variable (t), the symmetry of the ket $|\psi,t\rangle_S$ is maintained for all time t. In a similar manner, if the ket $|\psi,t\rangle_A$ is initially antisymmetric, then $H|\psi,t\rangle_A$ is antisymmetric, meaning that the ket at a later time must still be antisymmetric. Thus a state which is initially symmetric (antisymmetric) will always remain symmetric (antisymmetric).

Note that the properties (6.2) and (6.4) are maintained separately for fermions and bosons. Therefore the distinction between bosons and fermions is permanent.

As we saw in the last section, fermions are subject to Pauli's exclusion principle; the number of fermions occupying any particle-state is limited to one or zero. This is an important characteristic of fermions. It leads to a special statistics called Fermi-Dirac statistics. Fermi (Enrico Fermi, 1901-54) first studied the statistics and Dirac (Paul Adrien Maurice Dirac, 1902-) established the fact that the particles subject to Pauli's principle are precisely those for which the antisymmetric states must be prescribed.

For bosons, no restriction on the occupation number is imposed. Thus, this number can be zero, one, two, or any positive integer. The statistics of bosons are not the same as the Boltzmann statistics of the classical theory because the quantum states are different from the classical dynamical states. The quantum statistics for bosons were first studied by Bose, and further explored by Einstein; they are called the Bose-Einstein statistics.

The question of whether a quantum particle is a boson or fermion is known to be related to the magnitude of its spin angular momentum in units of $\hbar$ (see §11.3). Particles with integer spin are bosons, while those with half-integer spin are fermions. This is known as the spin-statistics theorem, and was first proved by Pauli [1] within the framework of relativistic quantum field theory. Since the proof involves quite advanced physical arguments, we will assume the theorem with no further discussions. According to this theorem (and in agreement with all experimental evidence), electrons, protons, neutrons, and μ-mesons, which all have spins of magnitude $\frac{1}{2}\hbar$, are examples of fermions, while photons (quanta of electromagnetic radiation), with spins of magnitude $\hbar$, are bosons.

9.7 The Occupation Number Representation

Let us consider a collection of N bosons moving in one dimension. The momentum eigenvalues are

$$p_k = \frac{2\pi\hbar}{L} k, \quad k = \ldots,-2,-1,0,1,2,\ldots \tag{7.1}$$

and the corresponding normalized eigen-kets are denoted by $|p_k\rangle$. Using these eigen-kets, we can construct a symmetric state for the system:

$$(N!)^{-\frac{1}{2}} \sum_P P[|p_a^{(1)}\rangle|p_b^{(2)}\rangle\ldots|p_g^{(N)}\rangle] = |p_a p_b \ldots p_g\rangle_S \,. \tag{7.2}$$

This ket corresponds to a quantum statistical state in which N bosons occupy N momenta $\{p_a p_b \ldots p_g\}$. We now look at the same state from a different angle. Let us imagine a large number of boxes assigned to each of the momentum states $\{p_k\}$, as indicated in Figure 9.2. A particle is placed in box p_a, another in box p_b, and so on until all N particles are distributed in the boxes. The distribution of N bosons in the boxes represents the quantum statistical state for the system. If we denote the number of bosons in the box p_k by n_k, the set of occupation numbers $\{n_k\}$ represents the distribution. This is called the <u>occupation number representation</u>. Each occupation number n_k can be any non-negative integer subject to the overall restriction

$$\sum_k n_k = N \,. \tag{7.3}$$

The set of momentum-states $\{p_k\}$ is unbounded (of infinite order), so that most of the n_k are zero. If $p_a, p_b, \ldots, p_g$ are all different (i.e., if each n_k is either 0 or 1), the ket $|p_a p_b \ldots p_g\rangle_S$ is normalized, since in this case each term in the summation in eq.(7.2) is orthogonal to the others and contributes

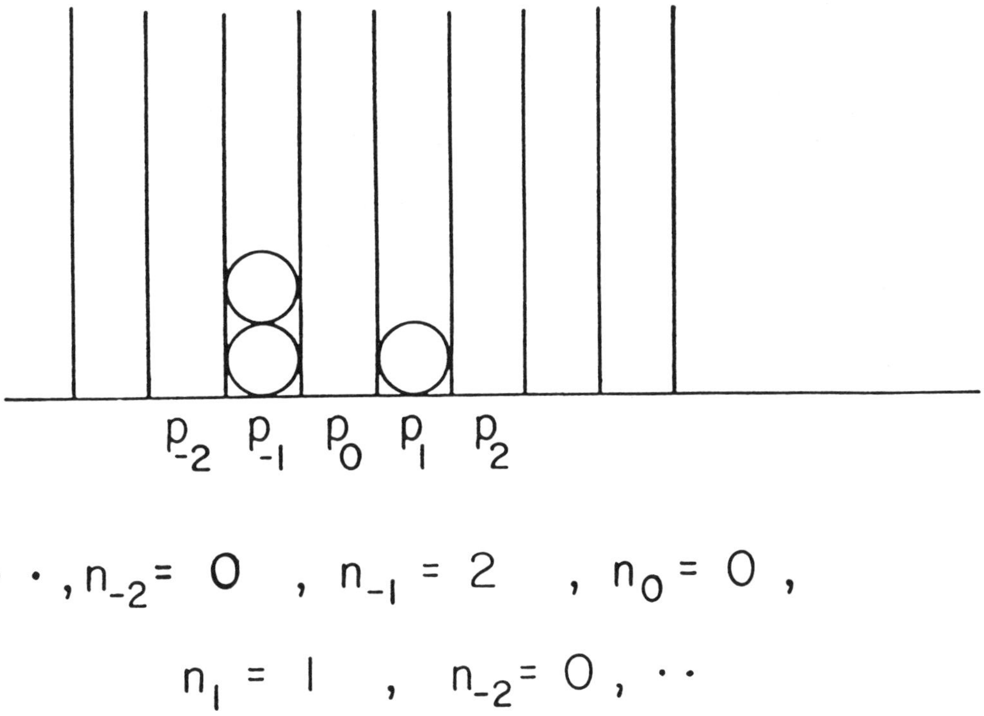

Fig. 9.2. The quantum state for bosons can be represented by the set of the numbers of bosons $\{n_k\}$ occupying the momentum states $p_k \equiv 2\pi\hbar k/L$.

$(N!)^{-1}$ to the squared length of the ket. However, if $p_a, p_b, \ldots, p_g$ are not all different, those terms generated by permutations which merely interchange bosons in the same momentum state will be equal. The number of equal terms will be $n_1!\; n_2! \; \ldots$, so that the squared length of the ket will be (see Problem 4.1)

$$_S\langle p_a p_b \cdots p_g | p_a p_b \cdots p_g \rangle_S = \ldots n_{-1}!\; n_0!\; n_1!\; n_2! \; \ldots \; . \tag{7.4}$$

We can therefore normalize the ket as follows :

$$(\ldots n_{-1}!\; n_o!\; n_1!\; n_2! \ldots)^{-\frac{1}{2}} \, |p_a p_b \cdots p_g\rangle_S \equiv |\{n\}\rangle \; . \tag{7.5}$$

A system of fermions can be treated in a similar manner. Because of the exclusion principle no two fermions can occupy the same momentum state. A quantum statistical state for the many fermion system can then be represented by a set of occupation numbers $\{n_k\}$, with the restriction that $n_k = 0$ or 1 for any k. A typical state is shown in Figure 9.3.

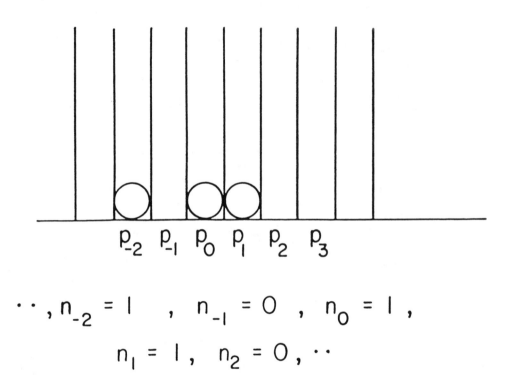

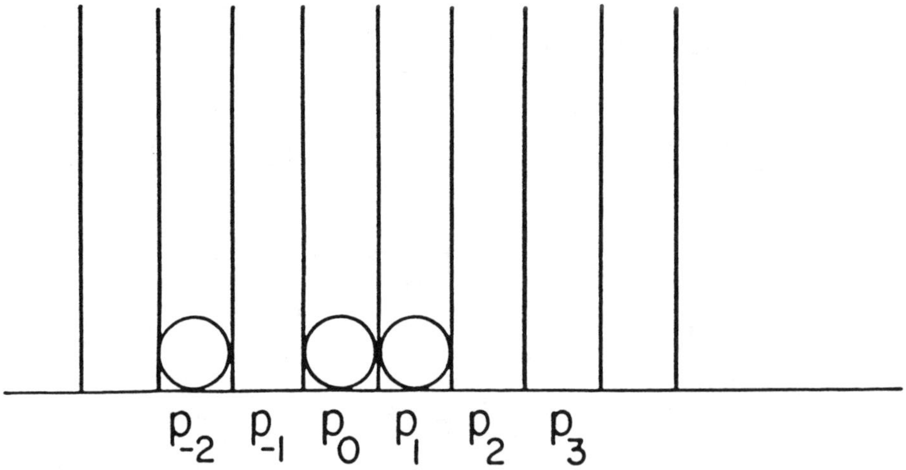

$$\cdot\cdot, n_{-2} = 1 \quad , \quad n_{-1} = 0 \quad , \quad n_0 = 1 \quad ,$$
$$n_1 = 1, \quad n_2 = 0, \cdot\cdot$$

Fig. 9.3. The quantum state for fermions can be represented by the set of occupation numbers $\{n_k\}$. Each occupation number is limited to zero or one.

In summary, we can represent quantum statistical states for bosons or fermions by the set of occupation numbers $\{n_k\}$. The important distinction is that the occupation numbers n_k for bosons can be any non-negative integers, while the numbers n_k for fermions are restricted to 1 or 0.

So far we have only taken up the case of one-dimensional momentum states. Clearly, we can extend the idea of the occupation number representation to <u>any</u> single-particle quantum state α_k.

We note that specification of the quantum statistical many-particle states by occupation numbers is a rather natural representation. For example, in the U.S. population census, we count the number of residents of each of the fifty states. The collection of head-counts, disregarding the individual, is reported as the population distribution over all of the states.

9.8 The Gibbs Ensemble of Many-Particle Systems. The Canonical Ensemble

The Gibbs ensemble of quantum mechanical single-particle systems was discussed earlier in §8.15. To extend the theory to a system of many particles, we require that the many-particle quantum state-vectors be symmetric for bosons and antisymmetric for fermions. Except for this requirement, all concepts and general formulations can be extended in a straightforward manner.

For completeness, we will enumerate important definitions and formulas below.

(a) Definition of the density operator

The Gibbs ensemble is characterized by the density operator (or quantum statistical operator) ρ, which is defined by

$$\rho \equiv \sum_m |m\rangle P_m \langle m| \; , \qquad (8.1)$$

where $|m\rangle$ and $\langle m|$ are normalized ket and bra vectors representing the quantum many-particle state m, and P_m represents the number of member-systems occupying the state m. Note that both definitions for the density operator, (8.15.21 and 22), have the form (8.1). For a system of many identical bosons (fermions), the many-particle state vectors, $|m\rangle$ and $\langle m|$, must be symmetric (antisymmetric). (If the system contains two or more kinds of particles, vectors should have proper symmetries.)

(b) The distribution of the systems in the ensemble

The number of systems in the ensemble occupying an arbitrary quantum state γ is given by

$$\langle\gamma|\rho|\gamma\rangle \; , \qquad (8.2)$$

where $\langle\gamma|$ and $|\gamma\rangle$ are normalized bra and ket vectors representing the state γ.

(c) The ensemble average of a dynamical function

The average value of an observable dynamical quantity ξ over the ensemble can be calculated from the formula

$$\langle\xi\rangle = \frac{\mathrm{Tr}\{\xi\rho\}}{\mathrm{Tr}\{\rho\}}, \tag{8.3}$$

where the symbol Tr means the trace (diagonal sum); for example,

$$\mathrm{Tr}\{\rho\} \equiv \sum_{\alpha} \langle\alpha|\rho|\alpha\rangle = N_G \tag{8.4}$$

where the summation is over all normalized bases $\{\alpha\}$. The quantity $\mathrm{Tr}\{\rho\}$ given by (8.4) represents the total number of systems, N_G in the ensemble.

(d) The evolution equation for the density operator

The density operator ρ changes with time, following the quantum Liouville equation [see (8.15.11)] :

$$i\hbar \frac{\partial\rho}{\partial t} = [H,\rho] . \tag{8.5}$$

We note that eqs. (8.1) - (8.5) have the same forms as those equations for the single-particle ensemble.

Earlier in §7.2, we discussed the canonical ensemble of classical mechanical systems. This is a stationary ensemble characterized by the canonical distribution function

$$\rho_c(q,p) = C \exp\left[-\frac{1}{k_B T} H(q,p)\right] . \tag{8.6}$$

We have seen in Chapter 7 that all equilibrium properties of a classical system can be calculated in terms of the canonical ensemble (with the bulk limit).

The canonical ensemble of quantum mechanical systems can be constructed in an analogous manner. This ensemble is characterized by the canonical density operator

$$\boxed{\rho_c \equiv C \exp\left[-\frac{1}{k_B T} H\right]}, \tag{8.7}$$

where H is the quantum Hamiltonian of the system, and C is a constant. Note that the operator ρ_c contains the absolute temperature T in the same form as the canonical distribution function.

Since this density operator ρ_c is a function of H only, it commutes with the Hamiltonian H:

$$[H, \rho_c(H)] = 0 \,. \tag{8.8}$$

Therefore from (8.5) we see that

$$\frac{\partial \rho_c}{\partial t} = \frac{1}{i\hbar}[H, \rho_c] = 0 \,. \tag{8.9}$$

Thus, the canonical density operator ρ_c does not change with time; that is, it is stationary. This is, of course, a necessary condition for the description of equilibrium properties.

The physical meaning of ρ_c may be further expounded in the following manner. The eigenvalue equation for the Hamiltonian H is

$$H|E_\nu\rangle = E_\nu |E_\nu\rangle \,, \tag{8.10}$$

where E_ν is the ν-th energy eigenvalue. In general, there may be more than one eigenstate corresponding to E_ν. For example, the eigenvalues of the Hamiltonian $H \equiv p^2/2m$ for a free particle moving in one dimension, obtained in §5.8 with the periodic boundary condition, are $\varepsilon_n \equiv \frac{p_n^2}{2m} = \frac{1}{2m}\left(\frac{2\pi\hbar}{L}\right)^2 n^2$, where n represents the set of all integers. Except for n = 0, there are two eigenstates corresponding to each ε_n. The degree of degeneracy tends to increase as the degree of freedom for the system increases. Therefore, as a general rule, several eigenstates correspond to a single energy-eigenvalue E_ν.

Let us assume that this degeneracy is removed, and that the set of the corresponding orthogonalized eigenstates are denoted by $\{\alpha_\nu\}$. (A simple case of such orthogonalization was discussed in §8.4.) These states will now be specified by the set (E_ν, α_ν), and be represented by normalized kets and bras $|E_\nu, \alpha_\nu\rangle$ and $\langle E_\nu, \alpha_\nu|$. Using these vectors as bases we may expand any ket $|\psi\rangle$ in the form

$$|\psi\rangle = \sum_{E_\nu, \alpha_\nu} |E_\nu, \alpha_\nu\rangle\langle E_\nu, \alpha_\nu|\psi\rangle . \tag{8.11}$$

Dropping $|\psi\rangle$ from both sides, we obtain the completeness relation:

$$1 = \sum_{E_\nu, \alpha_\nu} |E_\nu, \alpha_\nu\rangle\langle E_\nu, \alpha_\nu| . \tag{8.12}$$

Applying (8.12), we can re-express the canonical density operator ρ in the following form:

$$\rho_c = C \sum_{E_\nu, \alpha_\nu} \exp\left[-\frac{1}{k_B T} H\right]\left[|E_\nu, \alpha_\nu\rangle\langle E_\nu, \alpha_\nu|\right]$$

$$= C \sum_{E_\nu, \alpha_\nu} \exp\left[-\frac{1}{k_B T} E_\nu\right]\left[|E_\nu, \alpha_\nu\rangle\langle E_\nu, \alpha_\nu|\right] . \tag{8.13}$$

This has the standard form (8.1), and can be interpreted as follows: <u>The number of systems occupying the energy states</u> (E_ν, α_ν) <u>follow the Boltzmann distribution law</u>:

$$P(E_\nu, \alpha_\nu) = C \exp\left[-\frac{E_\nu}{k_B T}\right] . \tag{8.14}$$

Thus, in the energy-representation in which the Hamiltonian H is diagonal, the analogy between quantum and classical canonical ensembles is clearly seen.

In the high temperature limit $T \to \infty$, the canonical distribution becomes independent of the energy, as can be seen from (8.14). Then, the numbers of the systems occupying any two energy states become equal. Furthermore, since

$$
\begin{aligned}
&\langle \gamma | \rho(T \to \infty) | \gamma \rangle \\
&= \langle \gamma | \, C \sum_{E_\nu, \alpha_\nu} | E_\nu, \alpha_\nu \rangle \langle E_\nu, \alpha_\nu | \gamma \rangle && \text{[use of (8.13)]} \\
&= C \langle \gamma | \gamma \rangle && \text{[use of (8.12)]} \\
&= C \, , && (8.15)
\end{aligned}
$$

the number of systems occupying any quantum state is constant. (In the language of probability, this means that all quantum states are occupied with equal probability.) In short, the canonical ensemble approaches the uniform ensemble in the high temperature limit.

9.9 The Partition Function

In Chapter 7 we discussed the standard way of computing equilibrium properties of classical systems with the aid of the canonical ensemble. The main steps and results were:

(a) We can calculate all thermodynamic functions from the partition function defined by

$$Z^{(cl)} \equiv \mathrm{Tr}^{(cl)} \{ \exp [-H/k_B T] \} \, , \qquad (9.1)$$

where the symbol $\mathrm{Tr}^{(cl)}$ means the classical trace, that is, the phase-space integration with the symmetry factor arising from the indistinguishability of particles. The partition function

$$Z \equiv Z(T,V,N) \qquad (9.2)$$

for a pure gas depends on the temperature T, volume V, and the number of particles N.

(b) The Helmholtz free energy F of the system can be related to the partition function Z simply by

$$\boxed{F = -k_B T \ln Z(T,V,N) .} \tag{9.3}$$

(c) The internal energy E of the system, by postulate, is equal to the average energy over the ensemble:

$$E = \langle H \rangle \equiv \frac{Tr^{(cl)}\{H \exp(-H/k_B T)\}}{Tr^{(cl)}\{\exp(-H/k_B T)\}} . \tag{9.4}$$

Introducing the reciprocal temperature

$$\beta \equiv (k_B T)^{-1} , \tag{9.5}$$

we can rewrite (9.4) as

$$E = \frac{Tr^{(cl)}\{H e^{-\beta H}\}}{Tr^{(cl)}\{e^{-\beta H}\}} = -\frac{\partial}{\partial \beta} \ln Tr^{(cl)}\{e^{-\beta H}\}$$

or

$$\boxed{E = -\frac{\partial}{\partial \beta} \ln Z(\beta,V,N) ,} \tag{9.6}$$

where

$$Z(\beta,V,N) = Tr^{(cl)}\{e^{-\beta H}\} \tag{9.7}$$

is the partition function regarded as a function of β, V, and N.

(d) In order to insure the rigorous connection between the partition functions Z and thermodynamic functions, we must take the bulk limit (also called the thermodynamic limit):

$$N\to\infty,\ V\to\infty \text{ such that } N/V = n \text{ (number density)} . \tag{9.8}$$

In particular, the free energy per unit volume, $f(T,n)$, is given by

$$f(T,n) = \frac{F}{V} = \text{Lim} \frac{1}{V}\left[-k_B T \ln Z(T,N,V)\right] , \tag{9.9}$$

where the symbol Lim means the <u>bulk limit</u> (9.8).

All of these results can simply be carried over to the quantum statistical case. The only difference is that the partition function Z must now be defined by the quantum trace of the canonical density operator $\exp(-\beta H)$, where H is the quantum Hamiltonian operator; that is,

$$\begin{aligned} Z &\equiv \text{Tr}\{\exp(-\beta H)\} \\ &\equiv \text{diagonal sum of the matrix } \exp(-\beta H). \end{aligned} \tag{9.10}$$

In the energy-representation, we have

$$Z = \sum_{E_\nu,\alpha_\nu} \exp(-\beta E_\nu) , \tag{9.11}$$

where the summation is over all energy eigenstates. [The partition function is very often denoted by the capital letter Z because the sum-over-energy-states on the r.h.s. is called Zustandssumme (sum-over-states) in German.] To find Z from (9.11) we may solve the eigenvalue equation

$$H|E_\nu,\alpha_\nu\rangle = E_\nu|E_\nu,\alpha_\nu\rangle , \tag{9.12}$$

and find the energy E_ν and the <u>degree of degeneracy</u>, that is, the number of α_ν, and then compute the sum over the states as indicated in (9.11). Since the eigenvalues E_ν generally depend on the volume V, the quantum partition function Z depends on the temperature T, the number of particles N, and the volume V:

$$Z = Z(T,N,V) , \tag{9.13}$$

just as the classical counterpart $Z^{(cl)}$.

For illustration, let us take a simple harmonic oscillator. The Hamiltonian of this system is

$$H = \frac{p^2}{2m} + \frac{1}{2} kx^2 . \tag{9.14}$$

According to the study in §8.8, the corresponding energy eigenvalues are given in (8.8.13):

$$E_n = (\frac{1}{2} + n)\hbar\omega_o , \quad n = 0,1,2,\ldots , \tag{9.15}$$

where $\omega_o \equiv (k/m)^{\frac{1}{2}}$ is the angular frequency. We note no degeneracy of the eigenstates. The partition function Z can be calculated from (9.11) as follows:

$$\begin{aligned} Z &= \sum_n \exp(-\beta E_n) \\ &= \sum_{n=0}^{\infty} \exp\left[-\beta(\frac{1}{2} + n)\hbar\omega_o\right] \\ &= e^{-\frac{1}{2}\beta\hbar\omega_o}\left[1 + e^{-\beta\hbar\omega_o} + e^{-2\beta\hbar\omega_o} + \ldots\right] \\ &= \frac{e^{-\frac{1}{2}\beta\hbar\omega_o}}{1 - e^{-\beta\hbar\omega_o}} . \end{aligned} \tag{9.16}$$

For comparison, let us calculate the classical partition function,

$$Z^{(cl)} \equiv \frac{1}{2\pi\hbar} \int\int dx\, dp \exp\left\{-\beta\left[\frac{p^2}{2m} + \frac{1}{2} kx^2\right]\right\} . \tag{9.17}$$

The double integral can be carried out in a separate manner:

$$\int_{-\infty}^{\infty} dp\, e^{-p^2(\beta/2m)} = (\frac{2m\pi}{\beta})^{\frac{1}{2}}$$

$$\int_{-\infty}^{\infty} dx\, e^{-x^2(k\beta/2)} = (\frac{2\pi}{k\beta})^{\frac{1}{2}} . \tag{9.18}$$

We therefore obtain

$$Z^{(cl)} = \frac{1}{2\pi\hbar} \frac{2\pi}{\beta} \left(\frac{m}{k}\right)^{\frac{1}{2}} = \frac{1}{\hbar\beta\omega_o} . \tag{9.19}$$

Note that this expression is quite different from the corresponding quantum expression (9.17). Also, the actual steps of the calculations proceed in quite a different manner.

The average energy and heat capacity, calculated from the classical partition function $Z^{(cl)}$, are k_BT and k_B, respectively. The same quantities calculated from the quantum partition function Z are quite different (Problem 9.1).

The eigenvalue equation (9.12) can be solved for a very limited number of simple systems (most of which will be discussed in the text). For general and more complicated systems, the solution of the energy-eigenvalue equation is prohibitively difficult.

An important way of overcoming this difficulty is as follows. Since the trace can be calculated in any representation, a suitable representation with known bases can be used to compute the partition function Z from (9.10). For example, as the eigenvalues and eigenvectors for the momentum are known (as we saw in §8.11), we may use the momentum representation for certain problems.

Another way of circumventing the energy-eigenvalue problem is to check if the classical partition $Z^{(cl)}$ gives a reasonable approximation for the quantum partition function Z. As we will discuss later, the quantum partition function Z approaches the classical partition function $Z^{(cl)}$ in the <u>classical limit</u> in which Planck's constant h tends to zero. Fortunately, it is not difficult to see when and to what extent the classical approximation can be justified. This will be discussed in §9.12 and §9.13.

Problem 9.1 Consider a simple harmonic oscillator.

(a) Calculate the average energy and the heat capacity from the classical partition function $Z^{(cl)}$ given in (9.19).

(b) Calculate the same from the quantum partition function Z given in (9.16).

(c) Show that in the high temperature limit, the quantum expressions in (b) approach the classical expressions in (a).

9.10 The Grand Canonical Ensemble

Thermodynamic properties can also be calculated by means of the grand canonical ensemble. The main advantage of the grand canonical ensemble over the canonical ensemble lies in the ease of calculation. As we will see later, this is especially true for quantum statistical calculations.

The grand canonical ensemble of quantum many-particle systems is characterized by the grand canonical density operator

$$\rho_G \equiv \frac{1}{\Xi} \exp\left[\alpha N - \beta H_N\right] , \qquad \beta \equiv 1/k_B T , \qquad (10.1)$$

which has the same form as (7.11.11) for the grand canonical distribution function. Only the Hamiltonian H_N is a quantum operator; the quantity Ξ is the grand partition function given by

$$\Xi\,(\alpha,\beta,V) \equiv \sum_{N=0}^{\infty} \mathrm{Tr}_{,N}\{\exp[\alpha N - \beta H_N]\}$$

$$\equiv \mathrm{TR}\,\{\exp[\alpha N - \beta H]\}\,, \qquad (10.2)$$

where the symbol TR means the sum of all N-particle traces. We note that the grand partition function Ξ depends on α, β, and V, of which the first two appear explicitly in the definition equation (10.2). The volume dependence arises through the energy eigenvalues.

The average of a physical quantity A over the grand canonical ensemble is given by

$$\boxed{\langle A\rangle \equiv \frac{\mathrm{TR}\{A\exp[\alpha N-\beta H]\}}{\mathrm{TR}\{\exp[\alpha N-\beta H]\}}\,.} \qquad (10.3)$$

In particular, the average energy is

$$E = \langle H\rangle = \frac{\mathrm{TR}\{H\exp[\alpha N-\beta H]\}}{\mathrm{TR}\{\exp[\alpha N-\beta H]\}}\,. \qquad (10.4)$$

The operator exp $(-\beta H)$ contains the parameter β. The derivative with respect to the parameter β can be defined in a standard manner by

$$\frac{\partial}{\partial\beta}\exp(-\beta H) \equiv \lim_{\Delta\beta\to 0}\frac{1}{\Delta\beta}\left[e^{-(\beta+\Delta\beta)H} - e^{-\beta H}\right]. \qquad (10.5)$$

We can calculate the r.h.s. as follows:

$$\lim_{\Delta\beta\to 0}\frac{1}{\Delta\beta}\{[1 - (\beta+\Delta\beta)H + \frac{1}{2}(\beta+\Delta\beta)^2H^2 - \ldots]$$

$$- [1 - \beta H + \frac{1}{2}\beta^2H^2 - \ldots]\}$$

$$= -H[1 - \beta H + \frac{1}{2}\beta^2H^2 -]\}$$

$$= -H\exp(-\beta H).$$

Thus, we obtain

$$\frac{\partial}{\partial\beta} \exp(-\beta H) = -H \exp(-\beta H) . \qquad (10.6)$$

Using this expression, we can re-express (10.4) as follows:

$$\langle H\rangle = -\frac{\partial}{\partial\beta} \ln \mathrm{TR} \{\exp [\alpha N - \beta H]\} \qquad (10.7)$$

or

$$\boxed{\langle H\rangle = -\frac{\partial}{\partial\beta} \ln \Xi (\alpha, \beta, V) ,} \qquad (10.8)$$

which has the same form as the corresponding classical expression (7.11.19)

In a similar manner, we can show that the average number of particles in the given volume V, represented by

$$\langle N\rangle \equiv \frac{\mathrm{TR} \{N \exp [\alpha N - \beta H]\}}{\mathrm{TR} \{\exp [\alpha N - \beta H]\}} , \qquad (10.9)$$

is related to the grand partition function Ξ by

$$\boxed{\langle N\rangle = \frac{\partial}{\partial\alpha} \ln \Xi (\alpha, \beta, V) .} \qquad (10.10)$$

The relation between the grand partition function Ξ and thermodynamic functions has the same form for both the quantum and classical cases. Some important relations are

$$n \text{ (number density)} = \mathrm{Lim} \frac{\langle N\rangle}{V} = \mathrm{Lim} \frac{1}{V} \frac{\partial}{\partial\alpha} \ln \Xi (\alpha, \beta, V) \qquad (10.11)$$

e (internal energy density)

$$= \mathrm{Lim} \frac{\langle H\rangle}{V} = \mathrm{Lim} \frac{(-1)}{V} \frac{\partial}{\partial\beta} \ln \Xi (\alpha, \beta, V) \qquad (10.12)$$

$$\alpha \equiv \beta\mu \equiv \mu/k_B T , \qquad (10.13)$$

where μ is the chemical potential.

Eq.(10.11) gives the functional relationship between the absolute activity α and the number density n. Solving this with respect to α, and substituting the result into eq.(10.12), we can, in principle, obtain the internal energy e as a function of β and n.

Other important relations are

$$P = \beta^{-1} \frac{\partial}{\partial V} \ln \Xi (\alpha,\beta,V) , \tag{10.14}$$

$$S = k_B \left[1 - \beta \frac{\partial}{\partial \beta} - \alpha \frac{\partial}{\partial \alpha}\right] \ln \Xi (\alpha,\beta,V) . \tag{10.15}$$

For illustration, let us take a system of free bosons moving in one dimension. The Hamiltonian H of the system is the sum of the kinetic energies:

$$H = \sum_{j\ (\text{particle})\ =\ 1}^{N} \frac{p_j^2}{2m} . \tag{10.16}$$

According to the study in §8.8, the possible momentum eigenvalues are given by [see (8.8.21)]

$$p_k = \frac{2\pi\hbar}{L} k , \qquad k = \ldots,-1,0,1,2,\ldots .$$

The single-particle Hamiltonian $h \equiv p^2/2m$ depends on the momentum p only, and therefore is diagonal in the momentum representation:

$$h(p)|p_k> = \varepsilon_k|p_k> , \qquad \varepsilon_k \equiv p_k^2/2m . \tag{10.17}$$

The total Hamiltonian H for the system will then be diagonal in the momentum representation in which many-particle basis-vectors are constructed from the momentum states $\{p_k\}$, as prescribed in §§9.4 and 9.5. This means that the many-particle Hamiltonian is also diagonal in the momenum-state-occupation-number representation:

$$H|\{n\}> = E(\{n\})|\{n\}> , \tag{10.18}$$

where the eigenvalue $E(\{n\})$ is given by

$$E(\{n\}) \equiv \sum_{k\ (\text{integers})} \varepsilon_k n_k . \tag{10.19}$$

In the last expression, the n_k represent the numbers of particles occupying the momentum states p_k. The ket vector $|\{n\}>$ represents the many-particle state characterized by the set of the occupation numbers $\{n_k\}$.

The total number of particles, N, is the sum of the occupation numbers:

$$N = \sum_k n_k . \tag{10.20}$$

We now wish to calculate the grand partition function Ξ. As we saw above, the total Hamiltonian H is diagonal in the momentum-state occupation-number representation. Therefore the grand canonical density operator $\exp[\alpha N - \beta H]$ is also diagonal:

$$\exp[\alpha N-\beta H]\,|\{n\}> = \exp[\alpha N-\beta E(\{n\})]\,|\{n\}> . \tag{10.21}$$

To obtain the grand partition function through the formula (10.2), we simply sum the diagonal elements $\exp[\alpha N-\beta E(\{n\})]$ over all possible states $\{n\}$. Since

$$\begin{aligned} \exp[\alpha N-\beta E(\{n\})] &= \exp\Big[\sum_k(\alpha-\beta\varepsilon_k)n_k\Big] && [\text{use of (10.19) and (10.20)}] \\ &= \prod_k \exp[(\alpha-\beta\varepsilon_k)n_k] , && [e^{a_1+a_2} = e^{a_1}e^{a_2}] \end{aligned} \tag{10.22}$$

we obtain

$$\begin{aligned} \Xi &= \sum_{\{n\}} \exp[\alpha N-\beta E(\{n\})] \\ &= \sum_{\substack{n \\ (\text{all states})}} \prod_k \exp[(\alpha-\beta\varepsilon_k)n_k] . \end{aligned} \tag{10.23}$$

According to the rules explained in §9.7, the occupation number for bosons can run on any non-negative integers. Therefore, we obtain

$$\sum_{n_k} e^{(\alpha-\beta\varepsilon_k)n_k} = 1 + e^{\alpha-\beta\varepsilon_k} + e^{2(\alpha-\beta\varepsilon_k)} + \ldots$$
$$= \left[1 - e^{\alpha-\beta\varepsilon_k}\right]^{-1} . \tag{10.24}$$

Using this result for each of the momentum states p_k, we can carry out the summation in (10.23) and obtain

$$\Xi = \prod_k \left[1 - e^{\alpha-\beta\varepsilon_k}\right]^{-1} . \quad \text{(for free bosons)} \tag{10.25}$$

As a second example, let us consider a system of free fermions. All of the arguments proceed in a parallel manner up to eq.(10.23). The occupation number for fermions is however restricted to 0 or 1. We therefore obtain

$$\sum_{n_k=0,1} e^{(\alpha-\beta\varepsilon_k)n_k} = 1 + e^{\alpha-\beta\varepsilon_k} . \tag{10.26}$$

Using this result, we obtain

$$\Xi = \prod_k (1 + e^{\alpha-\beta\varepsilon_k}) . \quad \text{(for free fermions)} \tag{10.27}$$

Physical discussions of expressions (10.25) and (10.27) will be given in the following section.

9.11 The Bose and Fermi Distribution Functions

In the present section we consider free identical particles only.

Let us take free particles moving in one dimension. The grand partition functions Ξ for free bosons and free fermions were computed in the last section. The results for both cases given in (10.25) and (10.27), can be written in a single line:

$$\Xi = \prod_k \left[1 \mp e^{\alpha - \beta\varepsilon_k}\right]^{\mp 1} , \tag{11.1}$$

where the upper signs correspond to the case of bosons and the lower signs to that of fermions; this convention will be used in the present section wherever convenient.

Taking the logarithm of (11.1) and differentiating it with respect to α, we obtain

$$\frac{\partial}{\partial\alpha} \ln \Xi \equiv \frac{\partial}{\partial\alpha} \ln \left(\prod_k \left[1 \mp e^{\alpha-\beta\varepsilon_k}\right]^{\mp 1} \right)$$

$$= \frac{\partial}{\partial\alpha} \sum_k \ln \left[1 \mp e^{\alpha-\beta\varepsilon_k}\right]^{\mp 1} \qquad [\ln (ab) = \ln a + \ln b]$$

$$= \sum_k \frac{e^{\alpha-\beta\varepsilon_k}}{1 \mp e^{\alpha-\beta\varepsilon_k}} . \tag{11.2}$$

According to (10.10), this is equal to the average particle number $\langle N \rangle$. We thus obtain

$$\boxed{\langle N \rangle = \sum_k f(\varepsilon_k) ,} \tag{11.3}$$

where

$$f(\varepsilon_k) \equiv \frac{e^{\alpha-\beta\varepsilon_k}}{1 \mp e^{\alpha-\beta\varepsilon_k}} = \frac{1}{e^{\beta\varepsilon_k-\alpha} \mp 1} . \tag{11.4}$$

The functions $f_B(\varepsilon)$ and $f_F(\varepsilon)$, defined here corresponding to the upper and lower signs, are called the Bose and Fermi distribution functions, respectively.

Comparing (11.3) and (10.20), we observe that

$$\langle n_k \rangle = f(\varepsilon_k) . \tag{11.5}$$

This means that the Bose and Fermi distribution functions represent the relative probabilities with which the states p_k are occupied. We note that these probabilities depend only on the energy ε_k.

The relation (11.5) can be established directly with the aid of (10.3). In fact,

$$\langle n_k \rangle \equiv \frac{\mathrm{TR}\,\{n_k \exp\,[\alpha N-\beta H]\}}{\mathrm{TR}\,\{\exp\,[\alpha N-\beta H]\}}$$

$$= \frac{\sum_{\{n\}} \langle\{n\}|n_k \exp\,[\alpha N-\beta H]\,|\{n\}\rangle}{\sum_{\{n\}} \langle\{n\}|\exp\,[\alpha N-\beta H]\,|\{n\}\rangle}$$

$$= \frac{\sum_{\{n\}} n_k \prod_{\ell} \exp\,[(\alpha-\beta\varepsilon_\ell)n_\ell]}{\sum_{\{n\}} \prod_{\ell} \exp\,[(\alpha-\beta\varepsilon_\ell)n_\ell]} .$$

Examining the sum closely at the states k and k+1 (both integers), we obtain

$$\langle n_k \rangle = \frac{\ldots[\sum_{n_k} n_k\, e^{(\alpha-\beta\varepsilon_k)n_k}]\,[\sum_{n_{k+1}} e^{(\alpha-\beta\varepsilon_{k+1})n_{k+1}}]\ldots}{\ldots[\sum_{n_k} e^{(\alpha-\beta\varepsilon_k)n_k}]\,[\sum_{n_{k+1}} e^{(\alpha-\beta\varepsilon_{k+1})n_{k+1}}]\ldots}$$

$$= \frac{\sum n_k e^{(\alpha - \beta\varepsilon_k) n_k}}{\sum\limits_{n_k} e^{(\alpha - \beta\varepsilon_k) n_k}} \qquad \text{[dropping common factors]}$$

$$= \frac{\partial}{\partial\alpha} \ln \{ \sum_{n_k} e^{(\alpha - \beta\varepsilon_k) n_k} \}$$

$$= \frac{\partial}{\partial\alpha} \ln \{ [1 \mp e^{\alpha - \beta\varepsilon_k}]^{\mp 1} \} \qquad \text{[use of (10.24) and (10.26)]}$$

$$= \frac{e^{\alpha - \beta\varepsilon_k}}{1 \mp e^{\alpha - \beta\varepsilon_k}}$$

$$\equiv f(\varepsilon_k) \ . \qquad \text{Q.E.D.} \qquad (11.7)$$

We thus find that the grand-canonical-ensemble average of the occupation number n_k is given by the Bose or Fermi distribution function. In this ensemble all system-states with unlimited numbers of particles are involved in the averaging process. Therefore, the occupation numbers n_k may run over all non-negative integers for bosons and over 0 and 1 for fermions. However, if the canonical ensemble average is considered, the occupation number states {n} involved in the averaging process will be restricted to those states for which eq.(10.20): $\sum_k n_k = N$, is satisfied. This restriction will make the canonical ensemble average of n_k a complicated function of N.

The principal result eq.(11.5) can simply be extended to the three dimensional case. The momentum states are now characterized by three quantum numbers:

$$\left(\frac{2\pi\hbar}{L} j, \frac{2\pi\hbar}{L} k, \frac{2\pi\hbar}{L} \ell \right) \equiv (p_x', p_y', p_z') \equiv \vec{p}' , \qquad (11.8)$$

corresponding to the three components (p_x, p_y, p_z) of the momentum operator $\vec{p}$. All of the steps leading to eq.(11.7) can be followed in a parallel manner.

We then obtain

$$\langle n_{\vec{p}} \rangle \equiv \text{(average number of particles occupying the momentum state } \vec{p}\text{)}$$

$$= \frac{e^{\alpha-\beta\varepsilon_p}}{1 \mp e^{\alpha-\beta\varepsilon_p}} = \frac{1}{e^{\beta\varepsilon_p-\alpha} \mp 1} \equiv f(\varepsilon_p) \tag{11.9}$$

$$\varepsilon_p \equiv (p_x^{\,2} + p_y^{\,2} + p_z^{\,2})/2m \equiv p^2/2m \;, \tag{11.10}$$

where we dropped the prime on $\vec{p}$.

Again, we note that the average occupation number depends on the state $\vec{p}$ only through the energy ε_p. We further note that the energy dependence is characterized by the <u>same</u> Bose and Fermi distribution functions.

The Bose and Fermi distribution functions are markedly different from each other. They are also different from the Boltzmann distribution function:

$$f_c(\varepsilon) = e^{\alpha-\beta\varepsilon_p} \;. \tag{11.11}$$

This point will be discussed further in §§9.12 and 9.13.

In three dimensions, eqs.(10.20) and (10.21) can be extended to

$$N = \sum_{\vec{p}} n_{\vec{p}} \tag{11.12}$$

$$E(\{n\}) = \sum_{\vec{p}} \varepsilon_p \, n_{\vec{p}} \;. \tag{11.13}$$

Their averages over the grand canonical ensemble are given by

$$\langle N \rangle \equiv \sum_{\vec{p}} \langle n_{\vec{p}} \rangle = \sum_{\vec{p}} f(\varepsilon_p) \tag{11.14}$$

$$E \equiv \langle H \rangle \equiv \sum_{\vec{p}} \varepsilon_p \langle n_{\vec{p}} \rangle = \sum_{\vec{p}} \varepsilon_p \, f(\varepsilon_p) \;. \tag{11.15}$$

Many formulas such as (11.14) and (11.15) derived in this section will be used to discuss simple systems in later chapters.

Problem 11.1 Prove that the entropy S of an ideal quantum gas is given by

$$S = -k_B \sum_{\vec{p}} [f \ln f \mp (1 \pm f) \ln (1 \pm f)] ,$$

where f represents the Bose or Fermi distribution function.

Hint: Start with the general formula;

$$S = k_B [1 - \beta \frac{\partial}{\partial \beta} - \alpha \frac{\partial}{\partial \alpha}] \ln \Xi$$

$$\Xi = \prod_k [1 + e^{\alpha - \beta \varepsilon_k}]^{+1}.$$

9.12 Quantum Statistics in the Classical Limit

The dynamics of microscopic particles such as electrons and protons are known to be described by the quantum mechanical laws rather than the classical mechanical laws. Therefore, the statistical mechanics must in principle be founded on the basis of many-body quantum mechanics. However, in some cases the statistical mechanical theory based on classical mechanics leads to results indistinguishable from those of the theory based on quantum mechanics. For example, a dilute gas of monatomic molecules at high temperatures can be adequately treated by classical statistical mechanics. In such a case, classical statistical mechanics has advantages over quantum statistical mechanics. It appeals more directly to common sense, and it is also easier in computation. In other cases, the classical theory may yield useful approximate results which may not be obtained as easily from the quantum theory.

In some cases however, classical statistical mechanics leads to results essentially different from the correct results of quantum statistical mechanics. Fortunately, the question of when we can (or cannot) use the classical theory to obtain good results can be answered relatively simply, often without involving detailed calculations. We will investigate this question from a general point of view in the present section.

Let us recall that the quantum theory of a many-particle system is based on six fundamental postulates: the set of five quantum mechanical postulates enumerated in §8.13 and the quantum statistical postulate in §9.4. In §8.14 we saw that the quantum mechanical laws reduce to the classical mechanical laws when Planck's constant h tends to zero:

$$h \to 0 . \tag{12.1}$$

This is called the classical limit. We now state a very useful theorem:

Theorem: All results of quantum statistical mechanics reduce to those of classical statistical mechanics in the classical limit.

We will demonstrate this by taking simple examples. Let us consider a particle moving along a straight line of length L. In classical mechanics, a dynamical state of the system, (x,p), can be represented by a point in the phase space. In quantum mechanics, the dynamical state cannot be represented by a point because of Heisenberg's uncertainty principle [see §8.9]. The set of momentum eigenstates, $p_k \equiv 2\pi\hbar k/L$, $k = \ldots -1,0,1,2,\ldots$, however, may be represented by the set of quantum cells shown in Figure 9.4. We note that

(a) each cell extends from 0 to L in position, implying that a particle can be found with an equal probability everywhere along the line from 0 to L, and

(b) the area of each cell equals $2\pi\hbar$.

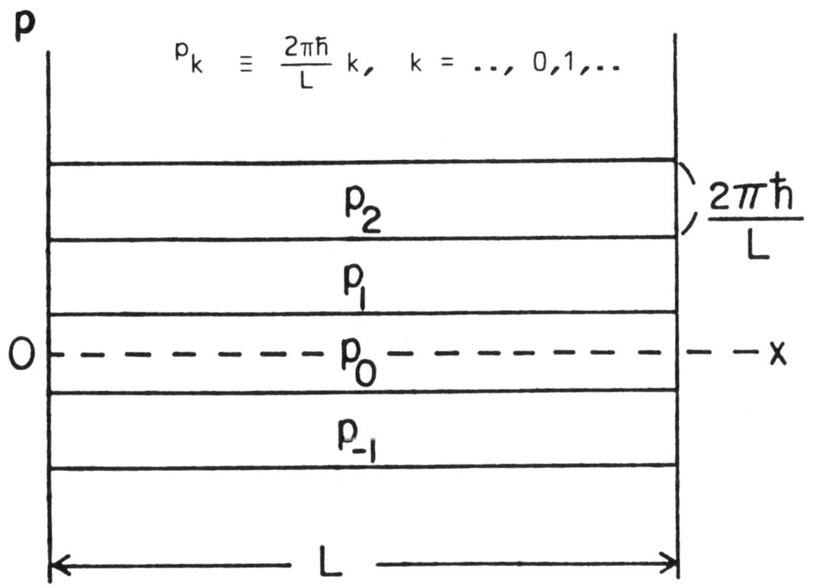

Fig. 9.4 The momentum states $\{p_k\}$ for linear motion are represented by rectangular cells of equal area $(2\pi\hbar)$ in the phase space.

Such representation of quantum states in phase space is not restricted to this special case. The quantum states for a simple harmonic oscillator can be represented by the elliptical shells as shown in Figure 9.5. Each cell has an area equal to $2\pi\hbar$. Whether or not every discrete quantum state can be represented by a cell of area $2\pi\hbar$ is not known, but is plausible.

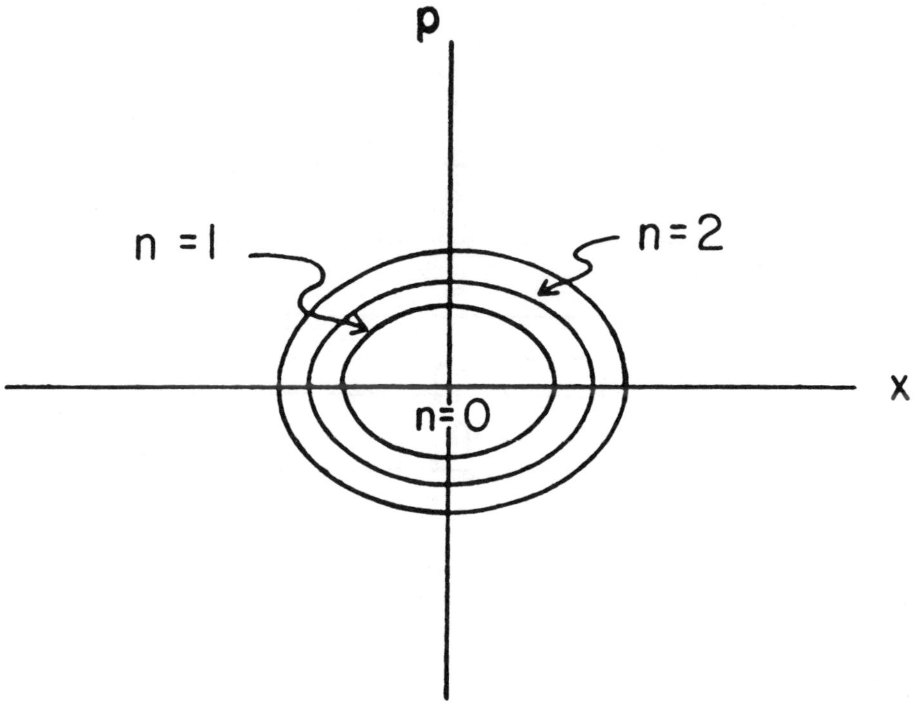

Fig. 9.5 The quantum mechanical eigenstates for a simple harmonic oscillator are represented by the quantum cells of phase-space area $2\pi\hbar$.

Let us now imagine that the phase space is divided into cells of area $2\pi\hbar$. See Figure 9.6. Each cell corresponds to a maximum specification of a quantum state in accordance with the uncertainty principle and should correspond to a distinct quantum state. In fact a typical cell located at the interval $(x-\frac{1}{2}\Delta x, x+\frac{1}{2}\Delta x)$ in position and $(p-\frac{1}{2}\Delta p, p+\frac{1}{2}\Delta p)$ in momentum may be thought to represent a quantum state with a momentum in $(p-\frac{1}{2}\Delta p, p+\frac{1}{2}\Delta p)$ and a position in

$(x-\frac{1}{2}\Delta x, x+\frac{1}{2}\Delta x)$. That is, this cell corresponds to a quantum state with approximate eigenvalues (x,p) with uncertainties Δx and Δp.

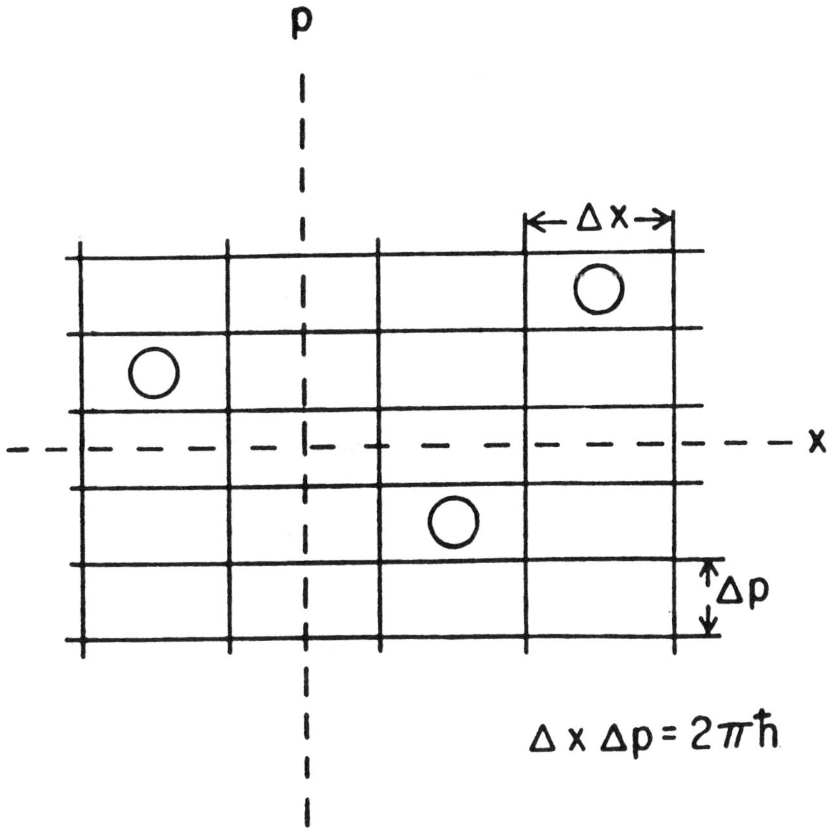

Fig. 9.6 The quantum cell of phase-space area $\Delta x \Delta p = 2\pi\hbar$ at (x,p) represents the quantum state with the momentum in $(p-\frac{1}{2}\Delta p, p+\frac{1}{2}\Delta p)$ and the position in $(x-\frac{1}{2}\Delta x, x+\frac{1}{2}\Delta x)$.

We may now regard the classical limit: $h \to 0$ as a process in which the unit cell area is reduced to zero. This means that we specify the dynamical state of the particle, (x,p), more and more accurately.

Let us now take a system of identical particles which may interact with each other and/or be subjected to arbitrary conservative forces. As noted earlier, the concept of identical (or indistinguishable) particles is the same for both quantum and classical cases. As we discussed in §§9.4 - 9.7, a dynamical state for a system of identical particles can best be described by stating that certain particle-states, that is, certain phase cells, $\{a,b,\ldots,g\}$ are occupied (without specifying which particles are in which cells).

For definiteness let us take a system of three identical particles. The Hamiltonian $H(x_1,p_1,x_2,p_2,x_3,p_3)$ by assumption is symmetric. The partition function Z for the quantum system is given by

$$Z \equiv \mathrm{Tr}\,\{e^{-\beta H}\} \,. \tag{12.2}$$

In calculating the quantum trace we may use the quantum phase cells just described. A typical state indicated in Fig.9.6 where the cells (a,b,c) are occupied by particles can be represented by the ket $|a,b,c\rangle$. From the recognition that the states a, b and c are quantum states with approximate eigenvalues (x_a,p_a), (x_b,p_b) and (x_c,p_c), it follows that the ket $|a,b,c\rangle$ is an approximate eigenket of the Hamiltonian H with the approximate eigenvalue $H(x_a,p_a,x_b,p_b,x_c,p_c) = H(a,b,c)$. We can therefore express the partition function Z in the form:

$$Z = \frac{1}{3!} \sum_{\substack{a \\ a\neq b}} \sum_{\substack{b \\ a\neq c}} \sum_{\substack{c \\ b\neq c}}{}' \, e^{-\beta H(a,b,c)}$$

$$+ \text{ (terms corresponding to the possible multiple occupancy of states by bosons)} \,. \tag{12.3}$$

The factor $(3!)^{-1}$ is needed to take care of the redundant repetition in the summation with respect to a, b and c. It is observed that the first sum is the same irrespective of whether the particles are bosons or fermions. Other contributions must be added corresponding to the cases in which states are occupied doubly or triply by bosons. Such contribution is zero for fermions.

Let us now take the classical limit. We can easily see that in the small cell area limit (a) the summation in (12.3) is replaced by a phase-space integral, and (b) the second term becomes negligibly small compared to the first term [Problem 12.1]. Thus, we obtain

$$Z \to \frac{1}{3!} \int \cdots \int \frac{dx_1 dp_1}{(2\pi\hbar)} \frac{dx_2 dp_2}{(2\pi\hbar)} \frac{dx_3 dp_3}{(2\pi\hbar)} e^{-\beta H(x_1, p_1, x_2, p_2, x_3, p_3)}$$

Here, the factor $(2\pi\hbar)^{-1}$ arose from the fact that each cell has an area equal to $2\pi\hbar$.

The generalization to the case of an arbitrary number N of particles moving in three dimensions is obvious. We obtain

$$Z \equiv \mathrm{Tr}_{,N}\{e^{-\beta H}\}$$

$$\to \frac{1}{(2\pi\hbar)^{3N} N!} \int \cdots \int d^3r_1 d^3p_1 \cdots d^3r_N d^3p_N \, e^{-\beta H} \equiv Z_{c\ell} \tag{12.5}$$

as $h \to 0$.

We thus found that the quantum mechanical partition function approaches the classical expression including $(N!)^{-1}$ and $(2\pi\hbar)^{-3N}$ in the classical limit: $h \to 0$. The importance of the factor $(N!)^{-1}$ was discussed in §7.8 and §7.10. The factor $(2\pi\hbar)^{-3N}$ was introduced earlier in §4.3. Let us recall that this was done primarily to make quantities of interest such as the distribution function and partitition function dimensionless. The choice of the factor $(2\pi\hbar)^{-3N}$ can now be justified from the classical limit relation (12.5).

If we have a binary mixture of N_1 and N_2 particles, the Hamiltonian is symmetric under the permutations of indices referring to identical particles only. In this case, the classical limit of the quantum partition function will be as follows:

$$Z \rightarrow \frac{1}{(2\pi\hbar)^{3N_1+3N_2} N_1!N_2!} \int\cdots\int d^3r_1 d^3p_1 \cdots d^3r_{N_1} d^3p_{N_1} \iint d^3R_1 d^3P_1 \cdots$$
$$d^3R_{N_2} d^3P_{N_2} \quad \exp\left[-\beta H(\vec{r}_1, \vec{p}_1, \ldots, \vec{r}_{N_1}, \vec{p}_{N_1}, \vec{R}_1, \vec{P}_1, \ldots, \vec{R}_{N_2}, \vec{P}_{N_2})\right] \tag{12.6}$$

It was shown in §7.10 that the factor $(N_1!N_2!)^{-1}$ is essential in accounting for the entropy of mixing.

In some statistical mechanics textbooks, the statement is made that the classical particles are distinguishable in principle. If one takes such a viewpoint, one is forced to correct the expression for the classical partition function by a factor like $(N_1!N_2!)^{-1}$ in an ad hoc manner. The formulation presented here in the present text is straightforward and more logical.

Problem 12.1

(a) Let us assume that two free particles are in a cell (box). When the cell is divided into n subcells of equal size, what is the probability of finding two particles in one of the subcells ? What is the probability of finding them in separate subcells ?

(b) Solve similar problems when there are three particles in the original cell.

9.59

(c) Show that the probability of finding m particles in m separate subcells, approaches unity as the number of division, n, tends to infinity. First check if your answers in (a) and (b) confirm this statement for the cases of $m = 2$ and 3. Then treat the general case.

Problem 12.2

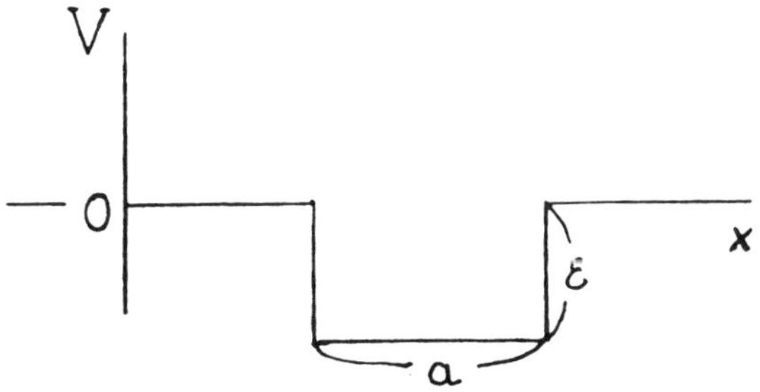

Consider a square-well potential of width a and energy depth ε, as shown on the left. If the well is large enough, it will allow a number of quantum mechanical bound states with discrete energies. A way of finding the number of states in an approximate manner is as follows.

(a) Compute the phase-space volume (area) available to a particle of mass m with a negative energy.

(b) Divide the phase-space volume by $2\pi\hbar$. The resulting number will give an estimate for the number of bound states. Carry out steps (a) and (b) for the square well indicated.

Note that the results depend on the particle mass m as well as on the potential parameters (a, ε).

Problem 12.3* Refer to the preceding problem.

(a) Apply (a) and (b) to a harmonic potential $V = \frac{1}{2} kx^2$ and show that the number of bound states with energies below the energy parameter E, increases linearly with that parameter E.

(b) If the enery levels are separated in equal distance, what is the energy separation ? Express your result in terms of the particle mass m and the force constant k.

9.13 Applicability of Classical Statistical Mechanics

Earlier in Chapter 7, we calculated the properties of a molecular gas by means of classical statistical mechanics. Since molecules actually move in accordance with quantum mechanics, this calculation was only an approximation. We now wish to study the validity of such calculations in the present section.

Let us consider a gas of helium molecules at one atmosphere of pressure. The masses M_3 and M_4 of He^3 and He^4 are known:

$$M_3 = 5.01 \times 10^{-24} \text{ g}$$

$$M_4 = 6.68 \times 10^{-24} \text{ g} . \tag{13.1}$$

The intermolecular potential V(r) between any He^4-He^4, He^3-He^3 and He^4-He^3 pairs can be expressed by the same Lennard-Jones potential

$$V(r) = 4\varepsilon\left[\left(\frac{\sigma}{r}\right)^{12} - \left(\frac{\sigma}{r}\right)^{6}\right] , \tag{13.2}$$

where ε and σ are, respectively, the depth and the zero point of V(r), as indicated in Figure 9.7. They are given numerically by

$$\varepsilon = 14.04 \times 10^{-16} \text{ ergs} = 1.404 \times 10^{-22} \text{ J} \tag{13.3}$$

$$\sigma = 2.56 \times 10^{-8} \text{ cm} = 2.56 \times 10^{-10} \text{m}. \tag{13.4}$$

The range of force R may be defined by

$$R = a\sigma \tag{13.5}$$

with a being a numerical factor of about 2.

According to the idea introduced first by de Broglie before the advent of the quantum theory, every microscopic particle in motion carries a "material wave," whose wavelength is related to its momentum p by

$$\lambda = \frac{h}{p} . \tag{13.6}$$

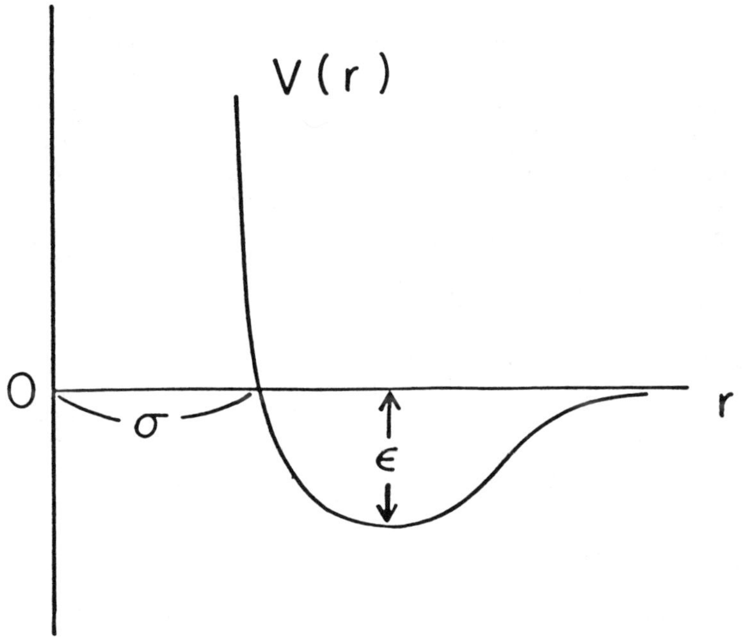

Fig. 9.7 The Lennard-Jones potential (13.2) representing the interaction between inert-gas atoms is characterized by the potential depth and the linear size σ .

A molecule in a dilute gas at temperature T has an average kinetic energy

$$\langle \frac{p^2}{2m} \rangle = \frac{3}{2} k_B T . \tag{13.7}$$

From this, we can obtain

$$\langle p_x^2 \rangle = \langle p_y^2 \rangle = \langle p_z^2 \rangle = m k_B T . \tag{13.8}$$

We now define the thermal de Broglie wavelength λ_T by

$$\lambda_T \equiv \frac{h}{\sqrt{mk_BT}} \equiv \frac{2\pi\hbar}{\sqrt{mk_BT}}. \tag{13.9}$$

Loosely speaking, this wavelength λ_T is a measure of the uncertainty associated with the linear position of a molecule in the gas at temperature T.

Consider now a binary encounter of molecules. The quantum mechanical diffraction effect is important when the de Broglie wavelength λ_T becomes comparable to, or greater than, the molecular dimension characterized by the range of force R,

$$\lambda_T \geqq R . \tag{13.10}$$

Using (13.1), (13.4), (13.5) and (13.9), we can re-express this inequality by

$$T \leqq \begin{cases} 24.2 \text{ K} & \text{for He}^3 \\ 18.2 \text{ K} & \text{for He}^4 . \end{cases} \tag{13.11}$$

In this temperature region one must treat the collision process quantum mechanically. Furthermore, if both colliding particles are identical, the indistinguishability requires that we take into account the direct and exchange collision simultaneously. This is important in the same temperature region as specified in (13.11).

If the temperature is further lowered so that the de Broglie wavelength λ_T becomes comparable to, or greater than, the average molecular separation characterized by $(N/V)^{-1/3}$:

$$\lambda_T > \left(\frac{N}{V}\right)^{-1/3} , \tag{13.12}$$

the quantum statistical effects become important. This condition (13.12) is satisfied by the entire liquid-state region :

$$0 < T < \begin{cases} 3.2\text{ K} & \text{for He}^3 \\ 4.2\text{ K} & \text{for He}^4 \,. \end{cases} \qquad (13.13)$$

In this temperature region we must treat the many-body problem on the basis of quantum statistics, that is, the Bose-Einstein statistics for liquid He^4, and the Fermi-Dirac statistics for liquid He^3. See further discussions in §10.7.

As we see in (13.9), the thermal de Broglie wavelength λ_T depends on the mass of a molecule as well as on the temperature T. At a fixed T, the heavier the mass, the smaller is the λ_T, which means that heavier molecules behave more like classical particles than lighter molecules. Our numerical estimate (13.11) indicates that a gas of He molecules as well as that of heavier molecules can be treated classical mechanically at room temperature, as we assumed in the earlier classical treatment in Chapter 7.

Problem 13.1 The intermolecular potential V(r) for monoatomic molecules is often represented by the Lennard-Jones (6, 12) function

$$V(r) = 4\varepsilon\left[\left(\frac{\sigma}{r}\right)^{12} - \left(\frac{\sigma}{r}\right)^{6}\right].$$

(a) Find the distance at which V(r) = 0.

(b) Find the distance at which dV/dr = 0.
What is the value of V at this distance?

(c) Draw a curve representing V(r).

References

[1] Pauli, W., Phys.Rev. 58, 716 (1940)

Review Questions

1. Write down mathematical symbols, expressions and/or figures and explain briefly :

(a) Four group properties

(b) Even and odd permutations. Cite examples

(c) The sign of the parity of a permutation

(d) Symmetric states for two identical bosons

(e) Antisymmetric states for fermions

(f) Pauli's exclusion principle

(g) Occupation number representation of a many-body quantum state

(h) Quantum Gibbs ensemble. Its characterization in terms of the density operator

(i) The canonical ensemble and canonical partition function

(j) The grand canonical ensemble

(k) The Bose, Fermi and Boltzmann distribution functions

(l) Classical limit

(m) The thermal de Broglie wavelength.

General Problems

1. State all four group properties. Check if the following sets satisfy the group properties:

(a) All integers including zero with respect to the addition. Point out the identity, and the inverses, if any.

(b) All integers including zero with respect to the multiplication.

(c) All rational number excluding zero with respect to the multiplication.

2. Consider a system of two identical fermions and boson, all moving independently along a straight line.

(a) Construct a ket representing a state of the system in which the two fermions and boson occupy different momentum states (p_a, p_b, p_c). Explain the restriction imposed on the symmetry of the state.

(b) Can the two fermions occupy the same momentum state? Explain your answer.

(c) Can one of the fermions and the boson occupy the same momentum state? Explain your answer.

(d) Does your ket constructed in (a) help explain your answers in (b) and (c) in any way?

(e) Do your answers in (b) and (c) depend on whether or not there are interactions between the particles?

3. In general, the quantum statistical results should approach the classical ones in the classical limit: $\hbar \to 0$. In particular the Bose distribution function $f(\varepsilon;\beta,\mu) = e^{\beta(\mu-\varepsilon)}\,[1 - e^{\beta(\mu-\varepsilon)}]^{-1}$ should approach the Boltzmann distribution function $f_c = e^{\beta(\mu-\varepsilon)}$.

This is certainly so if

$$e^{\beta(\mu-\varepsilon)} \ll 1 \text{ for all } \varepsilon > 0, \tag{1}$$

which seems to mean that μ must be negative and large in magnitude, and $\beta \equiv (k_B\ T)^{-1}$ is large and therefore the temperature is low. On the other hand, we know that the classical approximation should hold in the high temperature limit. Resolve this apparent paradox. Apply your arguments to the similar case of the Fermi distribution function.

Hint : Use the normalization condition for a start.

4. Let us consider a quantum mechanical simple harmonic oscillator, characterized by the Hamiltonian $H = \frac{1}{2}\, m^{-1} p^2 + \frac{1}{2}\, kx^2$.

(a) Write down the energy eigenvalue equation, using the ket vector notation and the wave mechanical notation.

(b) Show that $\varphi_0 = e^{-\alpha x^2}$ is an eigenfunction with the energy $E_0 = \frac{1}{2}\hbar\,\omega_0 = \frac{1}{2}\,\hbar\,(k/m)^{1/2}$.

(c) Write down expressions for all of the energy eigenvalues.

(d) Write down the equation of motion for the quantum state $|\psi,t\rangle$.

(e) Write down the Liouville equation for the density operator $\rho(t)$.

(f) Prove that the canonical density operator $c\ e^{-\beta H}$ is stationary, using the Liouville equation.

(g) Evaluate the partition function $Z \equiv \mathrm{Tr}\ \{e^{-\beta H}\}$.

(h) Evaluate the average energy $\langle H\rangle \equiv \mathrm{Tr}\ \{\ H\ e^{-\beta H}\}\ /\mathrm{Tr}\ \{e^{-\beta H}\}$ and the heat capacity, explicitly.

(i) Evaluate the averages $\langle x\rangle$ and $\langle p\rangle$.

(j) Evaluate the averages $\langle \frac{1}{2}kx^2\rangle$ and $\langle \frac{1}{2}m^{-1}p^2\rangle$. Are they equal to each other ?

Chapter 10. CONDUCTION ELECTRONS AND LIQUID HELIUM

In this chapter two typical quantum statistical systems are discussed. In a metal, conduction electrons move almost freely. The electrons obey the Fermi-Dirac statistics, and the thermal properties of moving electrons are very different from what classical statistical mechanics predicts. This topic will be discussed in the first half of this chapter, §§10.1 - 10.6, mainly in terms of the free electron model.

Liquid He^4 undergoes the so-called superfluid transition at 2.2 K. This phenomenon is regarded as a manifestation of the fact that He^4 molecules obey the Bose-Einstein statistics. In the second half of this chapter,

§§10.7 - 10.9, the thermal properties of free bosons and in particular the Bose-Einstein condensation will be discussed.

10.1 Conduction Electrons in a Metal

A metal is a substance in which electrical current can flow with little resistance. It is known that such electrical current is generated by moving electrons. The electron has a mass m_e and an electrical charge, $-e$, which is negative by convention. Their numerical values are

$$m_e = 9.109 \times 10^{-28} \text{ g} = 9.109 \times 10^{-31} \text{ kg}$$
$$e = 4.8023 \times 10^{-10} \text{ esu} = 1.6021 \times 10^{-19} \text{ C (Coulomb)}.$$

The mass of an electron is smaller by a factor of about 1840 than the least massive atom, i.e. a hydrogen atom. This makes the electron extremely mobile. Also, it makes its quantum nature more pronounced.

The electrons participating in the transport of charge, called <u>conduction electrons</u>, are those which would have orbited in the outermost shells surrounding the atomic nuclei if the nuclei were separated from each other. Other electrons which are more tightly bound with the atomic nuclei form part of the metallic ions. In a pure crystalline metal, these metallic ions form a relatively immobile array of regular spacing, called a <u>lattice</u>. Loosely speaking, a metal can be pictured as a system of two components: mobile electrons and relatively immobile lattice ions.

Let us take a monovalent metal such as copper (Cu) or sodium (Na). The Hamiltonian H of the system can be represented by

$$H = \sum_{j=1}^{N} p_j^2 (2m)^{-1} + \sum_{j>k}\sum \frac{k_o e^2}{|\vec{r}_j - \vec{r}_k|} + \sum_{\alpha=1}^{N} P_\alpha^2 (2M)^{-1}$$

$$+ \sum_{\alpha>\gamma}\sum \frac{k_o e^2}{|\vec{R}_\alpha - \vec{R}_\gamma|} + \sum_j \sum_\alpha \frac{-k_o e^2}{|\vec{r}_j - \vec{R}_\alpha|} ; \qquad [k_o \equiv (4\pi\varepsilon_o)^{-1}] \qquad (1.1)$$

the sums on the r.h.s. represent respectively the kinetic energy of electrons, the interaction energy among electrons, the kinetic energy of ions, the interaction energy among ions, and the interaction energy between electrons and ions. The metal as a whole is electrically neutral, and therefore the number of electrons should equal the number of ions. Both numbers are denoted by N.

At very low temperatures, the ions will be almost stationary near the equilibrium lattice points. [Because of the zero-point motion, [see §8.9], the ions are not completely at rest even at 0 K. But this fact does not affect the following argument in a substantial manner.] Then, the system can be viewed as one in which the electrons move about in a periodic lattice potential. The Hamiltonian of this idealized system, which now depends on the electron variables only, can be written as

$$H = \sum_j p_j^2 (2m)^{-1} + \sum_{j<k}\sum \frac{k_o e^2}{|\vec{r}_j - \vec{r}_k|} + \sum_j V(\vec{r}_j) + C, \qquad (1.2)$$

where $V(\vec{r}_j)$ represents the lattice potential, and the constant energy C depends on the lattice configuration.

For the sake of argument, let us drop the Coulomb interaction energy from (1.2). We then have

$$H = \sum_j p_j^2 (2m)^{-1} + \sum_j V(\vec{r}_j) + \text{constant}, \qquad (1.3)$$

which now characterizes a system of non-interacting electrons in the lattice.

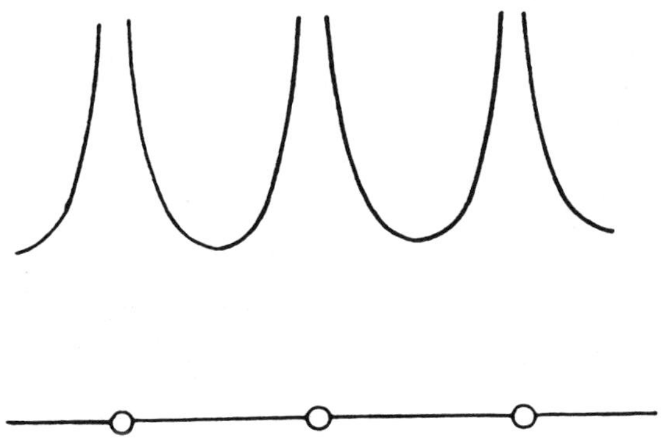

Fig. 10.1 A periodic potential in one dimension. The small circles represent the static lattice points occupied by ions.

In Figure 10.1, we draw a typical lattice potential (in one dimension). Quantum mechanical calculations show (Bloch's theorem) [1] that the wave function ψ, which satisfies the Schrödinger equation:

$$[- \frac{\hbar^2}{2m} \nabla^2 + V(\vec{r})] \psi(\vec{r}) = E\psi(\vec{r}), \tag{1.4}$$

is of the form

$$\psi_{\vec{k}}(\vec{r}) = e^{-i\vec{k}\cdot\vec{r}} u_{\vec{k}}(\vec{r}), \tag{1.5}$$

where $u_{\vec{k}}(\vec{r})$ has the same lattice periodicity as the potential $V(\vec{r})$. The wave vector $\vec{k}$ in (1.5) is real for an infinitely-extended lattice. The absolute square of the wave function,

$$|\psi_{\vec{k}}(\vec{r})|^2 = |u_{\vec{k}}(\vec{r})|^2, \tag{1.6}$$

then has the periodicity of the lattice. That is, the probability distribution function for the electron has the same periodicity as that of the lattice. The associated energy eigenvalues E have forbidden regions (energy gaps), and the energy eigenstates are characterized by the wave vector $\vec{k}$ and the zone number j, which enumerates the allowed energy bands:

$$E = E_j(\vec{k}). \tag{1.7}$$

A typical set of the energy bands is schematically drawn in Figure 10.2.

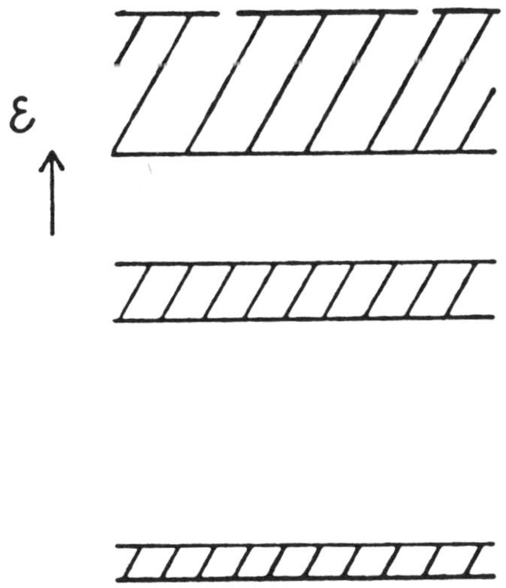

Fig. 10.2 A typical set of the energy bands for the electrons in a crystal.

At the absolute zero of temperature, these energy bands are filled with electrons from the bottom up, the upper limit being provided by the so-called Fermi energy. This is due to Pauli's exclusion principle: no two electrons can occupy the same quantum state.

If the uppermost energy band is completely filled with electrons, no electrons can gain energy in a continuous manner. In this case the electrons cannot be accelerated by an applied electric field, and the material having such an electron configuration behaves like an insulator. This is true, for example, for all crystals formed by inert gas atoms.

Consider now the case in which the uppermost energy band is partially filled with electrons, and the Fermi energy lies in the middle of the band. If a small electric field $\vec{E} = -\nabla\varphi$ is applied, those electrons with energies close to the Fermi energy can gain the electric energy $-e\,\varphi$ in a continuous manner and can be accelerated just as the classical electrons can be. This case corresponds to a metallic material. The detailed quantum mechanical calculations [1] show that those electrons in the partially filled band can flow in much the same manner as electrons in free space, whose energy-momentum relation,

$$E_{free} = \frac{p^2}{2m} \equiv \frac{\hbar^2}{2m} k^2, \tag{1.8}$$

is a simple function of the wave vector $\vec{k}$ with no energy gap. Only the response of the electric field is different. For most metals, the response of the conduction electrons to the field can be adequately described by the effective mass approximation. That is, the conduction electrons respond with a certain inertial mass m^*, whose value is slightly different from the free electron mass m_e, depending on the crystal structure and the electron orbital configuration of the metal in question. The dynamical behavior of the conduction electrons can then be characterized by the energy momentum relation,

$$E_{conduction\ electron} = \frac{p^2}{2m^*} \equiv \frac{\hbar^2}{2m^*} k^2. \tag{1.9}$$

The foregoing discussions in terms of the energy bands are based on the Hamiltonian in (1.3) with the neglect of the Coulomb interaction,

$$U \equiv \sum_{j>k}\sum \frac{k_o e^2}{|\vec{r}_j - \vec{r}_k|} . \tag{1.10}$$

This interaction is very important in the calculations of the actual properties of metals. However, inclusion of this interaction in the quantum mechanical calculations does not destroy such outstanding features as the energy bands with gaps and the sharply defined Fermi energy. In this sense, the Coulomb interaction plays a secondary role. Besides, this interaction, which acts on electrons in pairs, introduces a correlation among the electrons' motion and its proper treatment is quite difficult. In the present text, we will limit our discussions to those properties accountable in terms of free electrons with the effective mass m*.

10.2 Free Electrons. Fermi Energy

A system of free electrons is characterized by the Hamiltonian

$$H = \sum_{j=1}^{N} \frac{p_j^2}{2m} . \tag{2.1}$$

According to the study in §8.12, the momentum eigenstates for a quantum particle with a cube-shaped periodic boundary condition are characterized by three quantum numbers:

$$p_{x,j} \equiv \frac{2\pi\hbar}{L} j, \quad p_{y,k} \equiv \frac{2\pi\hbar}{L} k, \quad p_{z,\ell} \equiv \frac{2\pi\hbar}{L} \ell, \tag{2.2}$$

where j, k and ℓ are integers. For simplicity, we indicate the momentum states by a single Greek letter:

$$\vec{p}_\varkappa \equiv (p_{x,j}\ ,\ p_{y,k}\ ,\ p_{z,\ell}). \qquad (2.3)$$

The quantum state of the many-electron system can be specified by the set of occupation numbers $\{n_\varkappa\}$ with each $n_\varkappa \equiv n_{\vec{p}_\varkappa}$ taking on either one or zero. The ket vector representing such a state will be denoted by

$$|\{n\}\rangle \equiv |\{n_\varkappa\}\rangle. \qquad (2.4)$$

The corresponding eigenvalue is given by

$$E(\{n\}) = \sum_\varkappa \varepsilon_\varkappa n_\varkappa, \qquad (2.5)$$

where $\varepsilon_\varkappa \equiv (2m)^{-1} p_\varkappa^{\ 2}$ is the kinetic energy of the electron with momentum $\vec{p}_\varkappa$. The sum of the occupation numbers, $n_\varkappa$, equals the total number N of electrons:

$$\sum_\varkappa n_\varkappa = N. \qquad (2.6)$$

Let us suppose that the system is in thermodynamic equilibrium, which is characterized by temperature $T \equiv (k_B\beta)^{-1}$ and number density n. The thermodynamic properties of the system can then be computed in terms of the grand canonical ensemble with the density operator

$$\rho_G \equiv e^{\alpha N-\beta H}/TR\{e^{\alpha N-\beta H}\}. \qquad (2.7)$$

Earlier in §9.11 we saw that the average occupation number for the momentum state $p_\varkappa$ is characterized by the Fermi distribution function:

$$\langle n_\varkappa\rangle \equiv TR\{n_\varkappa e^{\alpha N-\beta H}\}/TR\{e^{\alpha N-\beta H}\}$$

$$= \frac{1}{e^{\beta\varepsilon_\varkappa-\alpha}+1} \equiv f(\varepsilon_\varkappa)\ . \qquad \text{[use of (9.11.7)]} \qquad (2.8)$$

The parameter α in this expression is to be determined from [see (9.11.14)]:

$$n = \text{Lim} \frac{\langle N \rangle}{V} = \text{Lim} \frac{1}{V} \sum_{\varkappa} \frac{1}{e^{\beta\varepsilon_{\varkappa} - \alpha} + 1} \equiv \text{Lim} \frac{1}{V} \sum_{\varkappa} f(\varepsilon_{\varkappa}). \tag{2.9}$$

We now investigate the behavior of the Fermi distribution function $f(\varepsilon)$ at very low temperatures. Let us put

$$\alpha \equiv \beta\mu. \tag{2.10}$$

As we discussed in §7.12, the quantity μ represents the chemical potential. In the limit of low temperature the chemical potential μ approaches a positive constant μ_0, as we will see presently ($\mu \to \mu_0 > 0$). We plot the Fermi distribution function $f(\varepsilon)$ at $T = 0$ ($\beta = \infty$) against the energy ε by a solid curve in Figure 10.3

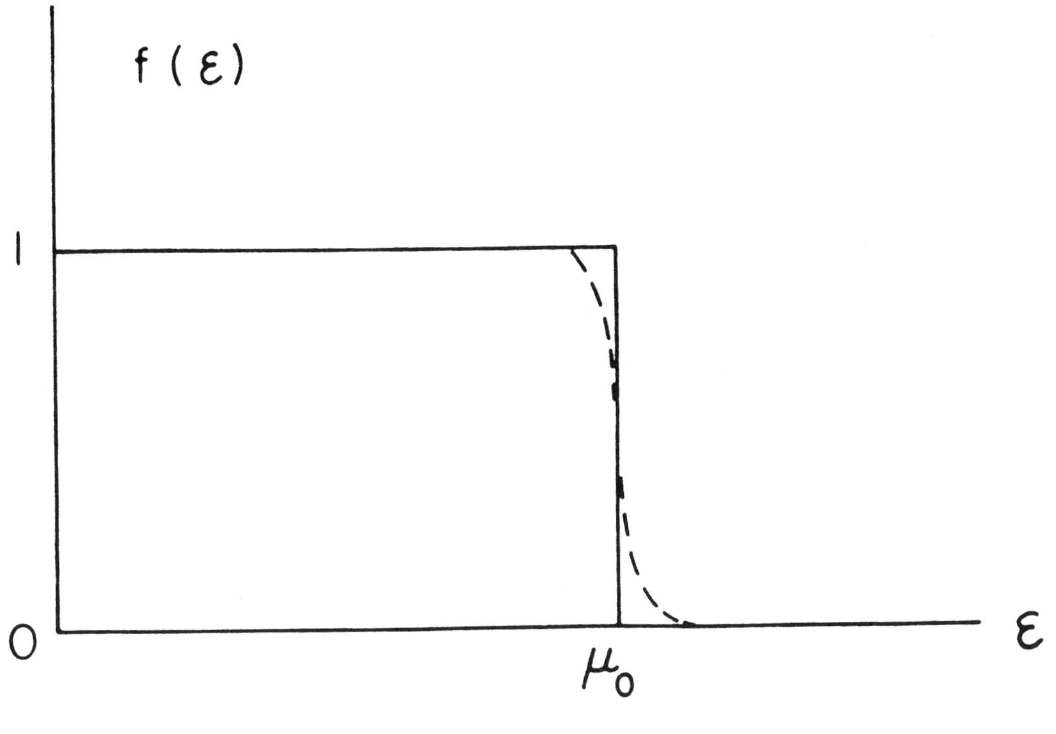

Fig. 10.3 The Fermi distribution function against the energy ε. The solid line is for $T = 0$ and the broken line for a small T.

It is a step function with the step at $\varepsilon = \mu_o$. This means that every momentum state $\vec{p}_\varkappa$ for which $\varepsilon_\varkappa \equiv p_\varkappa^{\,2}(2m)^{-1} < \mu_o$ is occupied with the probability 1, and any other momentum states are unoccupied. This special energy μ_o, called the Fermi energy, can be calculated in the following manner:

From (2.9), we have

$$n = \text{Lim} \frac{1}{V} \sum_{\varkappa} [f(\varepsilon_\varkappa)]_{T=0}$$

$$= \text{Lim} \frac{1}{V} \text{(the number of states } \varkappa \text{ for which } \varepsilon_\varkappa \leqq \mu_o). \qquad (2.11)$$

The momentum eigenstates in (2.2) can be represented by points in the three-dimensional momentum space as shown in Figure 10.4.

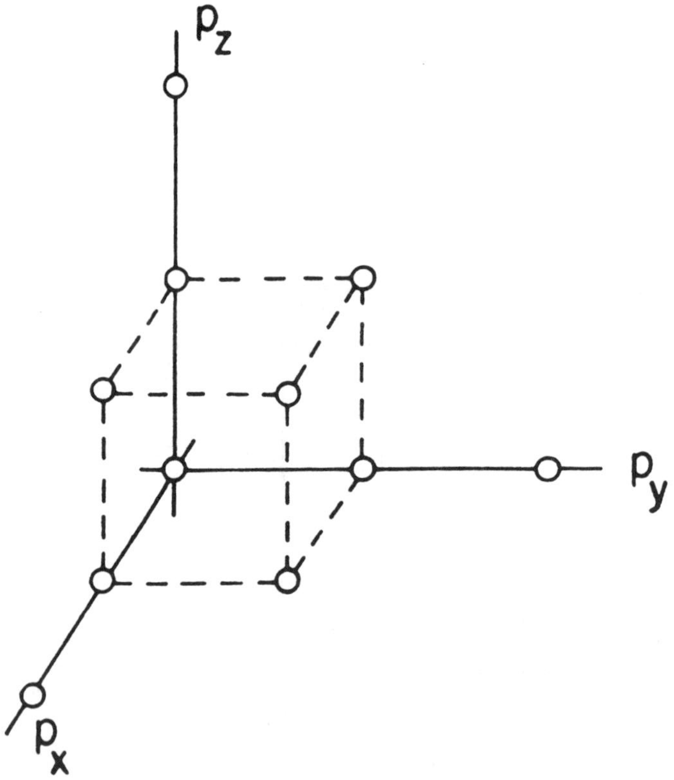

Fig. 10.4 The momentum states for a periodic cubic boundary condition are represented by points forming a simple cubic lattice with the lattice constant $2\pi\hbar/L$.

These points form a simple cubic lattice with the lattice constant 2 $\hbar/L$; the volume of the unit cell is equal to $(2\pi\hbar/L)^3$.

Let us define the Fermi momentum p_F by

$$\mu_o \equiv p_F^{\;2}\,(2m)^{-1} \;. \tag{2.12}$$

The number of occupied states will be equal to the number of lattice points within the sphere of radius p_F. Since one lattice point corresponds to one unit cell for the simple cubic lattice, this number is equal to the volume of the sphere, $(4\pi/3)p_F^{\;3}$, divided by the volume of the unit cell, $(2\pi\hbar/L)^3$. That is,

$$\text{the number of occupied states} \;=\; \frac{4\pi}{3}\, p_F^{\;3}/(2\pi\hbar/L)^3 \;. \tag{2.13}$$

Introducing this in (2.11), we obtain

$$n = \lim \frac{1}{L^3} \cdot \frac{4}{3}\,\pi p_F^{\;3} \cdot \frac{L^3}{(2\pi\hbar)^3} \;=\; \frac{1}{(2\pi\hbar)^3}\,\frac{4\pi}{3}\, p_F^{\;3} \;. \tag{2.14}$$

This result was obtained with the assumption of a periodic cubic boundary condition. The result obtained in the bulk limit, however, is valid independent of the shape of the boundary. [see Problem 2.1]

In our discussion so far, we have neglected the fact that an electron has a spin angular momentum (or simply spin) as an extra degree of freedom. It is known [see §12.4] that any quantum state for an electron must be characterized not only by the quantum numbers $(p_{x,j}\;,\; p_{y,k}\;,\; p_{z,\ell})$ describing its motion in the ordinary space, but also by the quantum number describing its spin. It is further known [see §12.5] that the electron has a permanent magnetic moment associated with its spin, and that the eigenvalues s_z of the z-component of the electronic spin are discrete and are restricted to the two values $\pm(\frac{1}{2})\hbar$. In the absence of the magnetic field, the magnetic potential energy is the same for both spin states. In the grand canonical ensemble, the states with the same energy are distributed with the same probability. In taking account of the spins, we must then multiply the r.h.s. of (2.14) by the factor 2, called the <u>spin degeneracy factor</u>. We thus obtain

$$n = \frac{8\pi}{3} \frac{p_F^{\,3}}{(2\pi\hbar)^3} . \qquad \text{[including the spin degeneracy]} \qquad (2.15)$$

After solving this equation for p_F, we obtain the Fermi energy as follows:

$$\mu_o \equiv \frac{p_F^{\,2}}{2m} = \frac{1}{2m} \left[\frac{3(2\pi\hbar)^3}{8\pi} n\right]^{2/3} \qquad \text{[use of (2.15)]}$$

or

$$\mu_o = \frac{\hbar^2}{2m} (3\pi^2 n)^{2/3} . \qquad (2.16)$$

Let us estimate the order of magnitude for μ_o by taking a typical metal, Cu. This metal has a specific weight of 9 g/cm^3 and a molecular weight of 63.5, yielding the number density $n = 8.4 \times 10^{22}$ electrons/cm^3 if one conduction electron is contributed by each Cu atom. Using this value for n, we find that

$$\mu_o \equiv k_B T_F \qquad (2.17)$$

$$T_F = 80{,}000 \text{ K} . \qquad (2.18)$$

The quantity T_F defined by (2.17) is called the <u>Fermi temperature</u>. The value found for the Fermi energy $\mu_o \equiv k_B T_F$ is very high compared with the thermal excitation energy of the order $k_B T$ per particle, which we will observe later in §10.5. This makes the thermodynamic behavior of the conduction electrons at room temperature drastically different from that of a classical gas.

The Fermi energy, by definition, is the chemical potential at the absolute zero. We may look at this connection in the following manner. For a box of a finite volume V, the momentum states are quantized as shown in Figure 10.4. As the volume V is raised, the unit cell volume in the momentum space, $(2\pi\hbar/L)^3$, decreases like V^{-1}. However, in the process of the bulk limit we must increase the number of electrons, N, in proportion to V. Therefore, the radius of the <u>Fermi sphere</u> within which all momentum states are filled with electrons neither grows nor shrinks. Obviously, this configuration corresponds to the lowest energy state for the system. The Fermi energy $\mu_0 \equiv p_F{}^2(2m)^{-1}$ represents the energy of an electron at the surface of the Fermi sphere. If we attempt to add an extra electron to the Fermi sphere, we must bring in an electron with an energy equal to μ_0.

Problem 2.1 The momentum eigenvalues for a particle in a rectangular box with sides of unequal length (L_1, L_2, L_3) are given by [see Problem 5.12.2]

$$p_{x,j} \equiv \frac{2\pi\hbar}{L_1} j, \quad p_{y,k} \equiv \frac{2\pi\hbar}{L_2} k, \quad p_{z,\ell} \equiv \frac{2\pi\hbar}{L_3} \ell.$$

Assuming that the particles are fermions, verify that the Fermi energy is the same in the bulk limit.

10.3 The Desity of States in the Momentum Space

In many applications of quantum statistical mechanics we meet the need for converting the sum over quantum states into an integral. This conversion becomes necessary when we first find discrete quantum states for a finite box, and then seek the sum over states in the bulk limit. [The necessity of such a conversion does not arise in the spin problem, which we will treat in Chapter 12.] This conversion is often a welcome procedure because the resulting integral, in general, is easier to handle than the sum. The conversion is of a purely mathematical nature, but it is an important step in carrying out statistical mechanical computations.

Let us first examine a sum over momentum states corresponding to a one-dimensional motion. Let us take the sum

$$\sum_k A(p_k), \tag{3.1}$$

where $p_k \equiv 2\pi\hbar k/L$; and $A(p)$ is an arbitrary function of p. The discrete momentum states are equally spaced, as shown by short bars in Figure 10.5. As the normalization length is made greater, the spacing (distance) between two successive states, $2\pi\hbar/L$, becomes smaller. This means that the number of states per unit momnetum interval increases as L increases. We denote the number of states within a small momentum interval Δp by Δn. We now take the ratio,

$$\frac{\Delta n}{\Delta p} \equiv \frac{\text{number of states in } \Delta p}{\Delta p}\,.$$

Dividing both the numerator and denominator by the number of states, we get

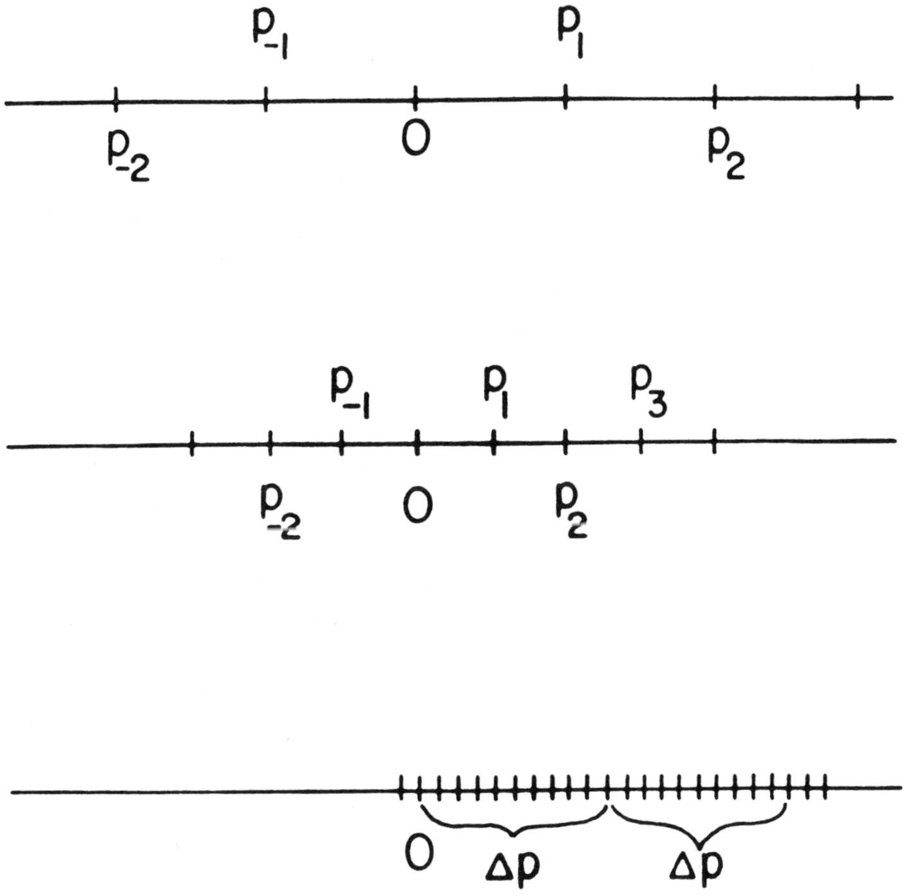

Fig. 10.5 The linear momentum states are represented by short bars forming a linear lattice with unit spacing equal to $2\pi\hbar/L$. As the normalization length L is made greater, the spacing becomes smaller.

$$\frac{\Delta n}{\Delta p} = \frac{1}{\text{momentum spacing per state}}$$

$$= \frac{1}{2\pi\hbar/L} = \frac{L}{2\pi\hbar} \,. \tag{3.2}$$

Note that this ratio $\Delta n/\Delta p$ increases linearly with the normalization length L.

Let us now consider a quantity

$$\sum_{\ell} A(p_\ell) \frac{\Delta n}{\Delta_\ell p} \Delta_\ell p \ , \tag{3.3}$$

where $\Delta_\ell p$ is the ℓ-th interval and p_ℓ represents a typical value of p within the interval $\Delta_\ell p$, say the p-value at the midpoint of $\Delta_\ell p$. Since

$$[\ \frac{\Delta n}{\Delta_\ell p} \Delta_\ell p\ ] = [\Delta n] = (\text{number}),$$

the two sums (3.1) and (3.3) have the same dimension. Furthermore, their numerical values will be close if (a) the function A(p) is a smooth function of p, and (b) there exist many states in $\Delta_\ell p$ so that $\Delta n/\Delta_\ell p$ can be regarded as the density of states. The condition (b) is satisfied for the momentum states $\{p_k\}$ when the length L is made sufficiently large. We can then expect that in the bulk limit, by choosing intervals infinitesimally small: $\Delta_\ell p \to dp$, expressions (3.1) and (3.3) will have the same value. That is,

$$\sum_{k(\text{states})} A(p_k) \to \sum_{\Delta_\ell p} A(p) \frac{\Delta n}{\Delta_\ell p} \Delta_\ell p \tag{3.4}$$

as $L \to \infty$. But by definition, the sum on the r.h.s. becomes the integral $\int dp\ A(p)\ dn/dp$ where

$$\frac{dn}{dp} \equiv \lim_{\Delta p \to 0} \frac{\Delta n}{\Delta p} = \frac{L}{2\pi\hbar} . \qquad [\text{use of } (3.2)] \tag{3.5}$$

In summary, we therefore have

$$\sum_k A(p_k) \to \int_{-\infty}^{\infty} dp\ A(p) \frac{dn}{dp} . \tag{3.6}$$

It is stressed that the condition (a) depends on the nature of the function A. Therefore, if A(p) is singular at some point, the condition (a) is not satisfied and this may invalidate the limit (3.6). Such exceptional cases do occur. See §10.8. We further note that the <u>density of states</u> $dn/dp = L(2\pi\hbar)^{-1}$ does not depend on the momentum.

The sum-to-integral conversion, which we have discussed, can easily be generalized for a multi-dimensional case. For example, in three dimensions, we have

$$\sum_{\vec{p}_k} A(\vec{p}_k) \to \int d^3p\, A(\vec{p})\, D(\vec{p}), \qquad (\text{as } V \equiv L^3 \to \infty) \tag{3.7}$$

where

$$D(\vec{p}) \equiv \frac{dn}{d^3p} \tag{3.8}$$

is the density of states per unit volume in the momentum space.

Let us choose the periodic cubic box of side length L for the normalization. The density of state dn/d^3p can then be calculated by extending the arguments leading to (3.2). The result is given by

$$D(p) = \frac{1}{(2\pi\hbar/L)^3} = \frac{L^3}{(2\pi\hbar)^3}\ . \tag{3.9}$$

For electrons, the spin degeneracy doubles this density. We therefore obtain

$$D(p) \equiv \frac{dn}{d^3p} = \frac{2L^3}{(2\pi\hbar)^3}\ . \qquad [\text{with spin degeneracy}] \tag{3.10}$$

Let us now use this result and simplify the normalization condition (2.9) which relates the chemical potential μ with the average density n:

$$n = \text{Lim}\, \frac{1}{V} \sum_{\varkappa} f(\varepsilon_{\varkappa}).$$

By choosing $A(p) = f(p^2/2m)$, we obtain

$$n = \text{Lim}\, \frac{1}{V} \int d^3p\; f(p^2/2m)\, \frac{dn}{d^3p}$$

$$= \text{Lim}\, \frac{1}{V} \int d^3p\; f(p^2/2m)\, \frac{2L^3}{(2\pi\hbar)^3} \qquad \text{[use of (3.9)]}$$

$$= \frac{2}{(2\pi\hbar)^3} \int d^3p\; f(p^2/2m). \qquad [L^3 = V] \qquad (3.11)$$

As a second example, we take the energy density of the system. From (2.5) and (2.6) we obtain

$$\bar{e} = \text{Lim}\, \frac{1}{V} \langle H \rangle = \text{Lim}\, \frac{1}{V} \sum_{\varkappa} \varepsilon_{\varkappa}\, f(\varepsilon_{\varkappa})$$

$$= \text{Lim}\, \frac{1}{V} \int d^3p\; \frac{p^2}{2m}\, f(p^2/2m)\, \frac{dn}{d^3p}$$

$$= \frac{2}{(2\pi\hbar)^3} \int d^3p\; \frac{p^2}{2m}\, f(p^2/2m)\;. \qquad (3.12)$$

Eqs. (3.11) and (3.12) were obtained starting with the momentum eigenvalues corresponding to the periodic boundary condition. The results in the bulk limit, however, do not depend on the choice of the boundary condition.

10.4 The Density of States in Energy

The concept of the density of states can also be applied to the energy domain. This is convenient when the sum over states has the form

$$\sum_{\varkappa} g(\varepsilon_{\varkappa}), \tag{4.1}$$

where $g(\varepsilon_{\varkappa})$ is a function of the energy $\varepsilon_{\varkappa}$ associated with the state $\varkappa$. The sums appearing in eqs. (2.5) and (2.9) are precisely of this form.

Let dn be the number of states within the energy interval $d\varepsilon$. In the bulk limit this number dn will be proportional to the interval $d\varepsilon$ so that

$$dn = D(\varepsilon)\, d\varepsilon. \tag{4.2}$$

Here, the proportionality factor

$$D(\varepsilon) \equiv \frac{dn}{d\varepsilon} \tag{4.3}$$

is called the density of states in the energy domain. This quantity $D(\varepsilon)$, in general, depends on the location of the interval $d\varepsilon$, and therefore on a typical energy ε within $d\varepsilon$, say the energy at the mid-point of the interval $d\varepsilon$.

If the set of the states $\{\varkappa\}$ becomes densely populated in the bulk limit, and the function g is smooth, then the sum may be converted into an integral of the form:

$$\sum_{\varkappa(\text{states})} g(\varepsilon_{\varkappa}) \to \int d\varepsilon\, g(\varepsilon)\, D(\varepsilon). \tag{4.4}$$

Let us now calculate the density of states $D(\varepsilon)$ for the system of free electrons.

The number of states, dn, in the spherical shell in momentum space shown in Figure 10.4 is obtained by dividing the volume of the shell, $4\pi p^2 dp$, by the unit cell volume $(2\pi\hbar/L)^3$ and multiplying the result by the spin degeneracy factor 2:

$$dn = \frac{2 \cdot 4\pi p^2 dp}{(2\pi\hbar/L)^3} = \frac{8\pi p^2 dp}{(2\pi\hbar)^3} V . \tag{4.5}$$

Since the energy ε and the momentum (magnitude) p are related by

$$\varepsilon \equiv p^2/2m, \quad p = (2m\varepsilon)^{\frac{1}{2}}, \tag{4.6}$$

we obtain the relation

$$dp = \frac{dp}{d\varepsilon} d\varepsilon = (m/2\varepsilon)^{\frac{1}{2}} d\varepsilon . \qquad \text{[use of (4.6)]} \tag{4.7}$$

Using the last three equations, we obtain

$$dn = V \frac{8\pi(2m\varepsilon)}{(2\pi\hbar)^3} \left(\frac{m}{2\varepsilon}\right)^{\frac{1}{2}} d\varepsilon = V \frac{8\sqrt{2}\pi\, m^{3/2}}{(2\pi\hbar)^3} \varepsilon^{\frac{1}{2}} d\varepsilon . \tag{4.8}$$

Dividing both sides by $d\varepsilon$, we get

$$D(\varepsilon) \equiv \frac{dn}{d\varepsilon} = V \frac{\sqrt{2}\, m^{3/2}}{\pi^2 \hbar^3} \varepsilon^{\frac{1}{2}} \tag{4.9}$$

for the density of states. The density of states, $D(\varepsilon)$, grows like $\varepsilon^{\frac{1}{2}}$. Its general behavior is shown in Figure 10.6.

We may now re-express the normalization condition (2.9) as follows:

$$n = \lim \frac{1}{V} \sum_{\kappa} f(\varepsilon_{\kappa})$$

$$= \lim \frac{1}{V} \int_0^{\infty} d\varepsilon\, f(\varepsilon)\, D(\varepsilon) \qquad \text{[use of (4.4)]}$$

$$= \lim \frac{1}{V} \int_0^{\infty} d\varepsilon\, f(\varepsilon)\, V \frac{\sqrt{2}\, m^{3/2}}{\pi^2 \hbar^3} \varepsilon^{\frac{1}{2}} \qquad \text{[use of (4.9)]}$$

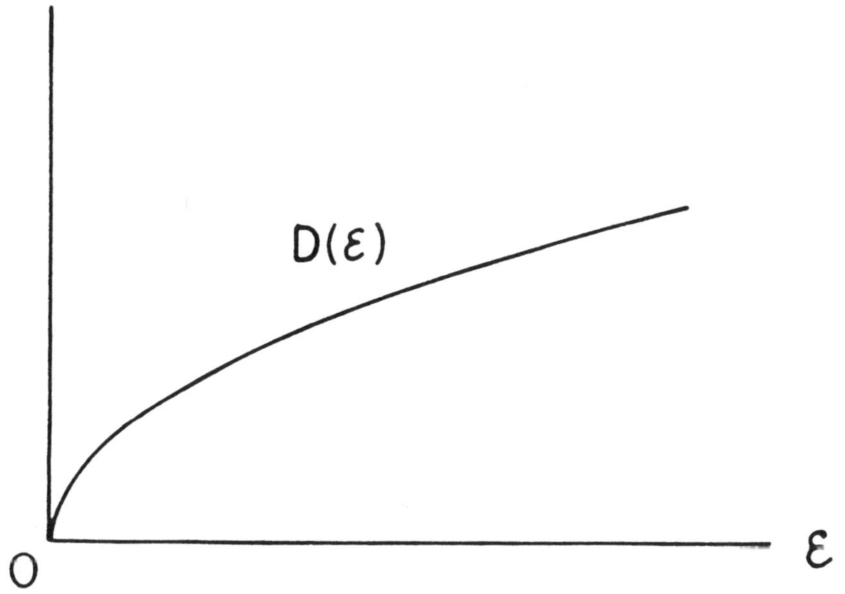

Fig. 10.6 The density of states in energy, $D(\varepsilon)$, for free electrons grows like $\sqrt{\varepsilon}$.

$$= \frac{\sqrt{2}\, m^{3/2}}{\pi^2 \hbar^3} \int_0^\infty d\varepsilon\ \varepsilon^{\frac{1}{2}}\ f(\varepsilon). \tag{4.10}$$

We note that the same result can be obtained from (3.11) in an alternative manner. In fact, let us use the spherical polar coordinates (p, ϑ, ϕ) for the momentum space integration. After integrating with respect to the angles (ϑ, ϕ) and rewriting the p-integral in terms of the ε-integral, we can obtain the same expression (4.10). [This procedure is carried out more explicitly in §10.8.] The density of states in the energy domain, defined through (4.2), is valid, however, even when we have states other than the momentum states. We will see such cases in later applications, see §§11.3 and 11.11.

Let us consider the zero temperature limit of (4.10). The Fermi distribution function $f(\varepsilon)$ then is a step function with the step at $\varepsilon = \mu_o$. We then obtain

$$n = \frac{\sqrt{2}\, m^{3/2}}{\pi^2 \hbar^3} \int_0^{\mu_o} d\varepsilon\, \varepsilon^{\frac{1}{2}} = \frac{\sqrt{2}\, m^{3/2}}{\pi^2 \hbar^3} \frac{2}{3} \mu_o^{3/2} .$$

Solving this equation for μ_o, we obtain

$$\mu_o = \frac{\hbar^2}{2m} (3\pi^2 n)^{2/3} . \qquad (4.11)$$

This result is in agreement with the previous calculation given in eq.(2.16).

The energy density given by $\bar{e} \equiv \text{Lim}\, V^{-1} \Sigma_\varkappa \varepsilon_\varkappa f(\varepsilon_\varkappa)$ can be expressed as

$$\bar{e} = \text{Lim} \frac{1}{V} \int_0^\infty d\varepsilon\, \varepsilon\, f(\varepsilon)\, D(\varepsilon) \qquad \text{[use of (4.4)]}$$

$$= \frac{\sqrt{2}\, m^{3/2}}{\pi^2 \hbar^3} \int_0^\infty d\varepsilon\, \varepsilon^{3/2} f(\varepsilon)\, D(\varepsilon) . \qquad \text{[use of (4.9)]} \qquad (4.12)$$

The zero-temperature limit of this expression can be calculated simply as follows:

$$\bar{e} = \frac{\sqrt{2}\, m^{3/2}}{\pi^2 \hbar^3} \int_0^{\mu_o} d\varepsilon\, \varepsilon^{3/2} = \frac{\sqrt{2}\, m^{3/2}}{\pi^2 \hbar^3} \frac{2}{5} \mu_o^{5/2}$$

$$= \frac{3}{5} n \mu_o . \qquad \text{[use of (4.11)]} \qquad (4.13)$$

This equation shows that the average energy per particle is equal to 3/5 of the Fermi energy μ_o.

Problem 4.1 Obtain (4.10) directly from (3.11) by using the spherical polar coordinates (p, ϑ, ϕ) in momentum space, integrating over the angles ϑ, ϕ and rewriting the p-integral in terms of the ε-integral.

10.5 The Heat Capacity of Degenerate Electrons. Qualitative Discussion

At room temperature most metals have molar heat capacities of about 3R just as for non-metallic solids. This experimental fact cannot be explained in terms of classical statistical mechanics. By applying the Fermi-Dirac statistics to conduction electrons, we can demonstrate the near absence of the electronic contribution to the heat capacity. In the present section, we will treat this topic in a qualitative manner.

Let us consider highly degenerate electrons with a large Fermi temperature T_F(80,000 K). At 0 K, the Fermi distribution function

$$f(\varepsilon;\mu,T) = \frac{1}{e^{(\varepsilon-\mu)/k_BT}+1} \tag{5.1}$$

is a step function, as indicated by the dotted line in the lower part of the diagram in Figure 10.7. At a temperature T of a few degrees, the abrupt drop near $\varepsilon = \mu_o$ becomes a smooth drop, as indicated by a solid line in the same diagram. In fact, the change in the distribution function $f(\varepsilon)$ is appreciable only in the neighborhood of $\varepsilon = \mu$. The function $f(\varepsilon;\mu,T)$ will drop from 1/2 at $\varepsilon = \mu$ to 1/101 at $\varepsilon - \mu = k_BT \ln 100$, [which can be directly verified from (5.1)]. This value $k_BT \ln 100 = 4.6\ k_BT$ is much less than the Fermi energy μ_o $k_B\ T_F$(80,000 K). This means that only those electrons with energies close to the Fermi energy μ_o are excited by the rise in temperature. In other words, the electrons with energies ε far below μ_o are not affected. There are

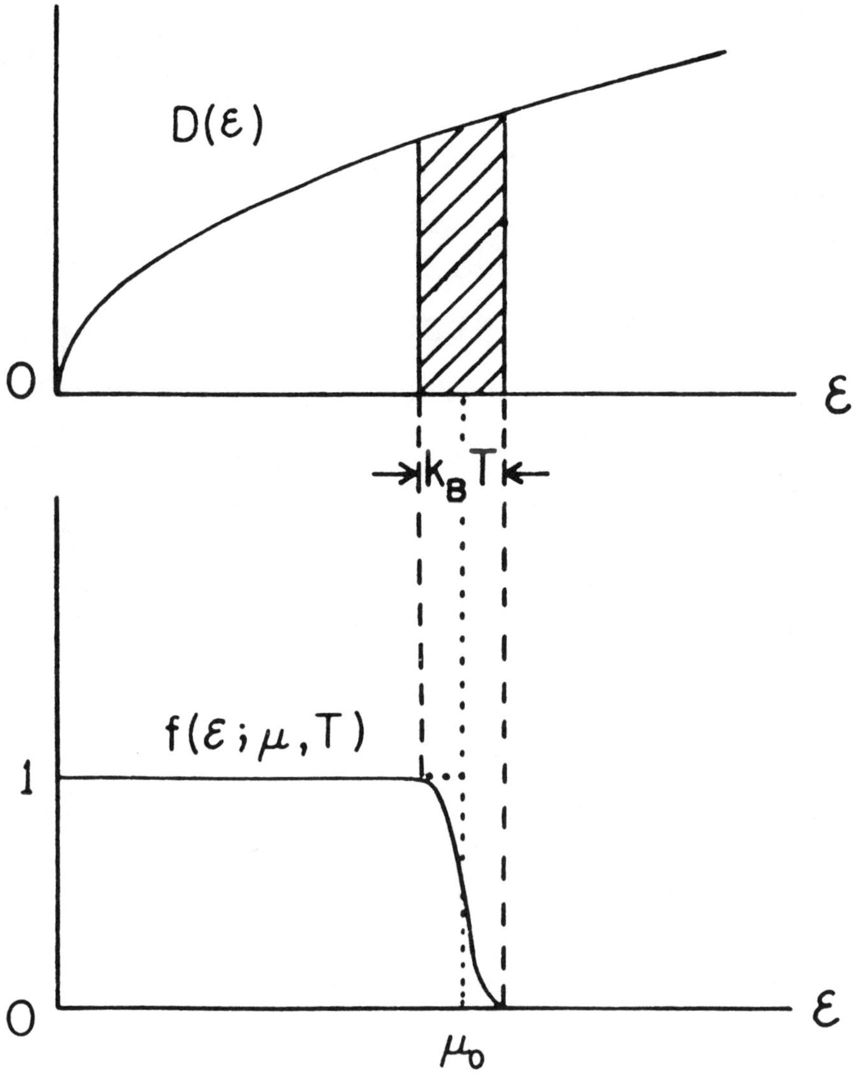

Fig. 10.7 The density of states in energy, D(ε), and the Fermi distribution function f(ε) are drawn as functions of the kinetic energy ε. The change in f is appreciable only near the Fermi energy μ_0 if $k_B T \ll \mu_0$. The shaded area in the upper part represents, approximately, the number of thermally excited electrons.

many such electrons, and in fact this group of unaffected electrons forms the great majority.

The number N_{ex} of electrons which are thermally excited can be found in the following manner. In the upper part of Figure 10.7, the density of state $D(\varepsilon)$ is drawn. Since $D(\varepsilon)d\varepsilon$ represents, by definition, the number of electrons within $d\varepsilon$, the integral of $D(\varepsilon)d\varepsilon$ over the interval in which the electron population is affected gives an approximation to the number of excited electrons, N_{ex}. This integral can be represented by the shaded area in the diagram. Since we know from the earlier arguments that the affected range of the energy is of the order k_BT, and $\mu_o >> k_BT$, we can estimate N_{ex} as follows:

$$N_{ex} = \text{the shaded area in the upper diagram}$$

$$= D(\mu_o)(k_BT) \ , \qquad (5.2)$$

where $D(\mu_o)$ is the density of states at $\varepsilon = \mu_o$ (Fermi energy). From (4.9) and (4.11), we obtain

$$D(\mu_o) = V \frac{\sqrt{2}\, m^{3/2}}{\pi^2 \hbar^3} \mu_o^{\frac{1}{2}} = \frac{3}{2} \frac{N}{\mu_o} \ . \qquad (5.3)$$

Using this expression, we get from (5.2)

$$N_{ex} \cong \frac{3}{2} N \frac{k_BT}{\mu_o} \ . \qquad (5.4)$$

The electrons affected will move up with extra energy of the order k_BT per particle. Therefore, the change in the total energy will be approximately given by

$$\Delta E = (N_{ex}) \times (k_BT) = \left(\frac{3}{2} N \frac{k_BT}{\mu_o}\right) k_BT = \frac{3}{2} N (k_BT)^2 \mu_o^{-1}. \qquad (5.5)$$

Differentiating this with respect to T, we obtain for the molar heat capacity:

$$C_V = \frac{\partial}{\partial T}\Delta E = 3N_o k_B \left(\frac{k_B T}{\mu_o}\right) \qquad [N_o = \text{Avogadro's number}]$$

$$= 3R\,\left(\frac{T}{T_F}\right)\,. \qquad [R = N_o k_B,\ \ \mu_o = k_B T_F] \tag{5.6}$$

This expression indicates that the heat capacity at room temperature (T = 300 K) is indeed small:

$$C_V = 3R\,(300/80{,}000) = \frac{3}{800}\,(3R) = 0.011\ R. \tag{5.7}$$

It is stressed that this result was obtained because the number of thermally excited electrons, N_{ex}, is much less than the total number of electrons, N. See (5.4). We also note that the electronic heat capacity is linear in the temperature.

In the next section, we will treat the same problem in a quantitative manner.

10.6* The Heat Capacity of Degenerate Electrons. Quantitative Calculation.

Historically, Sommerfeld (Arnold Sommerfeld, 1868 - 1951, German) first applied the Fermi-Dirac statistics to the conduction electrons and calculated the electronic heat capacity. His calculations resolved the heat capacity paradox mentioned in §8.1. In the present section we will carry out the calculations of the heat capacity in a quantitative manner.

The heat capacity at constant volume, C_V, can be calculated by differentiating the internal energy E with respect to the temperature T:

$$C_V = \frac{\partial E(T,V)}{\partial T} . \tag{6.1}$$

The internal energy density for free electrons was given in eq. (4.12),

$$\frac{E(T,V)}{V} = \frac{2^{1/2}m^{3/2}}{\pi^2\hbar^3} \int_0^\infty d\varepsilon\ \varepsilon^{3/2}\ f(\varepsilon;\mu,T). \tag{6.2}$$

Here μ is the chemical potential, and is related to the number density n by (4.10):

$$n = \frac{2^{1/2}m^{3/2}}{\pi^2\hbar^3} \int_0^\infty d\varepsilon\ \varepsilon^{1/2}\ f(\varepsilon). \tag{6.3}$$

The integrals on the r.h.s. of (6.2) and (6.3) may be evaluated in the following manner. Let us introduce

$$x \equiv \beta\varepsilon\ , \quad \alpha \equiv \beta\mu\ . \tag{6.4}$$

The function

$$\frac{1}{e^{x-\alpha}+1} \equiv F(x) \tag{6.5}$$

and the negative of its derivative

$$-F'(x) = \frac{e^{x-\alpha}}{(e^{x-\alpha}+1)^2} \quad (> 0) \tag{6.6}$$

are shown in Figure 10.8, where we assumed that

$$\alpha \equiv \beta\mu >> 1. \tag{6.7}$$

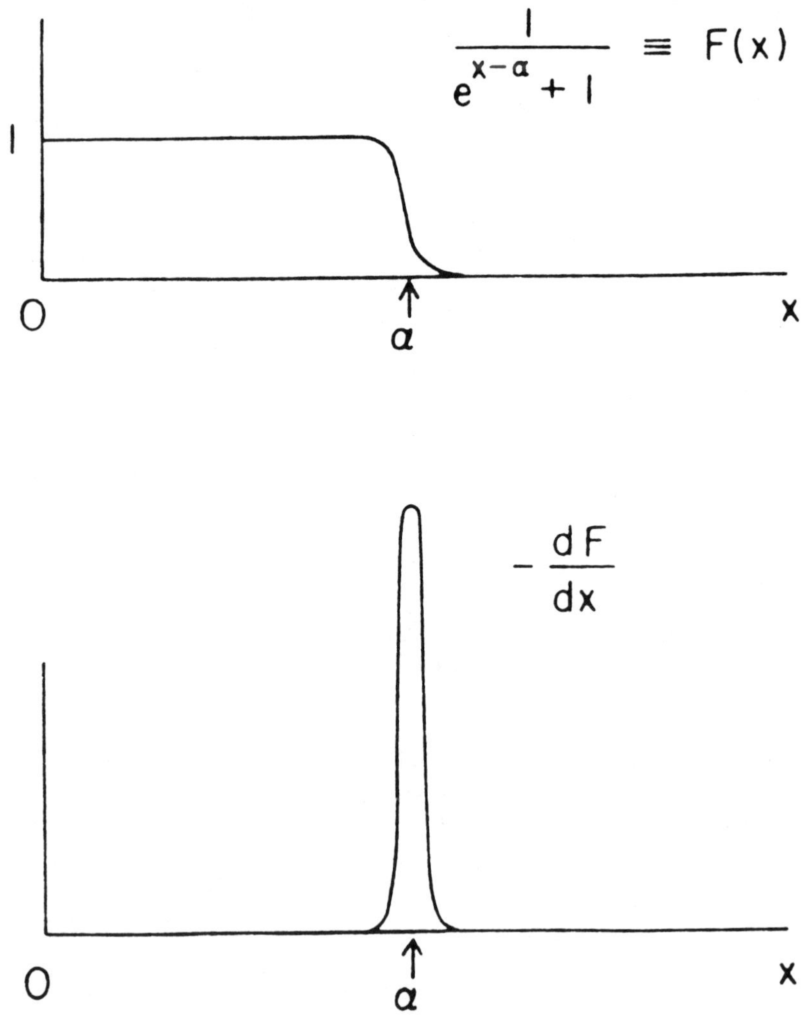

Fig. 10.8 The function $F(x) = (e^{x-\alpha} + 1)^{-1}$ and $-dF/dx$ are shown for positive x.

Let us take the integral

$$I \equiv \int_0^\infty dx\; F(x)\, \frac{d\varphi(x)}{dx}\,, \tag{6.8}$$

ere $\varphi(x)$ is a certain smooth function of x. By integrating by parts, we obtain

$$I = F(\infty)\;\varphi(\infty) - F(0)\;\varphi(0) - \int_0^\infty dx\; \frac{dF(x)}{dx}\;\varphi(x).$$

From (6.4) and (6.5), $F(\infty) = 0$ and $F(0) = 1$. We therefore obtain

$$I = -\varphi(0) - \int_0^\infty dx\; F'(x)\;\varphi(x). \tag{6.9}$$

We expand the function $\varphi(x)$ in a Taylor series at $x = \alpha$

$$\begin{aligned}\varphi(x) &\equiv \varphi[\alpha + (x-\alpha)]\\ &= \varphi(\alpha) + (x-\alpha)\;\varphi'(\alpha) + \tfrac{1}{2}(x-\alpha)^2\;\varphi''(\alpha) + \ldots\end{aligned} \tag{6.10}$$

If the function φ is smooth, the first few terms in this series should approximate φ in a reasonable manner. For example, the first three terms of the series (6.10):

$$\varphi(\alpha) + (x-\alpha)\varphi'(\alpha) + \tfrac{1}{2}(x-\alpha)^2\varphi''(\alpha) \equiv \psi(x) \tag{6.11}$$

have the same values in magnitude, gradient and curvature at $x = \alpha$ as the original function $\varphi(x)$ as indicated in Figure 10.9. Let us introduce the expansion of φ in eq. (6.9), and integrate term by term. To carry this out, let us consider

$$\begin{aligned}-\int_0^\infty dx\; F'(x)\;(x-\alpha)^n &= \int_0^\infty dx\; \frac{e^{x-\alpha}}{(e^{x-\alpha}+1)^2}\;(x-\alpha)^n \qquad \text{[use of (6.6)]}\\ &= \int_{-\alpha}^\infty dy\; \frac{e^y}{(e^y+1)^2}\;y^n \quad (\simeq J_n). \qquad [y = x-\alpha]\end{aligned} \tag{6.12}$$

By assumption, $\alpha \gg 1$ [see (6.7)]. We may then approximate the last integral by

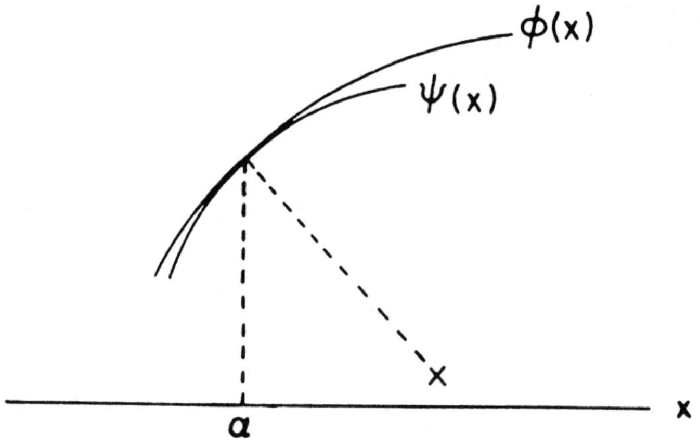

Fig. 10.9 The behavior of the function $\varphi(x)$ near $x = \alpha$ can be approximated by the function $\psi(x)$ which has the same magnitude, gradient and curvature as $\varphi(x)$.

$$\int_{-\infty}^{\infty} dy \frac{e^y}{(e^y+1)^2} y^n \equiv J_n \quad . \tag{6.13}$$

The definite integrals J_n vanish for odd n since the integrands are odd functions of y. For small even n, their numerical values are given by

$$J_0 = 1, \quad J_2 = \frac{\pi^2}{3} \quad . \tag{6.14}$$

Using the last several equations, we obtain

$$I \equiv \int_0^{\infty} dx\, F(x) \frac{d\varphi}{dx}$$

$$= \varphi(\alpha) - \varphi(0) + \frac{\pi^2}{6} \varphi''(\alpha). \tag{6.15}$$

We now apply this formula to evaluate the ε-integral in (6.3):

$$\int_0^\infty d\varepsilon\ \varepsilon^{\frac{1}{2}}\ f(\varepsilon)$$

$$= \beta^{-3/2} \int_0^\infty dx\ x^{\frac{1}{2}}\ F(x). \qquad [\beta\varepsilon = x]$$

We may choose

$$\varphi(x) = \frac{2}{3}x^{3/2}, \quad \frac{d\varphi}{dx} = x^{\frac{1}{2}}, \quad \frac{d^2\varphi}{dx^2} = \frac{1}{2}x^{-\frac{1}{2}}. \tag{6.16}$$

We then obtain

$$\int_0^\infty d\varepsilon\ \varepsilon^{1/2}\ f(\varepsilon) = \left(\frac{2}{3\beta^{3/2}}\right)(\beta\mu)^{3/2} + \frac{1}{\beta^{3/2}}\frac{\pi^2}{6}\frac{1}{2}(\beta\mu)^{-1/2}$$

$$= \frac{2}{3}\mu^{3/2}\left[1 + \frac{\pi^2}{8}\beta^{-2}\mu^{-2}\right]. \tag{6.17}$$

Using this, we obtain from (6.3)

$$n = \frac{2^{3/2}m^{3/2}}{3\pi^2\hbar^3}\mu^{3/2}\left[1 + \frac{\pi^2}{8}\beta^{-2}\mu^{-2}\right]. \tag{6.18}$$

In a similar manner, we can calculate the energy density from (6.2) and obtain

$$\bar{e} \equiv \text{Lim}\ \frac{E}{V} = \frac{2^{3/2}m^{3/2}}{5\pi^2\hbar^3}\mu^{5/2}\left[1 + \frac{5}{8}\pi^2\beta^{-2}\mu^{-2}\right]. \tag{6.19}$$

The chemical potential μ in general depends on density n and temperature T. This relation can be obtained from eq. (6.18). If we substitute the $\mu(n,T)$ so obtained into eq. (6.19), we get the internal energy density as a function of n and T. By subsequently differentiating this function with respect to T at fixed n, we can obtain the heat capacity at constant volume.

We can calculate the heat capacity by taking a route a little different from that prescribed above.

The Fermi energy μ_o by definition is the chemical potential at $T = 0$, and is given by $\hbar^2(3\pi^2 n)^{2/3}/2m$, see (2.16). This μ_o depends only on n. Solving eq.(6.18) for μ and expressing the result in terms of μ_o and T, we obtain [Problem 6.1]

$$\mu = \mu_o[1 - \frac{\pi^2}{12} \left(\frac{k_B T}{\mu_o}\right)^2]. \tag{6.20}$$

Introducing this expression into (6.19), we obtain [Problem 6.1]

$$\bar{e} = \frac{3}{5} n\mu_o \left[1 + \frac{5\pi^2}{12} \frac{k_B T}{\mu_o}\right]. \tag{6.21}$$

Differentiating this expression with respect to T at constant μ_o, we obtain the heat capacity as follows:

$$C_V = \frac{\pi^2}{2} N k_B^2 \, T/\mu_o. \tag{6.22}$$

Using the definition (2.17) of the Fermi temperature $T_F(k_B T_F \equiv \mu_o)$, we can then rewrite (6.22) as

$$C_V = N k_B \frac{\pi^2}{2} \left(\frac{T}{T_F}\right). \tag{6.23}$$

Here we see that the heat capacity C_V for degenerate electrons is greatly reduced by the factor T/T_F ($\ll 1$) from the ideal-gas heat capacity $\frac{3}{2} N k_B$. Also note that the heat capacity changes linearly with the temperature T. These findings are in agreement with the results of the qualitative calculations carried out in the preceding section. The result (6.22) is in agreement with the observed electronic contribution to the heat capacity of a metal. Normally the lattice contribution to the heat capacity, which will be discussed in Chapter 11, is much greater than the electronic contribution. Therefore, the experimental verification of the linear T-law as given above must be carried out at very low temperatures, where the lattice contribution becomes negligible. In this temperature region, the measured molar heat should rise linearly with the temperature T. By comparing the slope with

$$\frac{C_V}{T} = \frac{\pi^2}{2} \frac{R}{T_F}, \qquad \text{[use of (6.33)]} \tag{6.24}$$

we can find the Fermi temperature T_F numerically. Since this temperature is related to the effective mass m^* by $k_B T_F = \hbar^2(2m^*)^{-1}(3\pi^2 n)^{2/3}$, [use of (2.16) and (2.17)], we can obtain the numerical value for the effective mass m^* of the conduction electron. A more effective way of finding the m^*-value is through the transport and optical properties of conductors, and will be discussed in Chapter 13.

Problem 6.1 For free electrons at high degenracy, we have from (6.18) and (6.19).

$$n = \frac{2^{3/2} m^{3/2}}{3\pi^2 \hbar^3} \mu^{3/2} \left(1 + \frac{5}{8}\pi^2 k_B^2 T^2 \mu^{-2}\right). \tag{1}$$

$$\bar{e} = \frac{2^{3/2} m^{3/2}}{5\pi^2 \hbar^3} \mu^{5/2} \left(1 + \frac{5}{8}\pi^2 k_B^2 T^2 \mu^{-2}\right). \tag{2}$$

(a) Taking the zero-temperature limit of (1), we obtain

$$n = \frac{2^{3/2} m^{3/2}}{3\pi^2 \hbar^3} = \mu_o^{3/2}. \tag{3}$$

Postulate $\mu = \mu_o + a(k_B T/\mu_o)^2$, where a is a constant. Determine this constant a from eq.(1), and verify that

$$\mu = \mu_o \left\{1 - \frac{\pi^2}{12}\left(\frac{k_B T}{\mu_o}\right)^2 + O\left[\left(\frac{k_B T}{\mu_o}\right)^4\right]\right\}. \tag{4}$$

(b) Using (2) and (4) show that

$$\bar{e} = \frac{3}{5} n\mu_o \left\{1 + \frac{5\pi^2}{12}\left(\frac{k_B T}{\mu_o}\right)^2 + O\left[\left(\frac{k_B T}{\mu_o}\right)^4\right]\right\}.$$

10.7 Liquid Helium

Helium is the only substance in nature which does not solidify by a lowering of the temperature under normal atmospheric pressure. There exist two main isotopes, He^4, the most abundant, and He^3. Interactions between monoatomic molecules are practically identical for any pair, $He^4 - He^4$, $He^3 - He^3$ or $He^3 - He^4$. [These interactions are mainly determined by the electronic structure which both isotopes share. Numerical data regarding the masses and the interaction potential were given in §9.13.] The potential has a shallow attractive well such that two He^3 atoms may form a bound state with a very small binding energy. For He^4, the bound state has a 20% greater binding energy. [This difference is due to the difference in mass. Quantum mechanical calculations of the energy eigenvalues involve the total Hamiltonian, that is, the potential energy plus kinetic energy which contains the mass difference.] The boiling points for liquid He^3 and He^4 are 3.2 and 4.2 K respectively. These values do reflect the binding energies of the molecules.

Liquid He^4 undergoes the so-called λ-transition at 2.2 K into a superfluid phase whose flow properties are quite different from those of the ordinary fluid. For example, superfluid in a beaker creeps around the wall and drips down. See Figure 10.10. The reader interested in learning many fascinating properties of liquid He^4 should read the book by London, Super-fluid I [2]. On the other hand, liquid He^3 behaves quite differently. Very recently a phase transition was discovered for this liquid at a temperature of 0.002 K [3]. Its superfluid phase, however, is quite different from the superfluid phase of liquid He^4 [which is commonly called He II against He I which is the liquid phase above the λ-transition].

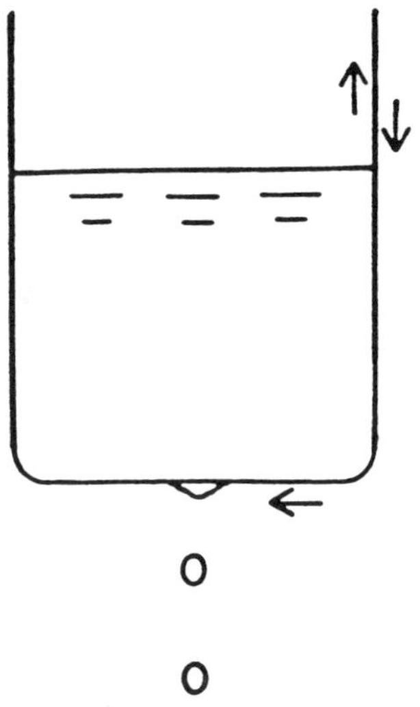

Fig. 10.10. Superfluid He in a beaker creeps around the wall and drips down.

Why does there exist such difference between liquid He^3 and He^4 ? Isotope He^4 has a nucleus (α-particle) consisting of two protons and two neutrons and possessing zero nuclear spin angular momentum, and two electrons orbiting around the nucleus with vanishing angular momentum in the ground state. He^4 molecules therefore are bosons according to the spin-statistics rule. Isotope He^3 has a nucleus, consisting of two protons and one neutron and possessing spin of magnitude $\frac{1}{2}\hbar$ due to the unpaired nucleon. The two electrons carried by He^3 have zero angular momentum in the ground state just as for He^4. Thus, He^3 molecules have a net spin of magnitude $\frac{1}{2}\hbar$, and accordingly are fermions. This difference in statistics generates a fundamental difference in the macroscopic properties of quantum fluids. To illustrate this last point, we will study the thermodynamic behavior of a system of free bosons in the following two sections.

10.8 Free Bosons. The Bose-Einstein Condensation

Let us consider a system of free bosons characterized by the Hamiltonian

$$H = \sum_{j=1}^{N} \frac{p_j^2}{2m} . \tag{8.1}$$

The system in equilibrium has the momentum distribution characterized by the Bose distribution function

$$f_B(\varepsilon_p) = \frac{1}{e^{\beta(\varepsilon_p - \mu)} - 1}, \qquad \varepsilon_p \equiv \frac{p^2}{2m}, \tag{8.2}$$

which is subject to the normalization condition :

$$\frac{1}{V} \sum_{\vec{p}} f_B(\varepsilon_p) = n \text{ (number density).} \tag{8.3}$$

Hereafter, we drop the subscript B.

By definition the Bose distribution function $f(\varepsilon_p)$ represents the relative probability of finding a particle with a momentum $\vec{p}$. Therefore, the function $f(\varepsilon_p)$ must be non-negative :

$$f(\varepsilon_p) = \frac{1}{e^{\beta(\varepsilon_p - \mu)} - 1} \geqq 0. \tag{8.4}$$

To insure this property for all $\varepsilon_p \geqq 0$ the chemical potential μ must be non-positive:

$$\mu \leqq 0 . \tag{8.5}$$

[or else the Bose distribution $f(\varepsilon_p)$ would become negative for $0 < \varepsilon_p < \mu$.] By the way, such restriction on the chemical potential μ does not apply to the Fermi or Boltzmann distribution function. [Why ? Problem 10.1].

In the bulk limit, the momentum eigenvalues form a continuous spectrum. For the moment, let us replace the sum over momentum states in (8.3) by the momentum integral :

$$n = \text{Lim}\, V^{-1} \int d^3p \, \frac{dn}{d^3p} \, f(\varepsilon_p). \tag{8.6}$$

The density of states dn/d^3p for free particles (without spin) was calculated in § 10.3, and is given by (3.10) :

$$\frac{dn}{d^3p} = \frac{V}{(2\pi\hbar)^3} . \tag{8.7}$$

Introducing this expression in (8.6), we obtain

$$n = \frac{1}{(2\pi\hbar)^3} \int d^3p \;\; f(\varepsilon_p). \tag{8.8}$$

The momentum-space integration may be carried out in the spherical polar coordinates (p, ϑ, φ) with the momentum - volume element $d^3p = p^2 \, dp \sin\vartheta \, d\vartheta \, d\varphi$. After performing the angular integration, we may rewrite the result in the form of an energy integral and obtain

$$n = \frac{M^{3/2}}{2^{1/2}\,\pi^2\,\hbar^3} \int_0^\infty d\varepsilon \;\; \varepsilon^{1/2} \, \frac{1}{e^{\beta(\varepsilon + |\mu|)} - 1} \equiv \frac{M^{3/2}}{2^{1/2}\,\pi^2\,\hbar^3} F(\beta, |\mu|), \tag{8.9}$$

[This result can also be obtained by finding the density of states in energy, D (ε), more directly as described in § 10.4,

$$D(\varepsilon) = V \frac{M^{3/2}}{2^{1/2}\pi^2\hbar^3} \varepsilon^{1/2}. \tag{8.10}$$

Note that because of no spin degeneracy, this expression is one half of expression (4.9)].

Let us now consider the ε- integral in (8.9) :

$$F(\beta, |\mu|) \equiv \int_0^\infty d\varepsilon\, \varepsilon^{\frac{1}{2}} \frac{1}{e^{\beta(\varepsilon+|\mu|)}-1}. \tag{8.11}$$

This quantity is a function of β and |μ|. For a fixed β, the function F is a decreasing function of |μ| with a maximum occurring at μ = 0 :

$$F(\beta, |\mu|) \leqq F(\beta, 0). \tag{8.12}$$

The maximum value for F can be evaluated as follows :

$$F(\beta, 0) = \int_0^\infty d\varepsilon\, \varepsilon^{1/2} \frac{1}{e^{\varepsilon\beta}-1}$$

$$= \beta^{-3/2} \int_0^\infty dx\, x^{1/2} \frac{1}{e^x-1}. \quad [\varepsilon\beta = x] \tag{8.13}$$

The x-integral has the numerical value

$$\int_0^\infty dx\, x^{\frac{1}{2}} \frac{1}{e^x-1} = \int_0^\infty dx\, x^{\frac{1}{2}} (e^{-x} + e^{-2x} + \ldots) = 1.306\sqrt{\pi}. \tag{8.14}$$

We then see that the integral on the r.h.s. of eq.(8.9) has an upper limit. The number density n on the l. h.s. could, of course, be increased without limit. Something must have gone wrong in our calculations.

A closer look at (8.2) shows that the function $f(\varepsilon)$ blows up in the neighborhood of $\varepsilon = 0$ if $\mu = 0$. This behavior therefore violates the validity condition (a), stated in § 10.3, for the conversion of the sum-over-states into an integral. In such a case, we must proceed more carefully.

Let us go back to eq.(8.3) and break the sum into two :

$$\frac{N}{V} = \frac{N_0}{V} + \frac{1}{V} \sum_{\vec{p}\,(\varepsilon_p > 0)} f(\varepsilon_p), \qquad (8.15)$$

where N_0 is the number of bosons with zero momentum, and is given by

$$N_0 \equiv \frac{1}{e^{\beta|\mu|} - 1} . \qquad (8.16)$$

This number can be made very large by choosing very small $\beta|\mu|$. For example, to have $N_0 = 10^{20}$, we may choose $\beta|\mu| = \ln(1 + N_0^{-1}) = \ln(1 + 10^{-20}) \simeq 10^{-20}$. In fact, N_0 can be increased without limit. We note further that, because the density of states in energy is proportional to $\varepsilon^{\frac{1}{2}}$, see (8.10), the contribution of zero momentum bosons is not included in the ε- integral. We therefore should write the normalization condition in the bulk limit as follows :

$$n = n_0 + \frac{M^{3/2}}{2^{1/2}\pi^2\hbar^3} \int_0^\infty d\varepsilon\, \varepsilon^{1/2} \frac{1}{e^{\beta(\varepsilon+|\mu|)} - 1} . \qquad (8.17)$$

The two terms in (8.17) [and also in (8.15)] represent the density $n_0 \equiv N_0/V$ of zero-momentum bosons and that of non-zero momentum bosons. The term

$n_0 \equiv N_0/V$ is important only when the number of zero-momentum bosons, N_0, is a significant fraction of the total number of bosons, N. The possibility of such an unusual state, called the Bose-Einstein condensation, was first recognized by Einstein. [Such possibilities exist neither for fermions nor classical particles.]

For low densities or high temperatures, the density of zero-momentum bosons n_0 is negligible against the total number density n. By raising the density or by lowering the temperature, the system undergoes a sharp transition into a state in which n_0 becomes a significant fraction of n. At absolute zero, all bosons have zero momentum. The sharp change in state resembles the gas-to liquid condensation but the Bose condensation occurs in the momentum space. Further characteristics of the Bose condensation will be discussed in the following section.

Problem 8.1 Let us consider a system of free fermions. Find, from the normalization, the range of the chemical potential μ which appears in the Fermi distribution function :

$$f_F(\varepsilon) \equiv [\, e^{\beta(\varepsilon-\mu)} + 1 \,]^{-1}.$$

Problem 8.2* Let us consider a system of free bosons moving in two dimensions.

(a) Does this system undergo transition into a condensed state at low temperatures as in the three dimensional case?

(b) Discuss the heat capacity of the system in all temperature ranges. Calculate explicitly the heat capacity at both low and high temperature limits.

10.9 Bosons in Condensed Phase

At absolute zero, all bosons have zero momentum. As the temperature is raised, bosons with non-zero momenta emerge. The number of those excited bosons, given by the volume V times the second term of the r.h.s. of eq.(8.15), can be represented by

$$N_{\varepsilon>0} \equiv \sum_{\vec{p}(\varepsilon_p>0)} f(\varepsilon_p)$$

$$= \frac{V\, M^{3/2}}{2^{1/2}\, \pi^2\, \hbar^3} \int_0^\infty d\varepsilon \quad \varepsilon^{1/2} \frac{1}{e^{\beta(\varepsilon+|\mu|)}-1} \quad \text{[use of (8.9)]}$$

$$= \frac{V\, M^{3/2}}{2^{1/2}\, \pi^2\, \hbar^3} F(\beta, |\mu|). \qquad \text{[use of (8.11)]} \tag{9.1}$$

Throughout the condensed phase, the chemical potential μ has a very small absolute value [Problem 9.1]. Since the function $F(\beta,|\mu|)$ is a slowly varying function of $|\mu|$ for $|\mu| \ll 1$, we may approximate (9.1) by its value at $\mu = 0$:

$$N_{\varepsilon>0} \cong V \frac{M^{3/2}}{2^{1/2}\, \pi^2\, \hbar^3} F(\beta,0)$$

$$= \frac{1.306\sqrt{\pi}\, V\, M^{3/2}}{\sqrt{2}\, \pi^2\, \hbar^3} (k_B T)^{3/2}. \quad \text{[use of (8.14)]} \tag{9.2}$$

Here we see that the number of bosons in the excited states with positive energies grows like $T^{3/2}$ as the temperature T is raised. This number may eventually reach the total number N , as T is raised to a critical temperature T_0. At and above the critical temperature T_0 practically all bosons are in excited states. This temperature T_0 can be obtained from

$$N = \frac{1.306\sqrt{\pi}}{\sqrt{2}\ \pi^2} \quad V \frac{M^{3/2}}{\hbar^3} \quad (k_B\ T_0)^{3/2}$$

$$= \frac{1}{6.032} \quad \frac{V\ M^{3/2}}{\hbar^3} \quad (k_B\ T_0)^{3/2}\ . \qquad (9.3)$$

Solving for T_0, we obtain

$$T_0 = 3.31 \frac{\hbar^2}{M\ k_B} \quad n^{2/3}. \qquad [(6.031)^{2/3} = 3.31] \qquad (9.4)$$

Using this relation, we can rewrite eq.(9.2) in the form :

$$\boxed{N_{\varepsilon>0} = N(T/T_0)^{3/2}, \quad T \lesseqgtr T_0}\ . \qquad (9.5)$$

The number of zero-momentum bosons, N_0, can be obtained by subtracting this number from the total number N :

$$N_0 \equiv N - N_{\varepsilon>0} = N\,[\,1 - (T/T_0)^{3/2}\,], \quad T \lesseqgtr T_0\ . \qquad (9.6)$$

Here we see that N_0 is in fact a finite fraction of the total number N. The number of bosons in excited states and in the ground state, $N_{\varepsilon>0}$ and N_0, are plotted against temperature in Figure 10.11.

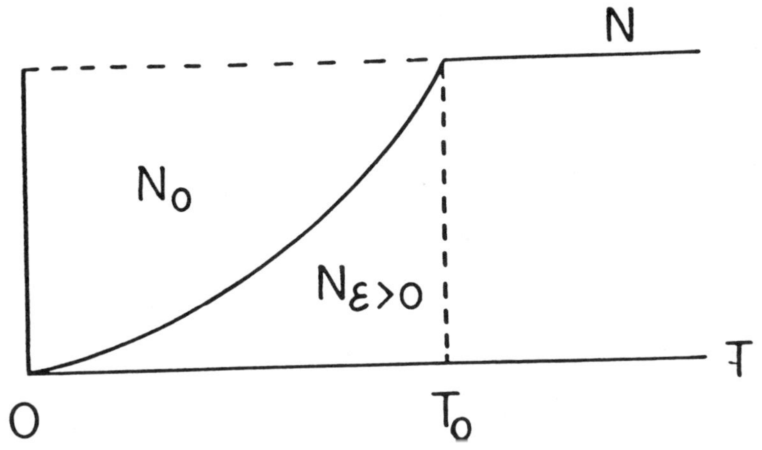

Fig. 10.11. The number of condensed bosons, N_o, below the critical temperature T_o, forms a finite fraction of the total number N.

The thermal energy E of the system comes from the excited bosons alone. The average energy per unit volume can be calculated as follows :

$$\frac{E}{V} = \mathrm{Lim}\,\frac{1}{V} \sum_{\vec{p}(\varepsilon_p>0)} \varepsilon_p\, f(\varepsilon_p)$$

$$= \frac{M^{3/2}}{2^{1/2}\,\pi^2\,\hbar^3} \int_0^\infty d\varepsilon\;\; \varepsilon^{3/2}\, \frac{1}{e^{\beta(\varepsilon+|\mu|)}-1}\,. \qquad (9.7)$$

The ε- integral here is also a slowly varying function of $|\mu|$ for $|\mu| \ll 1$. We may therefore approximate it by its value at $\mu = 0$:

$$\int_0^\infty d\varepsilon \quad \varepsilon^{3/2} \frac{1}{e^{\beta(\varepsilon + |\mu|)} - 1}$$

$$\cong \int_0^\infty d\varepsilon \quad \varepsilon^{3/2} \frac{1}{e^{\beta\varepsilon} - 1} \qquad [\ \mu \cong 0\ ]$$

$$= \beta^{-5/2} \int_0^\infty dx \quad x^{3/2} \frac{1}{e^x - 1}. \qquad [\ x = \beta\varepsilon\ ] \qquad (9.8)$$

The x - integral is numerically equal to 1.342 :

$$\int_0^\infty dx \quad x^{3/2} \frac{1}{e^x - 1} = 1.342\ . \qquad (9.9)$$

Using (9.8) and (9.9), we obtain

$$\frac{E}{V} = \frac{M^{3/2}}{2^{1/2}\pi^2\hbar^3} 1.342\ (k_B\ T)^{5/2}. \qquad (9.10)$$

This result can be rewritten with the aid of (9.3) as follows :

$$\frac{E}{V} = \frac{1.342}{2^{1/2}\ \pi^2} \cdot \frac{M^{3/2}}{\hbar^3}(k_B\ T_0)^{3/2} \cdot (k_B\ T)\ (T/T_0)^{3/2}$$

$$= 0.770\ n\ k_B\ T\ (T\ /\ T_0)^{3/2}. \qquad [\text{use of } (9.3)] \qquad (9.11)$$

Note that the internal energy density E/V grows like $T^{5/2}$ in the condensed phase.

Differentiating eq.(9.11) with respect to T and writing the result for a mole of of the gas, we obtain

$$C_V = \frac{\partial E}{\partial T}\Big)_V = \frac{\partial E}{\partial T}\Big)_{T_0}$$

$$= \frac{5}{2}\,\frac{E}{T} = 1.92\ N\ k_B\ (T\ /\ T_0)^{3/2}$$

$$= 1.92\ R\ (T\ /\ T_0)^{3/2}. \qquad (9.12)$$

The behavior of the heat capacity at constant volume, C_V, versus the temperature T is shown in Figure 10.12. The molar heat capacity C_V increases

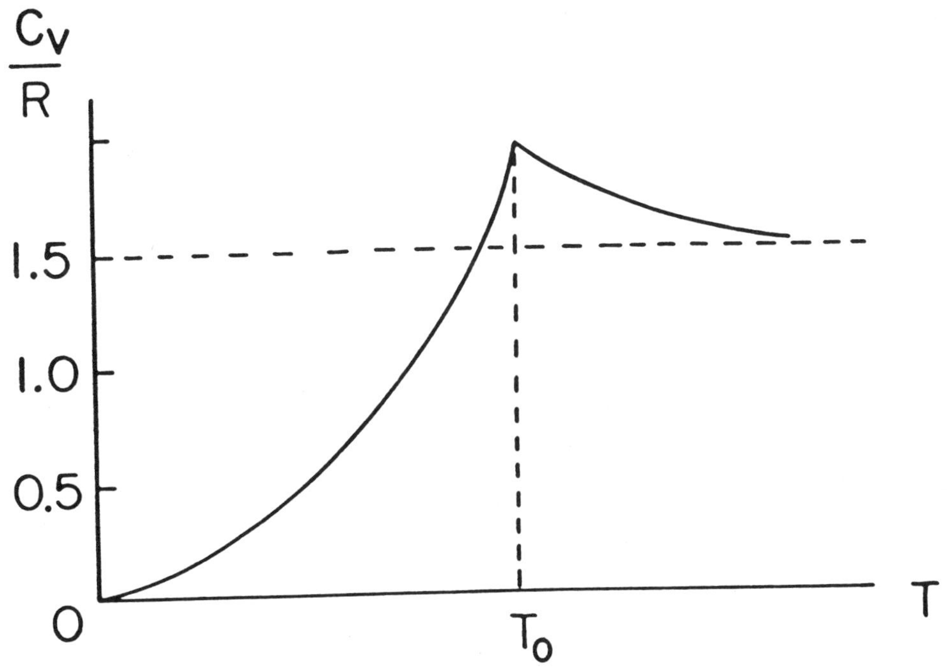

Fig. 10.12. The reduced heat capacity C_V/R exhibits a sharp cusp at the critical temperature T_0.

like $T^{3/2}$ throughout the condensed phase and reaches its maximum value 1.92 R at the transition temperature T_o. Above T_o, it gradually decreases and approaches the classical value 1.50 R at high temperatures. At the transition point, C_V is continuous in value but its derivative jumps from a positive to a negative value.

The Bose-Einstein condensation is similar to the gas-liquid transformation in the sense that it is a sudden macroscopic change of state. It is however, quite different in detail. As we just observed, the heat capacity is continuous at the transition point. In contrast, the heat capacity for the familiar gas-liquid transition has a discrete jump. Landau (Lev D. Landau, 1908-62, Russian) classified the Bose-Einstein condensation as a phase transition of second order as the second-order derivatives of the free energy such as heat capacity and compressibility, are discontinuous at the point of transition. For a phase transition of second order, there is no latent heat of condensation. In contrast, a phase transition of first order, for which the first derivatives of the free energy such as volume, entropy, internal energy, jump between the two phases, is accompanied by a latent heat. The readers interested in more about the general theory of the phase transition, are encouraged to read chapters in the classic book on Statistical Physics by Landau and Lifshitz [4].

To see relevance of the Bose-Einstein condensation to the actual liquid helium, let us look at a few properties of this substance. Figure 10.13 represents the P - T diagram, also called the phase diagram of pure He^4. Proceeding along the horizontal line at one atmospheric pressure the He^4 passes from gas to liquid at 4.2 K. This liquid, called the liquid He I , behaves like

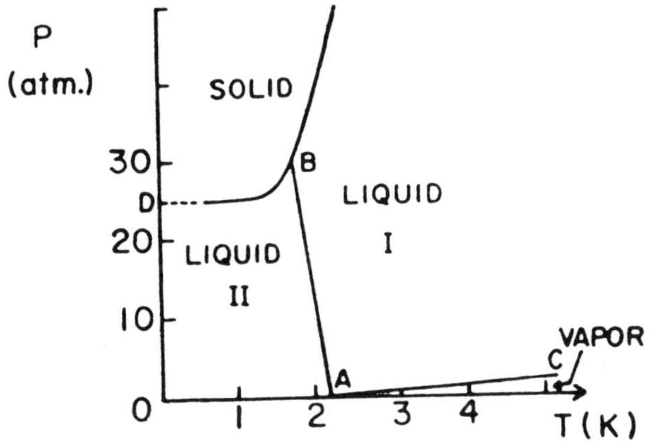

Fig.10.13 The phase diagram of pure He^4. The line A B, called the λ-line, separate liquid He II from liquid He I. The point C represents the critical point.

any other liquid, and has a finite viscosity. By cooling down further, the substance suffers a sudden change at 2.18 K. Below this temperature the liquid He II is a superfluid as briefly mentioned in §10.7. The heat capacity of liquid helium measured under its saturated vapor is shown in Figure 10.14.

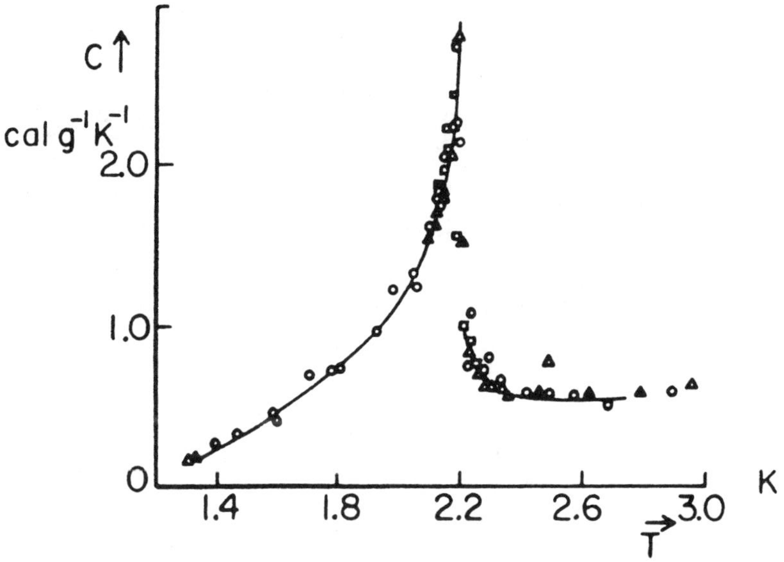

Fig. 10.14 Specific heat of liquid He^4 versus temperature. Because of the resemblance of this curve to the inverted Greek letter λ, the transition point is called the lambda point. From London's book, Superfluids [2].

Because of the similarity between this experimental curve and the inverted Greek letter λ, the transition is often called the λ-transition. Note that the heat capacity has an extremely sharp peak of a logarithmic type [5]. The resemblance between this curve and the heat capacity curve of the ideal Bose gas shown in Figure 10.12, is striking. If we calculate the value of the critical temperature T_0 from (9.4) with the use of the density of liquid helum = 0.145 g cm^{-3}, we find that T_0 = 3.14 K, which is remarkably close to the observed λ-transition temperature 2.18 K.

Problem 9.1 Assume that free bosons are in the condensed state below the critical temperature T_o. Calculate the pressure P of the system by means of formula

$$P = \lim \frac{1}{MV} \sum_{p(\text{states})} p_x^2 f(\varepsilon_p),$$

where $f(\varepsilon)$ is the Bose distribution function. Does P go to zero as T approaches zero?

References

[1] e.g. Haug, A., Theoretical Solid State Physics, vol.1, Pergamon, Oxford (1972); Kittel, C., Introduction to Solid State Physics, Wiley, New York, NY 10016, Fifth Edition (1976)

[2] London, F., Superfluid, vol.2, Wiley, New York, N.Y. 10016 (1954)

[3] Osheroff, D.D., Richardson, R.C. and Lee, D.M., Phys. Rev. Lett. 28 885 (1972); Osheroff, D.D., Gully, W.J., Richardson, R.C., and Lee, D.M., Phys. Rev. Lett. 29, 920 (1972)

[4] Landau, L.D. and Lifshitz, E.M., Statistical Physics, Addison - Wesley, Reading, Mass., Second Edition (1969)

[5] Buckingham, M.J. and Fairbank, W.M., The nature of the λ-transition, in Progress in Low Temperature Physics, edited by C.J. Gorter, vol. III, North-Holland Publ.Co., Amsterdam (1961)

Review Questions

1. Write down mathematical expressions, figures, etc, and explain briefly:

   (a) the Fermi temperature for Cu (order of magnitude)

   (b) the density of states in the momentum space

   (c) the density of states in the energy domain for free electrons

   (d) the spin degeneracy factor

   (e) the Bose-Einstein condensation.

General Problems

1. The electrons trapped on the surface of liquid helium can move in two dimensions only. A way of finding whether the electrons are statistically degenerate or not, is to calculate the Fermi energy of the corresponding free electrons. Find the Fermi temperature of the free electrons corresponding to the observed domain of concentration $n = 10^5 - 10^{10}\ cm^{-2}$.

   $m = 9.1 \times 10^{-28} g$, $\hbar = 1.05 \times 10^{-27}$ ergs sec, $k_B = 1.38 \times 10^{-16}$ erg deg^{-1}.

2. Calculate the ground state energy and low temperature heat capacity of free particles moving in one dimension with the assumption of

(a) the Maxwell-Boltzmann statistics

(b) the Fermi-Dirac statistics

(c) the Bose-Einstein Statistics.

3. Let us consider a system of free bosons moving in one dimension.

(a) Does this system undergo transition into a condensed state at low temperature as in the three dimensional case ?

(b) Discuss the heat capacity of the system in all temperature ranges. Calculate explicitly the heat capacity at both low and high temperature limits. Compare these results with those for classical free particles.

4. The heat capacity of free electrons moving in three dimensions follows a linear T - law at the lowest temperatures. Does this law still hold for systems of lower dimensions ? Guess an answer and substantiate it by the order-of-magnitude calculations.

5. Let us assume that a number N of free fermions having the energy-momentum relation : $\varepsilon = cp$, are contained in a colume V.

(a) Find an expression for the Fermi energy.

(b) Estimate the Fermi temperature when the number density $n \equiv N/V$ equals $10^{20}\ cm^{-3}$. Assume that $c = 3 \times 10^{10}\ m\ s^{-1}$.

(c) Find the heat capacity at extreme low temperatures.

(d) Find the heat capacity at extreme high temperatures.

6. Consider a system of free bosons having the energy-momentum relation $\varepsilon = cp$, where c is a constant. Investigate the possibility of the Bose-Einstein condensation in one, two and three dimensions.

Chapter 11. BLACK BODY RADIATION. LATTICE VIBRATIONS

In this chapter we will treat black body radiation and lattice vibrations. Quanta of electromagnetic radiation are called photons and quanta of lattice vibrations are called phonons. Photons and phonons both obey the Bose-Einstein statistics. They have great similarities, but also significant differences. The first five sections of this chapter deal with photons and the remaining seven with phonons.

11.1° Electric and Magnetic Fields in a Vacuum. the Wave Equation and its Plane-Wave Solutions

The basic equations governing electromagnetic fields are Maxwell's equations [1]. In a vacuum the electric field $\vec{E}(\vec{r},t)$ and the magnetic field $\vec{B}(\vec{r},t)$ obey the two vector equations:

$$\frac{\partial \vec{B}}{\partial t} + \nabla \times \vec{E} = 0 \tag{1.1}$$

$$\varepsilon_0 \frac{\partial \vec{E}}{\partial t} - \frac{1}{\mu_0} \nabla \times \vec{B} = 0, \tag{1.2}$$

and the two scalar equations:

$$\nabla \cdot \vec{B} = 0 \tag{1.3}$$

$$\nabla \cdot \vec{E} = 0. \tag{1.4}$$

It is known [1] that if we introduce a vector potential $\vec{A}(\vec{r},t)$ which satisfies

$$\nabla \cdot \vec{A} = 0 \qquad \text{(the Coulomb gauge condition)} \tag{1.5}$$

both $\vec{E}$ and $\vec{B}$ can be expressed in terms of A as follows:

$$\vec{B} = \nabla \times \vec{A}(\vec{r},t) \tag{1.6}$$

$$\vec{E} = -\frac{\partial}{\partial t} \vec{A}(\vec{r},t). \tag{1.7}$$

It is easy to verify that the fields $(\vec{B},\vec{E})$ defined by (1.5) - (1.7) satisfy Maxwell's equations (1.1) - (1.4) [Problem 1.1].

Introducing $\vec{E}$ and $\vec{B}$ from (1.7) and (1.6) in eq. (1.2), we obtain

$$-\varepsilon_0 \frac{\partial^2 \vec{A}}{\partial t^2} - \frac{1}{\mu_0} \nabla \times (\nabla \times \vec{A}) = 0.$$

Using the vector identity [see Appendix C(23)]

$$\nabla \times (\nabla \times \vec{C}) = (\nabla \cdot \vec{C}) - \nabla^2 \vec{C}, \tag{1.8}$$

we obtain

$$\frac{\partial^2 \vec{A}}{\partial t^2} = - \frac{1}{\varepsilon_o \mu_o} [\nabla(\nabla \cdot \vec{A}) - \nabla^2 \vec{A}]$$

$$= - \frac{1}{\varepsilon_o \mu_o} (-\nabla^2 \vec{A}). \qquad \text{[use of (1.5)]}$$

or

$$\frac{\partial^2 \vec{A}}{\partial t^2} = c^2 \nabla^2 \vec{A}, \tag{1.9}$$

where

$$c \equiv (\varepsilon_o \mu_o)^{-\frac{1}{2}} = 3 \times 10^{10} \text{ cm sec}^{-1} = 3 \times 10^8 \text{ m s}^{-1}. \tag{1.10}$$

Eq. (1.9) is called the <u>wave equation</u>, and characterizes the motion of the vector potential $\vec{A}$ in a vacuum. The constant c is the <u>speed of light</u>, and has the numerical value indicated.

Applying operators $\partial/\partial t$ and $\nabla\times$ to eqs. (1.1 - 1.2), we can verify that both $\vec{E}$ and $\vec{B}$ satisfy the same wave equation:

$$\frac{\partial^2 \vec{E}}{\partial t^2} = c^2 \nabla^2 \vec{E}, \quad \frac{\partial^2 \vec{B}}{\partial t^2} = c^2 \nabla^2 \vec{B}. \tag{1.11}$$

Earlier in §§3.11 - 3.12, we set up and solved the wave equation in one dimension, which is associated with the transverse oscillations of a stretched string. We can treat the three-dimensional wave equation (1.9) in an analogous manner.

We assume the periodic boundary condition:

$$A(x+L,y,z,t) = A(x,y+L,z,t) = A(x,y,z+L,t) = A(x,y,z,t), \tag{1.12}$$

where L is a length of periodicity. One can show that the exponential functions

$$e^{-i\omega_k t + i\vec{k}\cdot\vec{r}}, \tag{1.13}$$

where

$$\vec{k} = (k_x, k_y, k_z) \tag{1.14}$$

$$k_x = \frac{2\pi}{L} n_x \,, \quad k_y = \frac{2\pi}{L} n_y, \quad k_z = \frac{2\pi}{L} n_z$$

$$n_x, n_y, n_z = \ldots, -1, 0, 1, 2, \ldots \tag{1.15}$$

$$\omega_k \equiv c|\vec{k}| = ck \,, \tag{1.16}$$

are solutions of the wave equation (1.9) subject to the periodic boundary condition (1.12). [Problem 1.2]

For $\vec{k} = (k_x,0,0)$, $k_x = k > 0$, the function

$$e^{-\omega_k t + i\vec{k}\cdot\vec{r}} = e^{-ik(ct-x)} \tag{1.17}$$

represents a plane wave traveling along the positive x-axis with speed c. [See discussion in §3.12.] For a general $\vec{k} \equiv (k_x,k_y,k_z)$, the function $e^{-i\omega_k t + i\vec{k}\cdot\vec{r}}$ represents a plane wave traveling in the direction of $\vec{k}$, as indicated in Figure 11.1.

From (1.12), the vector potential $\vec{A}$ is periodic with the period L in the x, y, and z-directions. We may then expand $\vec{A}$ in the following form (Fourier's fundamental theorem): [2]

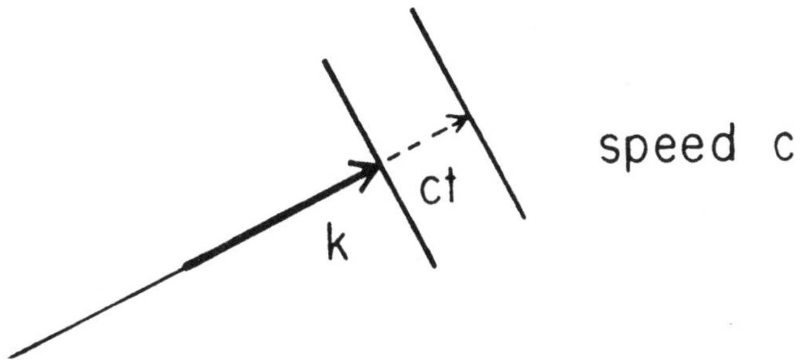

Fig. 11.1 A plane wave traveling in the direction of $\vec{k}$ and with the speed c.

$$\vec{A}(\vec{r},t) = \sum_{n_x=-\infty}^{+\infty} \sum_{n_y=-\infty}^{+\infty} \sum_{n_z=-\infty}^{+\infty}$$

$$\{a(n_x,n_y,n_z,t) \exp[i \frac{2\pi}{L} (xn_x+yn_y+zn_z)] + c.c.\}. \qquad (1.18)$$

By definition, the vector potential $\vec{A}$ is real. To insure this property, the complex conjugates (c.c.) are added on the r.h.s. Introducing the wave vectors $\vec{k}$ defined in (1.14) - (1.15), we can rewrite eq. (1.18) as

$$\vec{A}(\vec{r},t) = \sum_{\vec{k}} [\vec{a}_{\vec{k}}(t)e^{i\vec{k}\cdot\vec{r}} + c.c.]. \qquad (1.19)$$

Substitution of this expression into the wave equation (1.9) yields

$$\ddot{\vec{a}}_{\vec{k}}(t) + c^2k^2 \vec{a}_{\vec{k}}(t) = 0, \qquad (1.20)$$

which shows that the vector coefficient $\vec{a}_{\vec{k}}(t)$ is a periodic function of time with the angular frequency $\omega_k \equiv ck$. If we write

$$\vec{a}_{\vec{k}}(t) \equiv \vec{f}_{\vec{k}}\, e^{-i\omega_k t}, \qquad (1.21)$$

and substitute into (1.19), we obtain

$$\vec{A}(\vec{r},t) = \sum_{\vec{k}} (\vec{f}_{\vec{k}}\, e^{-i\omega_k t + i\vec{k}\cdot\vec{r}} + \text{c.c.}). \tag{1.22}$$

This expansion can be interpreted as follows: <u>The general solution for the wave equation with the periodic boundary condition is a superposition of plane-wave solutions</u>.

Applying the operator $\nabla\cdot$ to eq. (1.19), we obtain

$$\nabla\cdot\vec{A} = \sum_{\vec{k}} [\vec{a}_{\vec{k}}\cdot\nabla e^{i\vec{k}\cdot\vec{r}} + \text{c.c.}] \qquad [\vec{a}_{\vec{k}} \text{ independent of } \vec{r}]$$

$$= \sum_{\vec{k}} [(a_{\vec{k},x}\frac{\partial}{\partial x} + a_{\vec{k},y}\frac{\partial}{\partial y} + a_{\vec{k},z}\frac{\partial}{\partial z})e^{ixk_x+iyk_y+izk_z} + \text{c.c.}]$$

$$= \sum_{\vec{k}} [i(a_{\vec{k},x}\, k_x + a_{\vec{k},y}\, k_y + a_{\vec{k},z}\, k_z)e^{ixk_x+iyk_y+izk_z} + \text{c.c.}]$$

$$= \sum_{\vec{k}} [i\vec{k}\cdot\vec{a}_{\vec{k}}\, e^{i\vec{k}\cdot\vec{r}} + \text{c.c.}]$$

$$= 0. \qquad [\text{use of } (1.5)] \tag{1.23}$$

Since in general $e^{i\vec{k}\cdot\vec{r}}$ does not vanish, we must have

$$\vec{k}\cdot\vec{a}_{\vec{k}} = 0. \tag{1.24}$$

This means that the vector-coefficients $\vec{a}_{\vec{k}}$ are orthogonal to the wave vector $\vec{k}$.

Using (1.7), we obtain the electric field $\vec{E}$ as follows:

$$\vec{E} = -\frac{\partial}{\partial t}\vec{A}$$

$$= -\sum_{\vec{k}} [\dot{\vec{a}}_{\vec{k}}(t)\, e^{i\vec{k}\cdot\vec{r}} + \text{c.c.}]$$

$$= i\sum_{\vec{k}} \omega_k [\vec{a}_{\vec{k}}\, e^{i\vec{k}\cdot\vec{r}} - \vec{a}_{\vec{k}}^{\,*}\, e^{-i\vec{k}\cdot\vec{r}}]. \qquad [\text{use of } (1.21)] \tag{1.25}$$

In a similar manner, we obtain the magnetic field $\vec{B}$: [Problem 1.3]

$$\vec{B} = i \sum_{\vec{k}} [\vec{k}x\vec{a}_{\vec{k}} e^{i\vec{k}\cdot\vec{r}} - \vec{k}x\vec{a}_{\vec{k}}^{*} e^{-i\vec{k}\cdot\vec{r}}]. \tag{1.26}$$

The last three equations indicate that the electric and magnetic fields ($\vec{E}$,$\vec{B}$) are both perpendicular to the wave vector $\vec{k}$. In other words, the electromagnetic waves are transverse waves.

An electromagnetic traveling wave whose $\vec{E}$-field oscillates in a fixed plane is called a plane-polarized wave. The direction of the electric field $\vec{E}$, which is perpendicular to the wave vector $\vec{k}$, is customarily denoted by the polarization (unit) vector $\vec{\sigma}$. Then, we have by construction

$$\vec{k} \cdot \vec{\sigma} = 0. \tag{1.27}$$

From (1.26) we see that the associated magnetic field $\vec{B}$ oscillates in the direction perpendicular to both $\vec{k}$ and $\vec{E}$. The relationship between $\vec{E}$, $\vec{B}$ and $\vec{k}$ is shown schematically in Figure 11.2.

The electromagnetic wave which is emitted from a single molecule and which has traveled a long distance can be thought of as a plane-polarized wave. In a vacuum, electromagnetic waves of different wave vectors $\vec{k}$, frequemcoes ω_k and polarizations $\vec{\sigma}$ travel with the same speed c and without interaction. Each of these eaves carry certain energy. The sum of these energies represents the energy of the electromagnetic or radiation fields as we will see in the next section.

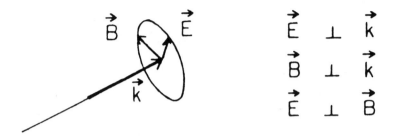

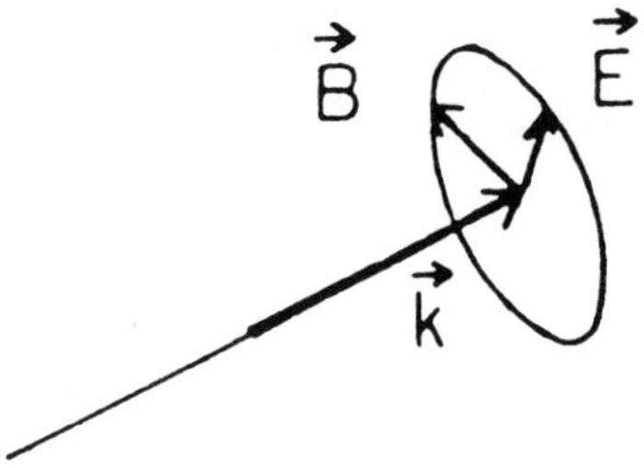

$$\vec{E} \perp \vec{k}$$
$$\vec{B} \perp \vec{k}$$
$$\vec{E} \perp \vec{B}$$

Fig. 11.2 An electromagnetic wave traveling in the direction $\vec{k}$ carries a pair of fields $(\vec{E},\vec{B})$, which are perpendicular to the wave vector $\vec{k}$, and which are also perpendicular to each other.

Problem 1.1 Verify that $\vec{E}$ and $\vec{B}$ given by $\vec{B} = \nabla \times \vec{A}(\vec{r},t)$, $\vec{E} = -\frac{\partial}{\partial t}\vec{A}(\vec{r},t)$, where A is a vector potential satisfying

$$\nabla\cdot\vec{A} = 0,$$

satisfy Maxwell's equations in a vacuum, eqs. (1.1) - (1.4).

Problem 1.2 Verify that the exponential functions

$$e^{-i\omega_k t \pm i\vec{k}\cdot\vec{r}},$$

with $\vec{k}$ and ω_k given by (1.15) and (1.16), satisfy the wave equation (1.9) and the boundary condition (1.12).

Problem 1.3 Calculate the magnetic field $\vec{B} \equiv \nabla \times \vec{A}$, when the vector potential $\vec{A}$ is given in the form of eq. (1.19):

$$\vec{A}(\vec{r},t) = \sum_{\vec{k}} [a_{\vec{k}}(t)\, e^{i\vec{k}\,\vec{r}} + a_{\vec{k}}^{*}(t)\, e^{-i\vec{k}\cdot\vec{r}}]$$

Answer:

$$\vec{B} = i \sum_{\vec{k}} [\vec{k} \times \vec{a}_{\vec{k}}\, e^{i\vec{k}\cdot\vec{r}} - \vec{k} \times \vec{a}_{\vec{k}}^{\,*}\, e^{-i\vec{k}\cdot\vec{r}}].$$

11.2 The Electromagnetic Field Theory

In the electromagnetic theory, the energy H of the radiation fields in the volume $V = L^3$ is defined by [1]

$$H = \frac{1}{2}\int_V d^3r\, (\varepsilon_o E^2 + \frac{1}{\mu_o} B^2) = \frac{1}{2}\varepsilon_o \int d^3r\, E^2 + \frac{1}{2\mu_o}\int d^3r\, B^2. \qquad (2.1)$$

We now wish to express this energy in terms of the wave amplitudes $\vec{a}_{\vec{k}}$.

The absolute square of (1.25) yields

$$E^2 = (i \sum_{\vec{k}} \omega_k [\vec{a}_{\vec{k}}\, e^{i\vec{k}\cdot\vec{r}} - \vec{a}_{\vec{k}}^{\,*}\, e^{-i\vec{k}\cdot\vec{r}}])\cdot(i \sum_{\vec{k}} \omega_{k'} [\vec{a}_{\vec{k}'}\, e^{i\vec{k}'\cdot\vec{r}} - \vec{a}_{\vec{k}'}^{\,*}\, e^{-i\vec{k}'\cdot\vec{r}}])$$

$$= - \sum_{\vec{k}} \sum_{\vec{k}'} \omega_k \omega_{k'} [\vec{a}_{\vec{k}}\cdot\vec{a}_{\vec{k}'}\, e^{i(\vec{k}+\vec{k}')\cdot\vec{r}} - \vec{a}_{\vec{k}}^{\,*}\cdot\vec{a}_{\vec{k}'}\, e^{-i(\vec{k}-\vec{k}')\cdot\vec{r}}$$

$$- \vec{a}_{\vec{k}}\cdot\vec{a}_{\vec{k}'}^{\,*}\, e^{i(\vec{k}-\vec{k}')\cdot\vec{r}} + \vec{a}_{\vec{k}}^{\,*}\cdot\vec{a}_{\vec{k}'}^{\,*}\, e^{-i(\vec{k}+\vec{k}')\cdot\vec{r}}]. \qquad (2.2)$$

Let us take the integral

$$I(\vec{k} + \vec{k}') \equiv \int_V d^3r\, e^{i(\vec{k}+\vec{k}')\cdot\vec{r}}$$

$$\equiv \int\int\int dxdydz\, e^{i(k_x+k'_x)x + i(k_y+k'_y)y + i(k_z+k'_z)z}, \qquad (2.3)$$

which can be factorized into three parts: the x-, y- and z- integrals. The x-integral

$$\int_0^L dx\, e^{i(k_x+k'_x)x} = \int_0^L dx\, e^{i2\pi x(n_x+n'_x)/L}$$

$$\left[k_x \equiv \frac{2\pi}{L} n_x, \quad k'_x \equiv \frac{2\pi}{L} n'_x\right]$$

vanishes, unless $n_x+n'_x = 0$, in which case the x-integral equals L. That is,

$$\int_0^L dx\, e^{i(k_x+k'_x)x} = \begin{cases} L & \text{if } k_x+k'_x = 0 \\ 0 & \text{otherwise;} \end{cases} \tag{2.4}$$

similar properties hold for the y- and z- integrals. We therefore obtain

$$I(\vec{k} + \vec{k}') = \begin{cases} L^3 \; (= V) & \text{if } \vec{k} + \vec{k}' = 0 \\ 0 & \text{otherwise.} \end{cases} \tag{2.5}$$

The same expression is valid if we replace $\vec{k} + \vec{k}'$ by $\vec{k} - \vec{k}'$ or $-\vec{k} - \vec{k}'$. In summary, we have

$$I(\vec{q}) \equiv \int_V d^3r\, e^{i\vec{q}\cdot\vec{r}} = V\, \delta^{(3)}_{\vec{q},0}$$

$$= \begin{cases} V & \text{if } \vec{q} = 0 \\ 0 & \text{otherwise ,} \end{cases} \tag{2.6}$$

where $\vec{q}$ is $\vec{k} + \vec{k}'$, $\vec{k} - \vec{k}'$, or $-\vec{k} - \vec{k}'$.

Integrating eq.(2.2) over the normalization volume V and using the property (2.6), we obtain

$$\int_V E^2\, d^3r = V \sum_{\vec{k}} \omega_k^{\,2}[2\, \vec{a}_{\vec{k}}^{\,*} \cdot \vec{a}_{\vec{k}} - \vec{a}_{\vec{k}} \cdot \vec{a}_{-\vec{k}} - \vec{a}_{\vec{k}}^{\,*} \cdot \vec{a}_{-\vec{k}}^{\,*}] \tag{2.7}$$

In a similar manner, we can show that [Problem 2.1]

$$\int_V d^3r\, B^2 = V \sum_{\vec{k}} [2(\vec{k} \times \vec{a}_{\vec{k}})\cdot(\vec{k} \times \vec{a}_{\vec{k}}^{\,*}) + (\vec{k} \times \vec{a}_{\vec{k}})\cdot(\vec{k} \times \vec{a}_{-\vec{k}})$$
$$+ (\vec{k} \times \vec{a}_{\vec{k}}^{\,*})\cdot(\vec{k} \times \vec{a}_{-\vec{k}}^{\,*})]. \tag{2.8}$$

Applying the general formula

$$(\vec{A} \times \vec{B})\cdot(\vec{C} \times \vec{D}) = \vec{A} \cdot [\vec{B} \times (\vec{C} \times \vec{D})] \qquad \text{[use of (C.14)]}$$
$$= \vec{A} \cdot [\vec{C}(\vec{B} \cdot \vec{D}) - \vec{D}(\vec{B} \cdot \vec{C})] \qquad \text{[use of (C.15)]}$$
$$= (\vec{A} \cdot \vec{C})(\vec{B} \cdot \vec{D}) - (\vec{A} \cdot \vec{D})(\vec{B} \cdot \vec{C}) \tag{2.9}$$

to $(\vec{k} \times \vec{a}_{\vec{k}})\cdot(\vec{k} \times \vec{a}_{\vec{k}}^{\,*})$, we obtain

$$(\vec{k} \times \vec{a}_{\vec{k}})\cdot(\vec{k} \times \vec{a}_{\vec{k}}^{\,*}) = k^2\, \vec{a}_{\vec{k}}^{\,*} \cdot \vec{a}_{\vec{k}} - (\vec{k} \cdot \vec{a}_{\vec{k}}^{\,*})(\vec{k} \cdot \vec{a}_{\vec{k}}) \qquad \text{[use of (2.9)]}$$
$$= k^2\, \vec{a}_{\vec{k}}^{\,*} \cdot \vec{a}_{\vec{k}}\, . \qquad \text{[use of (1.24)]} \tag{2.10}$$

In a similar manner, we can get

$$(\vec{k} \times \vec{a}_{\vec{k}})\cdot(\vec{k} \times \vec{a}_{-\vec{k}}) = k^2\, \vec{a}_{\vec{k}} \cdot \vec{a}_{-\vec{k}}$$
$$(\vec{k} \times \vec{a}_{\vec{k}}^{\,*})\cdot(\vec{k} \times \vec{a}_{-\vec{k}}^{\,*}) = k^2\, \vec{a}_{\vec{k}}^{\,*} \cdot \vec{a}_{-\vec{k}}^{\,*}\, . \tag{2.10a}$$

Using (2.10) and (2.10a), we can reduce eq. (2.8) to

$$\int_V d^3r\, B^2 = V \sum_{\vec{k}} k^2[2\, \vec{a}_{\vec{k}}^{\,*} \cdot \vec{a}_{\vec{k}} + \vec{a}_{\vec{k}} \cdot \vec{a}_{-\vec{k}} + \vec{a}_{\vec{k}}^{\,*} \cdot \vec{a}_{-\vec{k}}^{\,*}]. \tag{2.11}$$

Introducing the results (2.7) and (2.11) in eq. (2.1), and noting that $c^2 \equiv (\varepsilon_o\mu_o)^{-1}$, we obtain

$$H = 2\varepsilon_o V \sum_{\vec{k}} \omega_k^{\,2}\, \vec{a}_{\vec{k}}^{\,*} \cdot \vec{a}_{\vec{k}}\, . \tag{2.12}$$

Here we observe that the energy of radiation in a vacuum, which is positive by definition, is expressed in terms of the absolute squares of plane-wave amplitudes. Also note that the radiation energy is proportional to the volume V.

Let us now introduce real vectors:

$$\vec{Q}_{\vec{k}} \equiv V^{\frac{1}{2}} \varepsilon_o^{\frac{1}{2}} (\vec{a}_{\vec{k}} + \vec{a}_{\vec{k}}^{\,*}) \tag{2.13}$$

$$\vec{P}_{\vec{k}} \equiv -iV^{\frac{1}{2}} \varepsilon_o^{\frac{1}{2}} \omega_k (\vec{a}_{\vec{k}} - \vec{a}_{\vec{k}}^{\,*}) \equiv \dot{\vec{Q}}_{\vec{k}}. \tag{2.14}$$

The vectors $\vec{Q}_{\vec{k}}$ clearly obey the same equations of motion (1.20) as the amplitude vectors $\vec{a}_{\vec{k}}$:

$$\ddot{\vec{Q}}_{\vec{k}} = -\omega_k^{\,2}\, \vec{Q}_{\vec{k}}. \tag{2.15}$$

In terms of $\vec{Q}_{\vec{k}}$ and $\vec{P}_{\vec{k}}$, we can rewrite eq. (2.12) in the following form [Problem 2.2]

$$H = \sum_{\vec{k}} \frac{1}{2}(P_{\vec{k}}^{\,2} + \omega_k^{\,2}\, Q_{\vec{k}}^{\,2}). \tag{2.16}$$

Since $\vec{a}_{\vec{k}} \cdot \vec{k} = 0$, we obtain

$$\vec{Q}_{\vec{k}} \cdot \vec{k} = 0, \quad \vec{P}_{\vec{k}} \cdot \vec{k} = \dot{\vec{Q}}_{\vec{k}} \cdot \vec{k} = 0. \tag{2.17}$$

Thus, vectors $\vec{Q}_{\vec{k}}$ and $\vec{P}_{\vec{k}}$ are both perpendicular to the wave vector $\vec{k}$. Let us denote their components in the plane perpendicular to $\vec{k}$ by $(Q_{\vec{k},1}, Q_{\vec{k},2})$ and $(P_{\vec{k},1}, P_{\vec{k},2})$. We may then express eq. (2.16) in the form

$$H = \sum_{\vec{k}} \sum_{\sigma=1,2} \frac{1}{2} (P_{\vec{k},\sigma}^{\,2} + \omega_k^{\,2}\, Q_{\vec{k},\sigma}^{\,2}). \tag{2.18}$$

In this form, the radiation energy is the sum of the energies of harmonic oscillators, each oscillator corresponding to an electromagnetic plane-wave characterized by the wave vector $\vec{k}$, the angular frequency ω_k and the polarization index σ. Notice the similarity between (2.18) and (3.12.33), which represents the energy of a vibrating string written in terms of the normal coordinates and momenta.

We can look at eq. (2.18) as the definition equation for the Hamiltonian H with the set of canonical variables $\{Q_{\vec{k},\sigma}, P_{\vec{k},\sigma}\}$. In fact, one set of the Hamilton's equations,

$$\frac{d}{dt} Q_{\vec{k},\sigma} = \frac{\partial H}{\partial P_{\vec{k},\sigma}} = P_{\vec{k},\sigma} \tag{2.19}$$

coincides with the definition equations (2.14) for $P_{\vec{k},\sigma}$. The other set of equations,

$$\frac{d}{dt} P_{\vec{k},\sigma} = - \frac{\partial H}{\partial Q_{\vec{k},\sigma}} = -\omega_k^{\;2}\, Q_{\vec{k},\sigma}, \tag{2.20}$$

generates the equations of motion (2.15).

We now wish to move over to the quantum mechanical description of the same system. Before we take up this problem, we first review the case of a simple harmonic oscillator with mass m and spring constant k. The Hamiltonian H for this system is given by

$$\begin{aligned} H &= \frac{1}{2m} p^2 + \frac{1}{2} kx^2 \\ &= \frac{1}{2m} p^2 + \frac{1}{2} m\omega_o^{\;2}\, x^2, \end{aligned} \tag{2.21}$$

where $\omega_o \equiv (k/m)^{\frac{1}{2}}$. The position x and the momentum p satisfy the fundamental Poisson bracket relation: [see (3.6.10)]

$$\{x,p\} = 1. \tag{2.22}$$

In quantum mechanics, we postulate the quantum condition in the form of the commutation relation:

$$[x,p] = i\hbar, \tag{2.23}$$

where x and p are Hermitean operators representing position and momentum, respectively. The eigenvalues for the quantum Hamiltonian H, which has the same functional form as (2.21), were calculated earlier in §8.9, and are given by

$$E = (\tfrac{1}{2} + n)\hbar\omega_0 \,, \quad n = 0, 1, 2, \ldots \,. \tag{2.24}$$

We can obtain the same result in the following manner. Let us introduce a new set of canonical variables

$$Q \equiv \sqrt{m}\, x, \quad P \equiv \frac{p}{\sqrt{m}} \,, \tag{2.25}$$

in terms of which the Hamiltonian H is given by

$$H = \frac{1}{2}\,(P^2 + \omega_0{}^2\, Q^2). \tag{2.25}$$

The reader may verify by direct calculations [see Problem 2.3] that the correct equations of motion can be obtained from this Hamiltonian.

The new canonical variables (Q,P) are different from the old ones by constant factors. The corresponding quantum operators (Q,P) satisfy the commutation relation

$$[Q,P] = [\sqrt{m}\, x, \frac{p}{\sqrt{m}}\,] = [x,p] = i\hbar \tag{2.27}$$

The eigenvalues for the quantum Hamiltonian of the form (2.26) with the quantum condition (2.27) are also given by (2.24) [Problem 2.4].

We now return to the Hamiltonian for the radiation field given in (2.18). This Hamiltonian is similar to (2.26); only it has infinite degrees of freedom. In analogy with (2.27) we may postulate the following commutation relations:

$$[Q_{\vec{k},\sigma}, P_{\vec{k}',\sigma'}] = i\hbar\, \delta^{(3)}_{\vec{k},\vec{k}'}\, \delta_{\sigma,\sigma'}$$

$$[Q_{\vec{k},\sigma}, Q_{\vec{k}',\sigma'}] = [P_{\vec{k},\sigma}, P_{\vec{k}',\sigma'}] = 0. \tag{2.28}$$

The energy eigenvalues for the quantum Hamiltonian H in the form (2.18) can now be written as follows:

$$E = \sum_{\vec{k}} \sum_{\sigma} (\frac{1}{2} + n_{\vec{k},\sigma})\hbar\omega_k = E(\{n_{\vec{k},\sigma}\}), \tag{2.29}$$

where

$$n_{\vec{k},\sigma} = 0,1,2,\ldots \,. \tag{2.30}$$

The last expressions render the following interpretation: to each electromagnetic plane wave with the wave vector $\vec{k}$ and polarization σ, there corresponds a fictitious "particle" having the energy $\hbar\omega_k = \hbar ck$ and the momentum $\hbar\vec{k}$; the number of such particles may be zero, one, two and so on; the total electromagnetic energy is the sum of the energies of fictitious particles, called photons. In this picture, the quantum state of the radiation is characterized by the set of the numbers of photons, $\{n_{\vec{k},\sigma}\}$, in plane-wave states $(\vec{k},\sigma)$.

Problem 2.1 Prove eq. (2.8)
Hint: Use eq. (1.26) for the expression for $\vec{B}$ and the integration formula (2.6).

Problem 2.2 Derive eq. (2.16) from (2.12).

Problem 2.3 Using the Hamiltonian H in (2.26), derive the equation of motion. Compare the resulting equation with Newton's equation of motion.

Problem 2.4* The commutation relation $[Q, P] = i\hbar$ suggests the equivalence relation: $P = - i\hbar d/dQ$.

(a) Using this relation, write down the quantum differential operator corresponding to the Hamiltonian H in (2.26).

(b) Show that $\psi = e^{-\alpha Q^2}$, $\alpha = \frac{1}{2}\omega_0 \hbar^{-1}$, is an eigenfunction with the energy eigenvalue $\frac{1}{2}\hbar\omega_0$.

11.3 Black Body Radiation. Planck Distribution Function

Let us consider an oven maintained at a fixed temperature T. Inside the oven there are many moving molecules and electromagnetic radiation. We focus our attention on the latter. In the quantum picture, photons run with the speed of light. They may hit molecules moving within the oven or molecules which are stationary at the walls. They may be absorbed, re-emitted or reflected by these molecules. Photons do not interact with each other since the Maxwell equations which govern the electromagnetic fields are linear. We may therefore picture the oven filled with photons of all kinds of frequencies, wave-vectors, and polarizations. The population of photons is homogeneous within the oven, and also isotropic. Such states of photons in equilibrium are said to represent the black body radiation.

According to our earlier study in §§11.1 and 11.2, the energies for the radiation are given by (2.29):

$$E = \sum_{\vec{k}} \sum_{\sigma} (\tfrac{1}{2} + n_{\vec{k},\sigma}) \hbar\omega_k \equiv E(\{n_{\vec{k},\sigma}\}), \tag{3.1}$$

where $n_{\vec{k},\sigma}$ represents the number of photons with wave vector k and polarization σ. Assuming the canonical ensemble characterized by the distribution exp $(-\beta E)$, let us calculate the average number of photons, $\langle n_{\vec{k},\sigma}\rangle$:

$$\langle n_{\vec{k},\sigma}\rangle \equiv \mathrm{Tr}\{ n_{\vec{k},\sigma}\, e^{-\beta H}\}/\mathrm{Tr}\{e^{-\beta H}\}$$

$$= \frac{\sum \cdots \sum_{\{n_{\vec{k}',\sigma'}\}} n_{\vec{k},\sigma} \exp[-\beta E]}{\sum \cdots \quad \cdots \sum \exp[-\beta E]}$$

$$= \frac{\sum_{n_{\vec{k},\sigma}} n_{\vec{k},\sigma} \exp\left[-\beta n_{\vec{k},\sigma} \hbar\omega_k\right]}{\sum_{n_{\vec{k},\sigma}} \exp\left[-\beta\, n_{\vec{k},\sigma} \hbar\omega_k\right]}$$ [dropping common factors from numerator and denominator]

$$= -\frac{\partial}{\partial(\beta\hbar\omega_k)} \ln\Big[\sum_{n_{\vec{k},\sigma}=0} e^{-\beta\, n_{\vec{k},\sigma} \hbar\omega_k}\Big]$$ [d ln y/dx = y'/y]

$$= -\frac{\partial}{\partial(\beta\hbar\omega_k)} \ln\left[1 + e^{-\beta\hbar\omega_k} + e^{-2\beta\hbar\omega_k} + \ldots\right]$$

$$= -\frac{\partial}{\partial(\beta\hbar\omega_k)} \ln\left[1 - e^{-\beta\hbar\omega_k}\right]^{-1}$$

$$= \frac{e^{-\beta\hbar\omega_k}}{1 - e^{-\beta\hbar\omega_k}}$$

or

$$\langle n_{\vec{k},\sigma}\rangle = \frac{1}{e^{\beta\hbar\omega_k} - 1} \equiv n_0(\hbar\omega_k). \qquad (3.3)$$

Here we see that the average number of photons depends on the energy of the photon, $\hbar\omega_k$, only. The function $n_0(\varepsilon)$ is called the Planck distribution function. It is distinct from the Bose distribution function by the absence of the chemical potential.

The average energy of radiation per unit volume is

$$V^{-1}\langle E\rangle = V^{-1}\sum_{\vec{k}}\sum_{\sigma}\left[\frac{1}{2} + \langle n_{\vec{k},}\rangle\right]\hbar\omega_k$$

$$= V^{-1}\sum_{\vec{k}}\sum_{\sigma}\left[\frac{1}{2} + n_0(\hbar\omega_k)\right]\hbar\omega_k \qquad \text{[use of (3.3)]}$$

$$= u_0 + 2V^{-1}\sum_{\vec{k}} n_0(\hbar\omega_k)\,\hbar\omega_k\,, \tag{3.7}$$

where

$$u_0 \equiv \frac{1}{2}V^{-1}\sum_{\vec{k},\sigma}\hbar\omega_k = V^{-1}\sum_{\vec{k}}\hbar\omega_k \tag{3.8}$$

is the zero-point energy per unit volume, which is independent of temperature. Notice the polarization multiplicity factor 2, which appears after the summation over the polarization indices.

In the bulk limit, the sum $V^{-1}\sum_{\vec{k}} n_0(\hbar\omega_k)\,\hbar\omega_k$ can be converted into an integral in $\vec{k}$ space. By using (1.15), the density of states in $\vec{k}$ space, dn/d^3k, is calculated as follows:

$$\frac{dn}{d^3k} \equiv \frac{\text{no. of states in } d^3k}{d^3k} = \frac{1}{\text{k-volume per state}}$$

$$= \frac{1}{(2\pi/L)^3} = \frac{V}{(2\pi)^3}\,. \tag{3.9}$$

Using this, we obtain

$$u \equiv \lim V^{-1}\langle E\rangle$$

$$= u_0 + \frac{2}{(2\pi)^3}\int d^3k\,\hbar\omega_k\,n_0(\hbar\omega_k)\,. \tag{3.10}$$

The k-integral may be evaluated by using spherical polar coordinates. After carrying out angular integrations and expressing the result as an ω-integral ($\omega = ck$), we obtain [Problem 3.1]

$$u - u_0 = \frac{\hbar}{\pi^2 c^3} \int_0^\infty d\omega\, \omega^3 \frac{1}{e^{\beta\hbar\omega} - 1}$$

$$= \frac{\hbar}{\pi^2 c^3} \left(\frac{k_B T}{\hbar} \right)^4 \int_0^\infty dx \frac{x^3}{e^x - 1} . \qquad [x = \beta\hbar\omega] \qquad (3.11)$$

The x-integral is found to be

$$\int_0^\infty dx \frac{x^3}{e^x - 1} = \frac{\pi^4}{15} . \qquad (3.12)$$

We then obtain

$$u = \frac{\pi^2}{15} \frac{(k_B T)^4}{(c\hbar)^3} + u_0 . \qquad (3.13)$$

The average energy density apart from the zero-point energy is thus proportional to the fourth power of the temperature T. This is known as the Stephan-Boltzmann law of radiation.

Problem 3.1 Derive (3.11) from (3.10).

11.4 Experimental Verification of the Planck Distribution Function

The study of the black body radiation has played a most significant role in the development of quantum theory. In fact, the fundamental constant h, known today as Planck's constant, was first introduced in the interpretation of the energy distribution of the black body radiation. We will discuss the experimental study of this distribution in the present section.

Let us consider an oven maintained at a fixed temperature T. We make a small hole on the wall and measure the radiation emerging through the hole. If we view the radiation as the flow of photons, the situation here is similar to that of the effusing molecules through a small hole, which was discussed in §4.13.

Let us compute the photon flux escaping the small hole in the geometry shown in Figure 11.3. We ask how many photons arrive at the small surface element ΔA per unit time in the particular direction (ϑ,ϕ). The number density of photons moving in the "right" direction within the solid angle $d\Omega$ is given

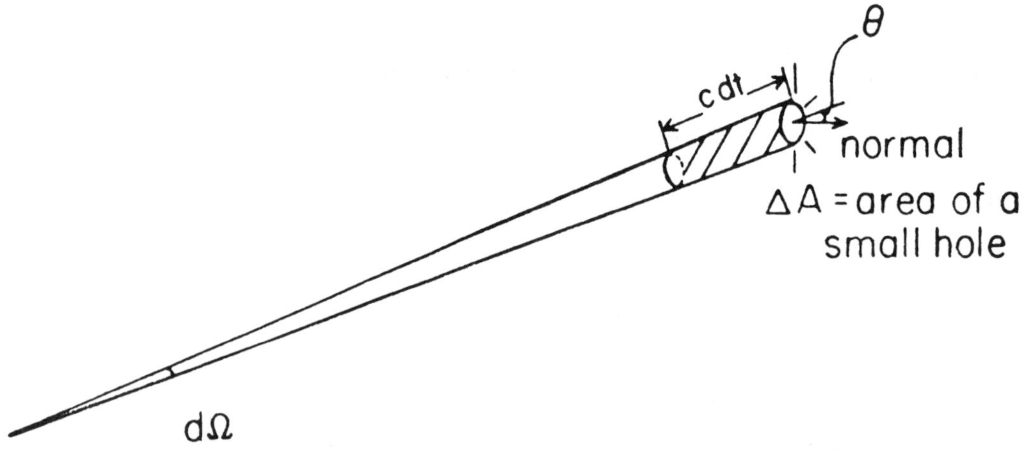

Fig. 11.3 A photon proceeding in the solid angle $d\Omega$ and located within the quasi-cylinder escapes through a small hole in the time dt.

by [see eq. (3.9)]

$$2n_o(\hbar\omega_k)\,\frac{k^2dk\;d\Omega}{(2\pi)^3} \equiv 2n_o\,\frac{d^3k}{(2\pi)^3}\,, \tag{4.1}$$

where n_o is the Planck distribution function and the factor 2 arises from the polarization multiplicity. Noting that all photons move with the same speed c, we obtain

photon flux

$$= 2(2\pi)^{-3}c\;n_ok^2\;dk\;d\Omega. \tag{4.2}$$

Since each photon carries the energy $\hbar\omega$, we multiply (4.2) by this factor, and obtain

energy flux

$$= 2(2\pi)^{-3}c\;\hbar\omega\;n_ok^2\;dk\;d\Omega. \tag{4.3}$$

Noting the relations

$$k = \frac{\omega}{c}\,,\quad dk = \frac{d\omega}{c}\,, \tag{4.4}$$

we can rewrite eq. (4.3) as

energy flux

$$= 2(2\pi)^{-3}\;\hbar\;c^{-2}\;n_o\;\omega^3\;d\omega d\Omega. \tag{4.5}$$

In optical spectroscopy, the frequency ν rather than the angular frequency $\omega \equiv 2\pi\nu$ is used most often; also the original Planck constant $h \equiv 2\pi\hbar$ is used. In these units, we may re-express (4.5) in the form:

$$\text{energy flux} = 2\, n_o \frac{h\nu^3}{c^2}\, d\nu\, d\Omega\ . \tag{4.6}$$

The energy flux per unit frequency per unit solid angle, $I(\nu)$, called the spectral energy density or the spectral intensity, is defined by the relation:

$$\text{energy flux} = I(\nu)\, d\nu\, d\Omega\ . \tag{4.7}$$

Comparing the last two expressions, we obtain the spectral intensity as

$$I(\nu) = 2\frac{h\nu^3}{c^2}\frac{1}{\exp(h\nu/k_BT) - 1}\ . \tag{4.8}$$

The spectral intensity I can directly be subjected to the experimental observation as sketched in Figure 11.4. The detector D and selector S are placed at large distances from the oven, and are arranged so that the detector

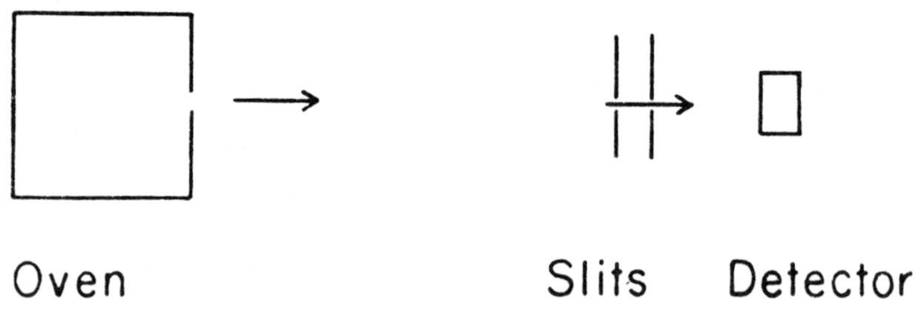

Fig. 11.4 An experimental arrangement for observing the spectral intensity.

can receive photons whose directions of travel are within a small solid angle d . If scattering, emission and absorption of photons outside the oven is neglected, the detected radiation intensity should clearly be proportional to the spectral intensity at the hole.

The <u>spectral intensity per unit wavelength</u>, $I'(\lambda)$, is defined by the relation

$$\text{energy flux} = I'(\lambda)\, d\lambda\, d\Omega$$
$$= I(\nu)\, d\nu\, d\Omega\ . \tag{4.9}$$

Noting thc relations

$$\lambda = \frac{c}{\nu}\ , \quad d\lambda = -\frac{c}{\nu^2}\, d\nu\ , \tag{4.10}$$

and comparing (4.6) and (4.9), we obtain

$$I'(\lambda) = 2\,\frac{hc^2}{\lambda^5}\,\frac{1}{\exp\,(hc/\lambda k_B T) - 1}\ . \tag{4.11}$$

Typical experimental data and theoretical curves for the spectral intensity $I'(\lambda)$ are shown in Figure 11.5. Generally, agreements are excellent. The position of the maximum intensity occurs at the wavelength λ_o, at which

$$\left.\frac{dI'(\lambda)}{d\lambda}\right]_{\lambda = \lambda_o} = 0\ . \tag{4.12}$$

Using the explicit form (4.11) for $I'(\lambda)$, we obtain

$$\lambda_o = 0.201\,\frac{hc}{k_B T}\ . \tag{4.13}$$

From this formula and also from Fig. 11.5, we observe that at T = 1000 K, the

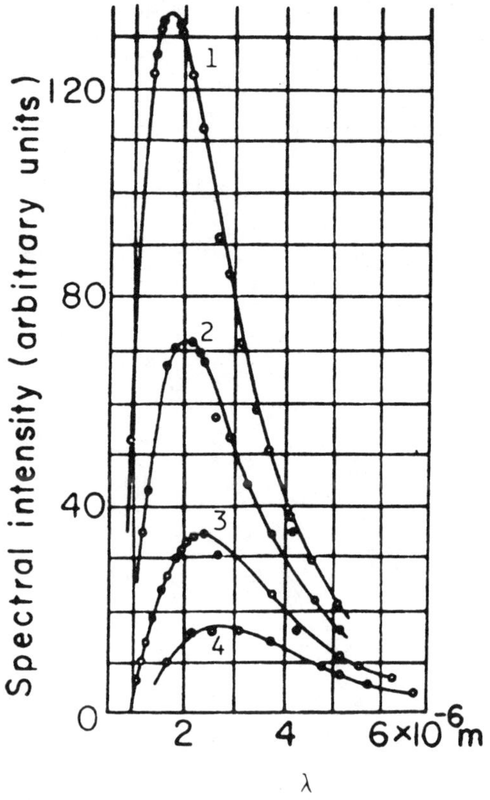

Fig. 11.5 Black body radiation for various temperatures as a function of wavelength.
1, 1646 K; 2, 1449 K
3, 1259 K; 4, 1095 K
[from P. Moon, [2]]

wavelength λ_0 is about 2800 A (ultraviolet region).

Due to the great importance of the Planck distribution function, the spectral intensity of the black body radiation has been studied repeatedly and with great care. We emphasize that the shape of the curve cannot be explained in terms of the classical statistics. It was first interpreted by Planck (Max Planck, 1858-1947, German) with the introduction of the fundamental constant h (Planck's constant). This was in 1900. This year is generally regarded as the year of birth of the quantum theory. In practice, the shpae of the distribution can give the temperature of the oven. Although it is not very accurate, observation of the wavelength λ_0 for the maximum intensity $I'(\lambda)$ can provide the numerical value for the ratio of fundamental constants, hc/k_B, as indicated in (4.13).

Problem 4.1 Using expression (4.11) for the spectral density per unit wavelength $I(\lambda)$, obtain the wavelength λ_o at which $I(\lambda)$ is greatest.

Answer: $\lambda_o = 0.201 \dfrac{hc}{k_B T}$

Problem 4.2 The spectral intensity per unit frequency, $I(\nu)$, is given by (4.8). Find the frequency ν_1, at which $I(\nu)$ is greatest. Compare the value of the corresponding wavelength $\lambda_1 = c/\nu_1$ with the wavelength λ_o obtained in Problem 4.1.

11.5 Radiation Pressure

In the quantum view of the electromagnetic radiation, the photon corresponding to the electromagnetic plane wave with the wave vector $\vec{k}$ has the momentum $\hbar\vec{k}$. If a number of photons is enclosed in a box of volume V, the walls of the box which reflect photons should experience pressure. This pressure for photons in equilibrium will be calculated and discussed in the present section.

The Hamiltonian H for the system and its eigenvalues are given by eqs.(2.18) and (3.1). The partition function Z is given by

$$Z(\beta, V) = \mathrm{Tr}\{\exp(-\beta H)\}$$

$$= \sum_{\text{states}} \exp\left(-\beta \sum_{\vec{k}\,\sigma} \left(\tfrac{1}{2} + n_{\vec{k},\sigma}\right)\hbar\omega_k\right), \qquad (5.1)$$

where the summation is over quantum states characterized by the occupation numbers $\{n_{\vec{k},\sigma}\}$ of photons in the wave modes $(\vec{k},\sigma)$. We can then transform (5.1) into

$$Z = \sum_{\{n_{\vec{k},\sigma}\}} \sum \cdots \sum \exp\left(-\beta \sum_{\vec{k},\sigma} \left(\tfrac{1}{2} + n_{\vec{k},\sigma}\right)\hbar\omega_k\right)$$

$$= \sum_{\{n_{\vec{k},\sigma}\}} \sum \cdots \sum \prod_{\vec{k},\sigma} \exp\left(-\beta\left(\tfrac{1}{2} + n_{\vec{k},\sigma}\right)\hbar\omega_k\right) \quad [e^{a+b} = e^a e^b]$$

$$= \prod_{\vec{k},\sigma} \left[\sum_{n_{\vec{k},\sigma}} \exp\left(-\beta\left(\tfrac{1}{2} + n_{\vec{k},\sigma}\right)\hbar\omega_k\right)\right]. \qquad (5.2)$$

For each mode,

$$\sum_{n_{\vec{k},\sigma}=0}^{\infty} \exp\left[-\beta \left(\tfrac{1}{2} + n_{\vec{k},\sigma}\right) \hbar\, \omega_k \right]$$

$$= e^{-\frac{1}{2}\beta\hbar\omega_k} \left[1 + e^{-\beta\hbar\omega_k} + e^{-2\beta\hbar\omega_k} + \ldots \right]$$

$$= e^{-\frac{1}{2}\beta\hbar\omega_k} \left[1 - e^{-\beta\hbar\omega_k} \right]^{-1} . \tag{5.4}$$

Applying the standard formula (7.3.13) : $P = \beta^{-1}\, \partial \ln Z(\beta, V)/\partial V$, we can calculate the pressure P in the following manner :

$$P = -2\, \beta^{-1} \frac{\partial}{\partial V} \sum_{\vec{k}} \left\{ \tfrac{1}{2}\beta\hbar\omega_k + \ln\left[1 - e^{-\beta\hbar\omega_k} \right] \right\}, \tag{5.5}$$

where the factor 2 resulted from the summation over the two possible polarizations (σ). The angular frequency ω_k defined in (1.16), depends on the normalization volume V as follows :

$$\omega_k = c\, k = 2\pi c \sqrt{n_x^2 + n_y^2 + n_z^2} \;/\; L$$

$$= 2\pi c \sqrt{n_x^2 + n_y^2 + n_z^2}\; V^{-1/3} . \tag{5.6}$$

Differentiating this expression with respect to V, we obtain

$$\frac{\partial \omega_k}{\partial V} = -\frac{1}{3} V^{-1}\, \omega_k . \tag{5.7}$$

Using this result, we obtain from (5.5)

$$P = -2\beta^{-1}\frac{\partial\omega_k}{\partial V}\frac{\partial}{\partial\omega_k}\left(\sum_{\vec{k}}\{\tfrac{1}{2}\beta\hbar\omega_k + \ln[1-e^{-\beta\hbar\omega_k}]\}\right)$$

$$= (-2\beta^{-1})\left(-\frac{\omega_k}{3V}\right)\left(\sum_{\vec{k}}\{\tfrac{1}{2}\beta\hbar + \frac{\beta\hbar e^{-\beta\hbar\omega_k}}{1-e^{-\beta\hbar\omega_k}}\}\right)$$

$$= \frac{2}{3V}\sum_{\vec{k}}\left\{\frac{1}{2} + \frac{1}{e^{\beta\hbar\omega_k}-1}\right\}\hbar\omega_k$$

$$= \frac{2}{3V}\sum_{\vec{k}}\left\{\frac{1}{2} + n_0(\hbar\omega_k)\right\}\hbar\omega_k . \text{ [use of (3.3)]} \tag{5.8}$$

Comparing this expression with eqs.(3.7) and (3.10), we obtain

$$\boxed{P = \frac{1}{3}u .} \tag{5.9}$$

That is, the pressure equals one third of the radiation energy density. Using (3.13), we then have

$$\boxed{P = \frac{\pi^2}{45}\frac{(k_B T)^4}{(c\hbar)^3} + \frac{1}{3}u_0,} \tag{5.10}$$

where u_0 represents the zero-point energy density and is independent of temperature.

Radiation pressure can be quite significant in astronomical problems. But it is very difficult to observe in the laboratory. In the usual operation of a radiometer, the black plates absorb radiation and warm the gas in their immediate vicinity. This generates molecular flow (heat convection) and a resulting change in the pressure. This effect is more than one order of magnitude greater than radiation pressure.

Problem 5.1 The pressure P of a system of non-interacting particles equals 2/3 of the energy density : $P = 2u/3$.

(a) Prove this for the classical mechanical system.

(b) Do the same for the quantum mechanical system.

11.6 Crystal Lattices

When a material forms a solid, the molecules align themselves in a special periodic structure called a crystal lattice. For example, in rock salt (NaCl), Na^+ and Cl^- ions occupy the sites of a simple cubic lattice alternately as shown in Figure 11.6. In copper (Cu), Cu^+ ions form a

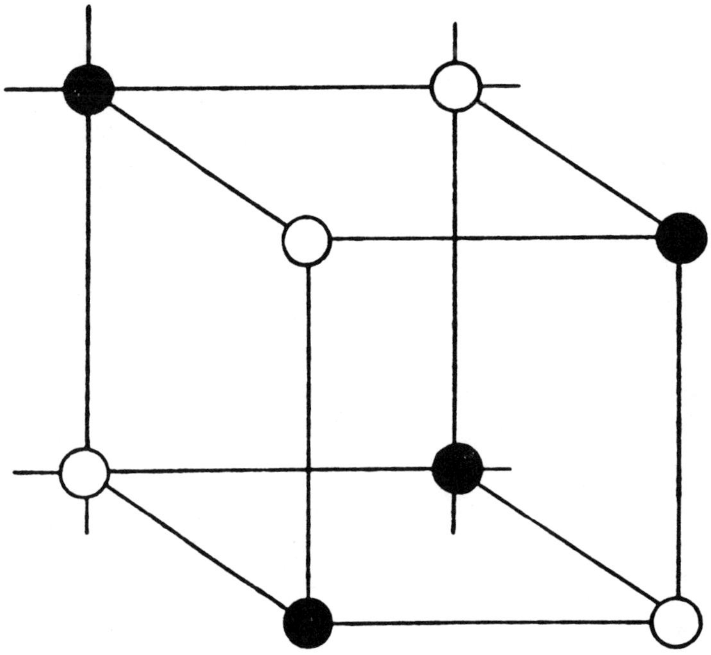

Fig.11.6. A simple cubic lattice. In rock salt, Na^+ and Cl^- occupy the lattice sites alternately.

body-centered cubic lattice as shown in Figure 11.7. In solid Ar, neutral

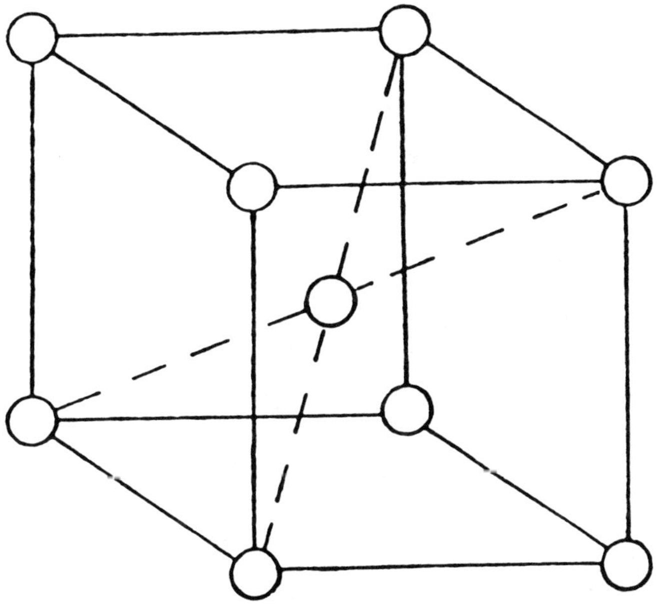

Fig. 11.7. A body-centered cubic lattice.

molecules form a face-centered cubic lattice, as shown in Figure 11.8. Diamonds and graphites are both composed of carbon molecules C. The arrangement of C molecules however, are different.

Why does a particular material exist in a particular crystalline state ? This is a good but hard question; the answer must involve in a complicated manner the composition and nature of the molecules constituting the material and the interaction between component particles (electrons, ions, neutral molecules).

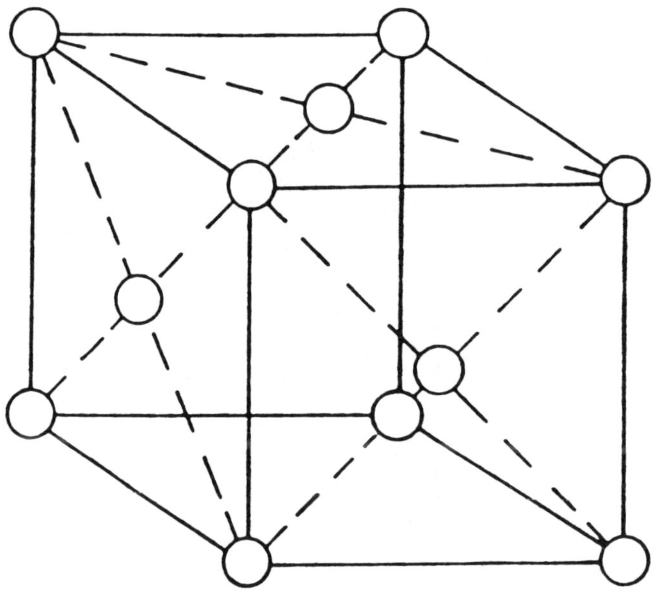

Fig. 11.8. A face-centered cubic lattice.

Solids can be classified into metals, insulators, semiconductors, etc., according to the electric transport properties. They can also be classified according to the modes of binding. The readers interested in this and other aspects of solids are encouraged to refer to elementary solid-state textbooks such as Kittel's Solid State Physics [3]. An overwhelming fact however is that almost all materials crystallize below a certain temperature. The exceptions are liquid He^3 and He^4, which do not freeze at any temperature [at the atmospheric pressure]. These quantum liquids were briefly discussed in § 10.7 of the last chapter.

11.33

The thermal properties of crystalline solids are remarkably similar for all materials. At high temperatures, Dulong-Petit's law for the heat capacity, which was discussed in §7.6, is observed. At moderate and low temperatures, deviations from this law become significant as pointed out in §8.1. We can treat the heat capacity of a solid without excessive mathematical complications. We will discuss this topics in the remainder of this chapter.

Problem 6.1 What is the number of the nearest neighbors for (a) the simple cubic lattice, (b) the body centered cubic lattice and (c) the face-centered cubic lattice? Study the lattice structures of diamond and graphite by referring to some solid-state physics textbooks. What are the number of the nearest neighbors for (d) the diamond and (e) the graphite lattice ?

[Answers : (a) 6 (b) 8 (c) 12 (d) 4 (e) 3.]

11.7 Lattice Vibrations. Einstein's Theory of the Heat Capacity

Let us consider a crystal lattice at low temperatures. We may expect that each molecule forming the lattice executes small oscillations around the equilibrium position. For illustration, consider a one-dimensional lattice shown in Figure 11.9. The motion of j-th molecule may be characterized by the Hamiltonian of the form

$$h_j = \frac{p_j^2}{2M} + \frac{1}{2} k_o u_j^2 , \qquad (7.1)$$

where u_j denotes displacement, and $p_j \equiv M\dot{u}_j$ represents the momentum. Here we assumed a parabolic potential which is reasonable for small oscillations. If the equipartition theorem is applied, the kinetic and potential energy parts contribute $\frac{1}{2} k_B T$ each to the average thermal energy. We then obtain

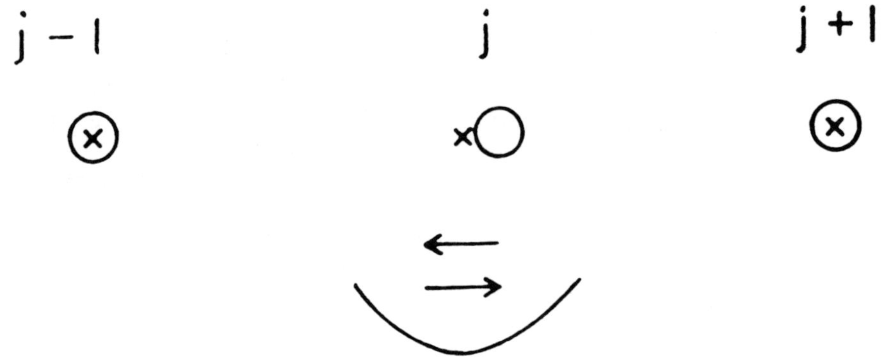

Fig. 11.9 The j-th molecule in the linear chain executes simple harmonic oscilations as characterized by eq.(7.1).

$$\langle h\rangle = \langle\frac{p^2}{2M}\rangle + \langle\frac{1}{2}k_o u^2\rangle = \frac{1}{2}k_B T + \frac{1}{2}k_B T = k_B T. \tag{7.2}$$

Multiplying this by the total number of molecules, N, we obtain $N k_B T$ for the total energy. By differentiating it with respect to T, we obtain the heat capacity $N k_B$ for the lattice.

In quantum mechanics, the eigenvalues of the Hamiltonian

$$h = \frac{1}{2M}p^2 + \frac{1}{2}k_o u^2 \tag{7.3}$$

are given by [see (8.8.13)]

$$\varepsilon_n \equiv (\frac{1}{2} + n)\hbar\omega_o, \quad n = 0,1,2,\ldots , \tag{7.4}$$

where

$$\omega_o \equiv (k_o/M)^{1/2} \tag{7.5}$$

is the angular frequency; and the quantum states are characterized by non-negative integers n.

If we assume a canonical ensemble of simple harmonic oscillators with the distribution exp $[-\beta\varepsilon]$, the average energy can be calculated as follows:

$$\langle h\rangle = \frac{\sum_n \varepsilon_n e^{-\beta\varepsilon_n}}{\sum_n e^{-\beta\varepsilon_n}} = -\frac{\partial}{\partial\beta}\,\ell n\,[\Sigma\, e^{-\beta\varepsilon_n}] = -\frac{\partial}{\partial\beta}\,\ell n\,[\sum_{n=0}^{\infty} e^{-\beta\hbar\omega_o(n+\frac{1}{2})}]$$

$$= -\frac{\partial}{\partial\beta}\,\ell n\,[e^{-(\frac{1}{2})\,\beta\hbar\omega_o}\,[1 - e^{-\beta\hbar\omega_o}]^{-1}]$$

$$= \frac{\partial}{\partial\beta}\,[\frac{1}{2}\,\beta\hbar\omega_o + \ell n\,(1 - e^{-\beta\hbar\omega_o})] = [\frac{1}{2} + n_o(\hbar\omega_o)]\,\hbar\omega_o, \qquad (7.6)$$

where

$$n_o(\eta) \equiv \frac{1}{e^{\beta\eta}-1} \qquad (7.7)$$

is the Planck distribution function. Notice that the average energy $\langle h\rangle$ is quite different from the classical average energy $k_B\,T$, see eq.(7.2).

For a three dimensional lattice with N molecules we should multiply expression (7.6) by the number of degrees of freedom, 3N. We then obtain

$$E = 3\,N\,[\frac{1}{2} + n_o\,(\hbar\omega_o)]\,\hbar\omega_o \qquad (7.8)$$

for the thermal energy of the lattice. Differentiating this expression with respect to T, we obtain the heat capacity: [Problem 7.1]

$$C_V = \frac{\partial E}{\partial T} = \frac{\partial E}{\partial\beta}\frac{d\beta}{dT} = \frac{3N\,(\hbar\omega_o)^2}{k_B\,T^2}\;\frac{e^{\beta\hbar\omega_o}}{(e^{\beta\hbar\omega_o} - 1)^2}\,.$$

We rewrite this expression in the form [Problem 7.1]

$$C_V = 3\,N\,k_B\left(\frac{\Theta_E}{T}\right)^2\,\frac{e^{\Theta_E/T}}{(e^{\Theta_E/T}-1)^2}\,, \qquad (7.9)$$

where

$$\Theta_E \equiv \frac{\hbar\omega_0}{k_B} = \frac{\hbar}{k_B}\left(\frac{k_0}{M}\right)^{\frac{1}{2}} \tag{7.10}$$

has the dimension of temperature, and is called the Einstein temperature.

The function

$$g(x) \equiv x^2 e^x (e^x - 1)^{-2} \qquad [x = \Theta_E/T] \tag{7.11}$$

has the asymptoic behavoir [Problem 7.2]

$$g \sim \begin{cases} x^2 e^{-x} & \text{for } x >> 1 \\ 1 & \text{for } x << 1. \end{cases} \tag{7.12}$$

At very high temperatures ($\Theta_E/T<<1$), the heat capacity, given by (7.9), approaches the classical value 3 Nk_B:

$$C_V = 3N k_B g(\Theta_E/T) \sim 3 N k_B. \tag{7.13}$$

At very low temperatures ($\Theta_E/T>>1$), the heat capacity C_V behaves like

$$C_V \sim 3N k_B (\Theta_E/T)^2 e^{-\Theta_E/T}, \tag{7.14}$$

and approaches zero exponentially as T tends to zero. The behavior of $C_V/3N k_B$ from (7.9) against T/Θ_E is plotted in Figure 11.10. For comparison, experimental data of C_V for diamond, are shown by points with the choice of Θ_E = 1320 K. The fit is quite reasonable for diamond. Historically, Einstein obtained the formula (7.9) in 1907 [4] before the advent of the quantum theory in 1925-6. At that time the discrepancy between experiment and theory (equipartition theorem) had been a mystery.

Einstein's theory explains very well why a certain solid like diamond exhibits a molar heat capacity substantially smaller than 3 R (the value in accordance with Dulong-Petit's law). It also predicts that the heat capacity for any solid should decrease as the temperature is lowered. This is in

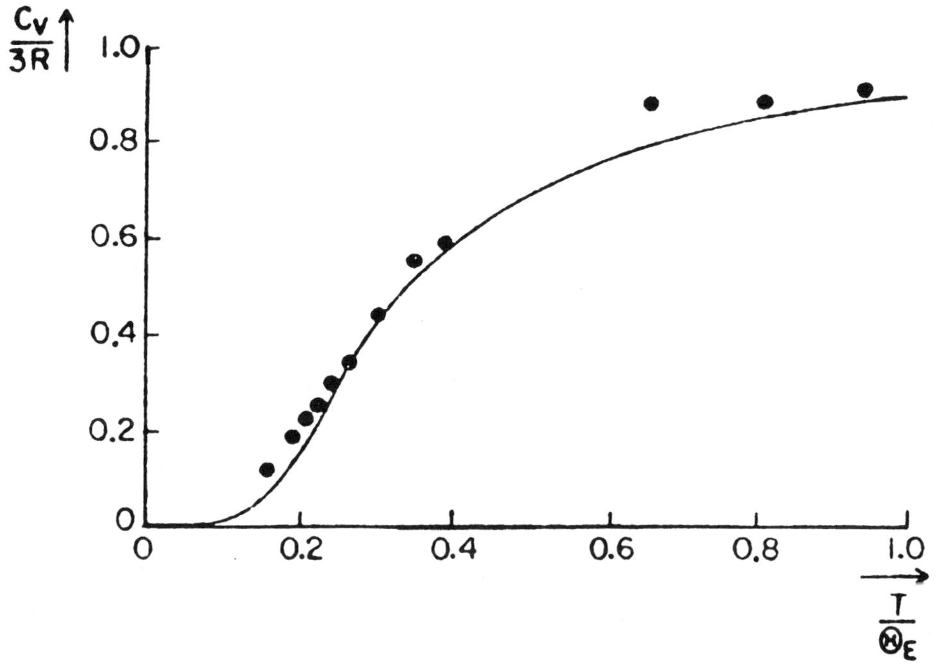

Fig. 11.10 The molar heat capacity for Einstein's model of a solid is shown in solid line. Experimental points are for a diamond with Θ_E = 1320 K. [After A. Einstein [4]]

qualitative agreement with experimental observations. However the temperature-dependence at the low temperature end indicates non-negligible discrepancies. Several years after Einstein's theory, Debye reported a very successful theory of the heat capacity over all temperature range. Before we present Debye's theory in §11.11, we will look at the solids from a macroscopic point of view in the following few sections.

Problem 7.1 Derive (7.9), using (7.7) and (7.8).

Problem 7.2 Verify the asymptotic behavior (7.12) from (7.11).

11.8° Elastic Properties.

In the present section, we will review elastic properties of a solid in an elementary manner.

Let us take a rectangular block of solid. If external tensions are applied to pull the block as shown in Figure 11.11, the length L of the block will extend a small amount ΔL. It is empirically

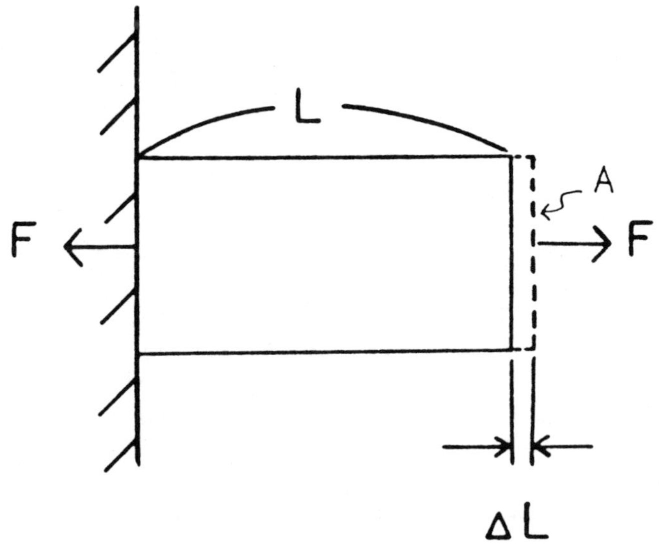

Fig. 11.11 A solid extends a small amount ΔL by applying a tensile stress F/A.

known that for small deformations, the tensile strain defined by $\Delta L/L$ is proportional to the tensile stress, that is, the tensile force per unit area:

$$Y \frac{\Delta L}{L} = \frac{F}{A}, \tag{8.1}$$

where A denotes the side area.

The proportionality factor Y is called the Young modulus. This factor depends on the nature of the material.

When the block elongates under tensile stress, the dimensions perpendicular to the direction of stress become smaller by an amount proportional to the tensile strain. If the original width and its small change are denoted by W and ΔW, respectively, then it is found that

$$\frac{\Delta W}{W} = -\sigma \frac{\Delta L}{L}, \tag{8.2}$$

where σ is a dimensionless constant, called Poisson's ratio, which is also a material constant. For many materials σ has a value between 0.1 and 0.4. See Table 11.1.

If an extra pressure ΔP is applied to the block, the volume should decrease. The volume strain defined by the ratio $\Delta V/V$ will be proportional to the change ΔP in pressure. This relation is represented by

$$-B \frac{\Delta V}{V} = \Delta P, \tag{8.3}$$

where the coefficient B is called the bulk modulus. The reciprocal of the bulk modulus is called the compressibility, and is given by

$$\kappa \equiv \frac{1}{B} = -\frac{1}{V}\frac{\Delta V}{\Delta P} . \tag{8.4}$$

For a solid or liquid, this quantity is small. The volume of a gas changes markedly under the applied pressure. This property therefore distinguishes a gas from a liquid (and solid) as pointed out earlier in §3.8.

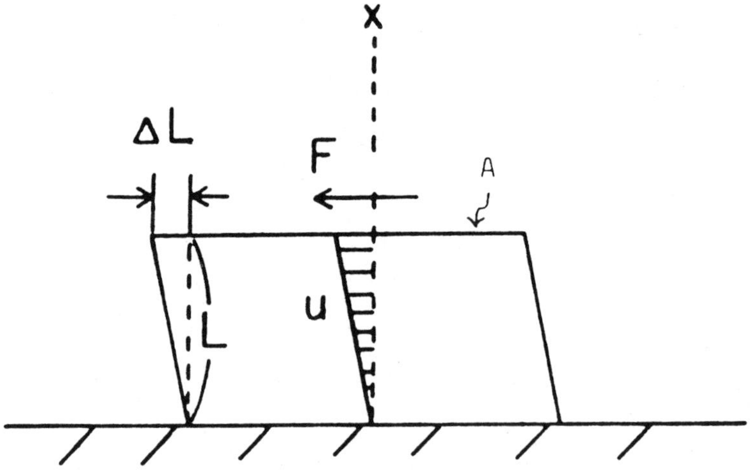

Fig. 11.12. Under a shearing force F, the solid block suffers a shear strain ΔL/L.

Let us now apply a shearing force on the block which is fixed at one end as shown in Figure 11.12. It is observed that for small deformations, the shear strain, defined by ΔL/L, is proportional to the shearing stress F/A:

$$S\frac{\Delta L}{L} = \frac{F}{A}. \tag{8.5}$$

The proportionality factor S is called the shear (or rigidity) modulus.

The response to the shearing force is a distinctive property of a solid. Such behavior is not known for a fluid (liquid or gas). We further note that all moduli, Y, B and S, have the dimension of stress (force per unit area). Table 11.1 shows the numerical values for typical solids.

Materials	Young modulus Y $10^{11} Nm^{-2}$	Bulk modulus B $10^{11} Nm^{-2}$	Shear modulus S $10^{11} Nm^{-2}$	Poisson's ratio σ
Aluminum	0.70	0.70	0.30	0.16
Copper	1.1	1.4	0.42	0.32
Iron	1.9	1.0	0.70	0.27
Steel	2.0	1.6	0.84	0.19
Tungsten	3.6	2.0	1.5	0.20

Table 11.1 Approximate values for elastic constants.

For a homogeneous and isotropic solid, the elastic constants are known to be interrelated such that only two of these are independent. For example, the shear modulus S can be expressed in terms of Young modulus Y and Poisson's ratio σ as follows:

$$S = \frac{Y}{2(1 + \sigma)}. \tag{8.6}$$

11.9° Elastic Waves

Elastic properties of a solid support elastic waves. There are two kinds of waves : transverse and longitudinal waves. We will discuss these waves in the present section.

Let us first take the case of a transverse wave. Consider a wave which proceeds in the x-direction and whose oscillations are in the y-direction. Such a wave is represented in Figure 11.13, where the displacement u(x,t) is shown

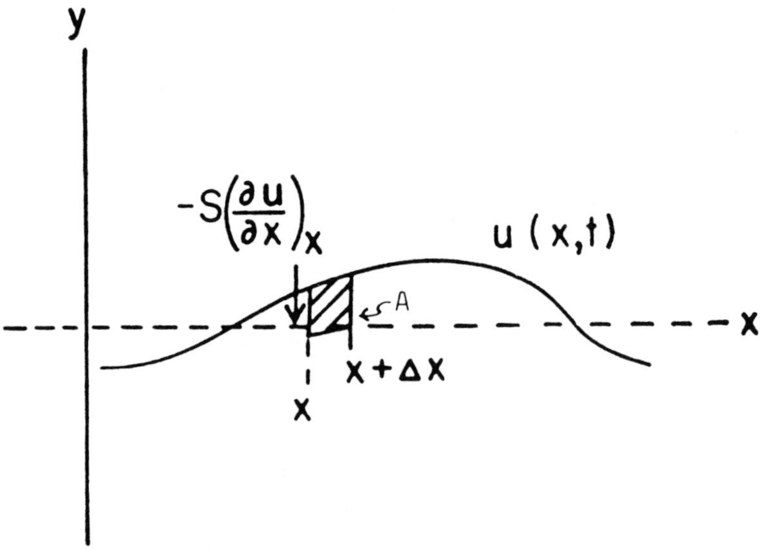

Fig. 11.13 A transverse elastic wave proceeding in the x-direction. The displacement is shown in full line.

by the solid line. Note that the displacement is measured from the straight line (the x-axis), corresponding to the equilibrium state of the solid.

[The transverse oscillations are similar to the transverse oscillations of a stretched string, which we discussed in §§3.11 and 3.12.]

As we will show below, the displacement u obeys the <u>wave equation</u>

$$\rho \frac{\partial^2 u}{\partial t^2} = S \frac{\partial^2 u}{\partial x^2} , \tag{9.1}$$

where ρ denotes the mass density.

Let us go back to Figure 11.12. Under the applied stress, the displacement u measured from the equilibrium position (dotted line) may vary as shown. Shear stresses acting at internal planes parallel to F are proportional to the shear strain so that

$$\text{the shear stress (magnitude) at } x = S \frac{u}{x} = S \frac{\partial u}{\partial x} . \tag{9.2}$$

We note that the shear stress of this magnitude acts on the opposite sides of the plane in such a direction that it may reduce the gradient (magnitude) of the displacement.

Let us now consider a portion of the solid within the volume element $A \, \Delta x$ where $A \equiv \Delta y \, \Delta z$ represents the rectangular area perpendicular to the x-axis. See Figure 11.13. This volume remains constant throughout during the transverse plane-wave motion. The shearing force acting on this portion at the plane $x = x_o$, is given by

$$- S \left(\frac{\partial u}{\partial x}\right)_{x_o} A, \tag{9.3}$$

where the negative sign arises from the fact that the force acts in the direction in which the gradient $\frac{\partial u}{\partial x}$ is reduced. In the case shown in Figure 11.13,

$[\partial u/\partial x]_{x_0} > 0$, and therefore the shearing force is directed downwards. In other words, at the plane $x = x_0$ where $[\partial u/\partial x]_{x_0} > 0$, the shearing forces act such that the left-hand part of the solid pulls down the right-hand part, and, by Newton's third law (action and reaction), the right-hand part pulls up the left-hand part. The direction (and magnitudes) of the forces may, in fact, be simulated in terms of a hypothetical stretched string lying on the line of the displacement which executes transverse oscillations.

At the plane $x = x_0 + \Delta x$, the shearing force acting on the portion with the correct direction, is given by

$$S \left(\frac{\partial u}{\partial x} \right)_{x_0 + \Delta x} A. \tag{9.4}$$

The sum of the shearing forces acting on the portion at the planes at x and $x + \Delta x$ (dropping the suffix o) is given by

$$- SA \left(\frac{\partial u}{\partial x} \right)_{x} + SA \left(\frac{\partial u}{\partial x} \right)_{x + \Delta x}$$

$$= SA \frac{\partial^2 u}{\partial t^2} \Delta x. \tag{9.5}$$

This must be equal to the mass ($\rho\ A\ \Delta x$) times acceleration $\dfrac{\partial^2 u}{\partial t^2}$. We therefore obtain

$$(\rho\ \Delta x\ A) \frac{\partial^2 u}{\partial t^2} = S \frac{\partial^2 u}{\partial x^2} \Delta x\ A,$$

which immediately yields eq.(9.1) after dropping the common factor $\Delta x\ A$.

Let us rewrite eq.(9.1) in the standard form of a wave equation :

$$\boxed{\frac{\partial^2 u}{\partial t^2} = c_t^2 \frac{\partial^2 u}{\partial x^2}\,,} \tag{9.6}$$

where

$$c_t \equiv (S/\rho)^{\frac{1}{2}} \tag{9.7}$$

represents the speed of the transverse wave.

Eq.(9.7) is valid for the transverse wave proceeding in the x-direction. For an isotropic solid, we can remove this restriction, and obtain the following wave equation :

$$\boxed{\frac{\partial^2 \vec{u}_t}{\partial t^2} = c_t^2 \nabla^2 \vec{u}_t .} \tag{9.8}$$

It is interesting to observe that the transverse elastic waves are similar to the electromagnetic waves discussed in §11.1. Eq.(9.8) corresponds to eq.(1.9). Only the speed of propagation is different.

A solid can also support a longitudinal compressional wave characterized by the wave equation

$$\frac{\partial^2 \vec{u}_\ell}{\partial t^2} = c_\ell^{\,2} \nabla^2 \vec{u}_\ell\,, \tag{9.9}$$

where c_ℓ is the propagation speed given by

$$c_\ell \equiv [(B + \tfrac{3}{4} S)/\rho]^{\frac{1}{2}} . \tag{9.10}$$

The restoring force for this case is more complex than the case of the transverse wave, and this is reflected in the expression (9.10) for the speed of propagation. In general, the speed c_ℓ for the longitudinal wave is greater than the speed c_t for the transverse wave:

$$c_\ell > c_t. \tag{9.11}$$

11.10* The Hamiltonian for Elastic Waves

In the last section, we set up the equations of motion (the wave equations) by the Newtonian method. We can derive the same equations by Lagrangian or Hamiltonian methods. In the present section, we will review these methods, and derive the Hamiltonians for the elastic waves.

Let us first look at the one-dimensional wave equation [see (9.1)]

$$\rho \frac{\partial^2 u}{\partial t^2} = S \frac{\partial^2 u}{\partial x^2} . \tag{10.1}$$

By analogy with the case of stretched string discussed in §§3.11-12, we may assume the Lagrangian of the form :

$$L = \int d^3r \left[\frac{1}{2} \rho \left(\frac{\partial u}{\partial t} \right)^2 - \frac{1}{2} S \left(\frac{\partial u}{\partial x} \right)^2 \right]$$

$$\equiv T - V. \tag{10.2}$$

The first integral corresponds to the kinetic energy T and the second to the negative of the potential energy V for the transverse wave proceeding in the x-direction. We further note that this expression is similar to (3.12.23); the main differences are that the quantities appearing in (10.2) refer to the three dimensional space.

The Lagrangian L in (10.2) is an integral (rather than a sum) of functions of field variables ($\partial u/\partial t$, $\partial u/\partial x$). In such a case, we cannot apply the usual form of Lagrange's equations of motion :

$$\frac{d}{dt} \left[\frac{\partial}{\partial \dot{q}_j} L(q_1, q_2, \ldots, \dot{q}_1, \dot{q}_2, \ldots) \right] - \frac{\partial L}{\partial q_j} = 0. \tag{10.3}$$

This difficulty can however be overcome in a general manner. The extension of the Lagrangian method to a field variable $u(\vec{r},t)$, can be found in textbooks of advanced dynamics [5,6]. According to this general theory, the equation of motion for the field u can be obtained in the following manner.

Let us suppose that the Lagrangian L for a general system is given in the form

$$L = \int d^3r \, \hat{L}\left(u, \frac{\partial u}{\partial t}, \frac{\partial u}{\partial x}\right), \tag{10.4}$$

where the integrand $\hat{L}$, called the Lagrangian density, is a function of the field u, and its x- and t-derivatives ($\partial u/\partial x$, $\partial u/\partial t$) . We then set up the Lagrange's field equation in terms of the Lagrangian density $\hat{L}$ as follows :

$$\boxed{\frac{\partial}{\partial t}\left[\frac{\partial \hat{L}}{\partial(\partial u/\partial t)}\right] + \frac{\partial}{\partial x}\left[\frac{\partial \hat{L}}{\partial(\partial u/\partial x)}\right] - \frac{\partial \hat{L}}{\partial u} = 0.} \tag{10.5}$$

This equation generates the equation of motion for the field u. By the way, much similarity exists between (10.3) and (10.5). The main differences are that (a) the discrete index j is replaced by the continuous space variable x, (b) the Lagrangian density $\hat{L}$ enters in the latter, and (c) an additional term involving the space derivative is present. Eq. (10.5) may be written down easily by remembering that each term must have the same dimension.

We now go back to our Lagrangian L in (10.2), from which we obtain the Lagrangian density as follows :

$$\hat{L} = \frac{1}{2}\rho\left(\frac{\partial u}{\partial t}\right)^2 - \frac{1}{2}S\left(\frac{\partial u}{\partial x}\right)^2. \tag{10.6}$$

Here we note that this particular Lagrangian density $\hat{L}$ does not depend on u.

Using (10.6), we obtain

$$\frac{\partial \hat{L}}{\partial u} = 0$$

$$\frac{\partial \hat{L}}{\partial\left(\frac{\partial u}{\partial t}\right)} = \frac{\partial}{\partial\left(\frac{\partial u}{\partial t}\right)} \left[\frac{1}{2}\rho\left(\frac{\partial u}{\partial t}\right)^2 - \frac{1}{2} S \left(\frac{\partial u}{\partial x}\right)^2 \right]\Bigg|_{u,\frac{\partial u}{\partial x}}$$

$$= \rho \frac{\partial u}{\partial t}$$

$$\frac{\partial \hat{L}}{\partial\left(\frac{\partial u}{\partial x}\right)} = - S \frac{\partial u}{\partial x} . \qquad (10.7)$$

Introducing these in eq.(10.5), we obtain

$$\frac{\partial}{\partial t}\left[\rho \frac{\partial u}{\partial t} \right] + \frac{\partial}{\partial x}\left[- S \frac{\partial u}{\partial x} \right] - 0 = 0$$

or

$$\rho \frac{\partial^2 u}{\partial t^2} - S \frac{\partial^2 u}{\partial x^2} = 0,$$

which is in agreement with eq.(10.1).

For a homogeneous isotropic solid, the general wave equation for a transverse wave was given in (9.8):

$$\frac{\partial^2 \vec{u}}{\partial t^2} = c_t^2 \, \nabla^2 \, \vec{u}, \qquad (10.8)$$

where we omitted the subscript t on the displacement vector $\vec{u}$. The Lagrangian density $\hat{L}$ which generates eq.(10.8), can be written down as follows :

$$\hat{L} = \frac{1}{2}\rho\left|\frac{\partial\vec{u}}{\partial t}\right|^2 - \frac{1}{2}s\left[\left|\frac{\partial\vec{u}}{\partial x}\right|^2 + \left|\frac{\partial\vec{u}}{\partial y}\right|^2 + \left|\frac{\partial\vec{u}}{\partial z}\right|^2\right]$$

$$= \frac{1}{2}\rho\left[\left(\frac{\partial u_x}{\partial t}\right)^2 + \left(\frac{\partial u_y}{\partial t}\right)^2 + \left(\frac{\partial u_z}{\partial t}\right)^2\right] - \cdots$$

$$= \hat{L}\left(\frac{\partial u_x}{\partial t}, \frac{\partial u_y}{\partial t}, \frac{\partial u_z}{\partial t}, \frac{\partial u_x}{\partial x}, \frac{\partial u_y}{\partial x}, \ldots, \frac{\partial u_y}{\partial z}, \frac{\partial u_z}{\partial z}\right). \tag{10.9}$$

This is an extension of expression (10.2); the generalization is twofold; (a) the displacement is characterized by the vector field $\vec{u} = (u_x, u_y, u_z)$ and (b) the direction of the wave propagation is unrestricted. The general Lagrange's field equations will be of the form :

$$\frac{\partial}{\partial t}\left[\frac{\partial\hat{L}}{\partial(\partial u_a/\partial t)}\right] + \frac{\partial}{\partial x}\left[\frac{\partial\hat{L}}{\partial(\partial u_a/\partial x)}\right] + \frac{\partial}{\partial y}\left[\frac{\partial\hat{L}}{\partial(\partial u_a/\partial y)}\right]$$

$$+ \frac{\partial}{\partial z}\left[\frac{\partial\hat{L}}{\partial(\partial u_a/\partial z)}\right] - \frac{\partial\hat{L}}{\partial u_a} = 0, \quad a = x,y,z. \tag{10.10}$$

The reader may verify the wave equation (10.8) from eqs.(10.9) and (10.10) [Problem 10.1].

Just as for the case of the string, we can express the Lagrangian density $\hat{L}$ in terms of normal coordinates and velocities. Let us assume a cubic-box fixed-end boundary condition with the side length L_0. The normal modes of oscillations are characterized by the wave vector $\vec{k} = (k_x, k_y, k_z)$,

$$k_a = \frac{\pi}{L_o} n_a,$$

$$n_a = 1, 2, \ldots, \quad a = x, y, z, \tag{10.11}$$

and the angular frequency

$$\omega_{t,k} \equiv c_t |\vec{k}| = c_t k. \tag{10.12}$$

These are generalizations of eqs.(3.11,22) and (3.11.23). The wave vector $\vec{k}$ along which the sinusoidal oscillations are repeated, is perpendicular to the displacement $\vec{u}$;

$$\vec{k} \cdot \vec{u} = 0. \qquad \text{(transverse wave)} \tag{10.13}$$

Let us follow the standard recipe described in §3.12, and introduce the normal coordinates as follows :

$$\vec{q}_{\vec{k}}(t) \equiv \left(\frac{2}{L_o}\right)^3 \int_0^{L_o} dx \int_0^{L_o} dy \int_0^{L_o} dz \; \vec{u}(\vec{r},t) \sin(x\, k_x) \sin(y\, k_y) \sin(z\, k_z).$$

[extension of (3.12.16)] (10.14)

After lengthy but straightforward calculations, we can then transform the Lagrangian density $\hat{L}$ into [Problem 10.2]

$$\hat{L} = \sum_{\vec{k}} \left\{ \frac{1}{4} \rho |\dot{\vec{q}}_{\vec{k}}|^2 - \frac{1}{4} S\, k^2 |\vec{q}_{\vec{k}}|^2 \right\}. \tag{10.15}$$

[extension of (3.12.28)]

From (10.13), the normal-coordinate vector $\vec{q}$ is perpendicular to the wave vector $\vec{k}$:

$$\vec{q}_{\vec{k}} \cdot \vec{k} = 0. \tag{10.16}$$

Let us denote the Cartesian components of $\vec{q}_{\vec{k}}$ by $q_{\vec{k},1}$ and $q_{\vec{k},2}$. We may then express the Lagrangian density $\hat{L}$ in the form

$$\hat{L} = \sum_{\vec{k}} \sum_{\sigma=1,2} [\frac{1}{4} \rho \, \dot{q}_{\vec{k},\sigma}^{2} - \frac{1}{4} S \, k^2 \; q_{\vec{k},\sigma}^{2}], \tag{10.17}$$

which is the sum of single-mode Lagrangian densities. From the Lagrangian field equations:

$$\frac{d}{dt} [\frac{\partial \hat{L}}{\partial \dot{q}_{\vec{k},\sigma}}] - \frac{\partial \hat{L}}{\partial q_{\vec{k},\sigma}} = 0, \tag{10.18}$$

we can easily verify that each normal coordinate $q_{\vec{k},\sigma}$ oscillates with the normal-mode frequency $\sqrt{S/\rho}\; k \equiv c_t k \equiv \omega_{t,k}$.

Multiplying (10.17) by the volume V, we obtain

$$\begin{aligned} L &\equiv V \hat{L} \\ &= \sum_{\vec{k}} \sum_{\sigma} (\frac{1}{4} V\rho \, \dot{q}_{\vec{k},\sigma}^{2} - \frac{1}{4} VS \, k^2 \, q_{\vec{k},\sigma}^{2}), \end{aligned} \tag{10.19}$$

which represents the Lagrangian L for the system. For later convenience, let us introduce new normal coordinates

$$Q_{\vec{k},\sigma} \equiv \left(\frac{V\rho}{2}\right)^{\frac{1}{2}} q_{\vec{k},\sigma}, \tag{10.20}$$

which are different from the old coordinates $q_{\vec{k},\sigma}$ by constant factors, and therefore change with time in the same manner as $q_{\vec{k},\sigma}$. Using this new set of normal coordinates, we can re-express the Lagrangian L as follows :

$$L = \sum_{\vec{k}} \sum_{\sigma} \frac{1}{2} [\dot{Q}^2_{\vec{k},\sigma} - \omega^2_{t,k} \; Q^2_{\vec{k},\sigma}] . \tag{10.21}$$

This is the most compact form of the Lagrangian written in the normal coordinates.

To derive the Hamiltonian , we define the canonical momenta $\{ P_{\vec{k},\sigma} \}$ by

$$P_{k,\sigma} \equiv \frac{\partial L}{\partial \dot{Q}_{\vec{k},\sigma}} = \dot{Q}_{\vec{k},\sigma} . \quad \text{[use of (10.21)]} \tag{10.22}$$

The Hamiltonian H_t can now be constructed by expressing $\sum_{\vec{k}} \sum_{\sigma} P_{\vec{k},\sigma} \; \dot{Q}_{k,\sigma} - L$ in terms of $\{ Q_{\vec{k},\sigma} , P_{\vec{k},\sigma} \}$. The result is given by

$$H_t = \sum_{\vec{k}} \sum_{\sigma} \frac{1}{2} [P^2_{\vec{k},\sigma} + \omega_{t,k}^2 \; Q^2_{\vec{k},\sigma}] . \tag{10.23}$$

We note that this Hamiltonian is the sum of simple-harmonic oscillator Hamiltonians over all normal modes. Further note that this expression is almost identical to the Hamiltonian (2.18) for the radiation fields. Only the angular frequencies are different; instead of $\omega_k = c\, k$ with c being the speed of light, we have $\omega_{t,k} = c_t \; k$ with c_t representing the speed of the transverse elastic wave.

In further detail, the normal modes which are characterized by the wave vector $\vec{k}$ and the frequency $\omega_{t,k}$ introduced here in this section represent standing waves [see (10.14)], while the normal modes introduced in §11.2 are running waves [see (1.19)]. This difference, which arises from different boundary conditions can be eliminated by using the same boundary condition.

[See the paragraph following eq.(11.13) for further comments.]

Let us now turn to the case of longitudinal waves as characterized by eq.(9.9). This case can be treated in a similar manner, and the Hamiltonian H_ℓ can be written as

$$H_\ell = \sum_{\vec{k}} \frac{1}{2}[P_{\vec{k}}^2 + \omega_{\ell,k}^2 \quad Q_{\vec{k}}^2]. \tag{10.24}$$

Note that the normal modes have different frequencies $\omega_{\ell,k} \equiv c_\ell k$. Also, since the displacement occurs only along the direction of the propagation of the wave, the single pair of the normal coordinates, $(Q_{\vec{k}}, P_{\vec{k}})$,is needed for each longitudinal mode.

In summary, we can write the total Hamiltonian for an elastic solid as

$$H = H_t + H_\ell . \tag{10.25}$$

Problem 10.1 Derive the wave equation (10.8) from the Lagrangian field equations (10.10) with the Lagrangian density (10.9).

Problem 10.2 Derive (10.15) from (10.9). Use (10.14).

11.11 Debye's Theory of the Heat Capacity

Earlier in §11.7, we discussed Einstein's theory of the heat capacity of a solid. This theory explains well why a certain solid like diamond exhibits a molar heat capacity substantially smaller than 3 R (the value in accordance with Dulong-Petit's law). It also predicts that the heat capacity for any solid should decrease as the temperature is lowered. This prediction is in qualitative agreement with experimental observations. However, as we will see later, the temperature - dependence at the low temperature end indicates non-negligible discrepancies. In 1912, Debye (Peter J.W. Debye, Dutch, 1884-1966)[7] reported a very successful theory of the heat capacity of solids. This theory will be discussed in the present section.

In Einstein's model, the solid is viewed as a collection of free harmonic oscillators with a common frequency ω_o. This is rather a drastic approximation since a molecule in a real crystal moves under the fluctuating potential field generated by its neighboring molecules. It is more reasonable to look at the crystalline solid as a collection of coupled harmonic oscillators. [We can set up a theory from this point of view as we will do so later in §11.12*.] It is known in dynamics (see §§3.9* and 3.10*) that the motion of a set of coupled harmonic oscillators can be described concisely in terms of the normal modes of oscillations. A major difficulty of this approach however is to find the set of the normal-modes frequencies. Debye overcame this difficulty by regarding a solid as a continuous elastic body, whose normal modes of oscillations are known as we saw in the last few sections.

Let us recall that a macroscopic solid can support transverse and longitudinal waves. We first look at the case of the transverse waves.

The Hamiltonian H_t for these waves is given in (10.23) :

$$H_t = \sum_{\vec{k}} \sum_{\sigma} \frac{1}{2} [P_{\vec{k},\sigma}^2 + \omega_{t,k}^2 Q_{\vec{k},\sigma}^2]. \tag{11.1}$$

Note that this Hamiltonian has the same form as (2.18) for the Hamiltonian of the electromagnetic radiation. We may now develop a quantum theory in an analogous manner to this case.

The energy eigenvalues for the corresponding quantum Hamiltonian of the form (11.1) are given by

$$E_t = \sum_{\vec{k}} \sum_{\sigma} (\frac{1}{2} + n_{\vec{k},\sigma}) \hbar\omega_{t,k} = E[\{n_{\vec{k},\sigma}\}], \tag{11.2}$$

where $n_{\vec{k},\sigma}$ are non-negative integers :

$$n_{\vec{k},\sigma} = 0,1,2, \ldots \tag{11.3}$$

Just as for the case of the quantum theory of radiation, it is convenient to interpret expression (11.2) in terms of the energies of fictitious particles called phonons. In this view, the quantum number $n_{\vec{k},\sigma}$ represents the number of phonons in the normal mode $(\vec{k},\sigma)$. The energy of the system is then specified by the set of the numbers of phonons in normal modes (states), $\{ n_{\vec{k},\sigma} \}$.

By taking the canonical-ensemble average of (11.2), we obtain

$$< E_t[\{n_{\vec{k},\sigma} \}] > = \sum_{\vec{k},} \sum_{\sigma} (\frac{1}{2} + <n_{\vec{k},\sigma}>) \hbar \omega_{t,k} \quad \text{[use of (11.2)]}$$

$$= \sum_{\vec{k}} \sum_{\sigma} [\frac{1}{2} + n_0(\hbar\omega_{t,k})] \hbar \omega_{t,k} \equiv E_t(\beta), \tag{11.4}$$

[use of (3.3)]

where

$$n_o(\varepsilon;\beta) = \frac{1}{e^{\beta\varepsilon} - 1} \equiv n_o(\varepsilon) \qquad (11.5)$$

denotes the Planck distribution function, [Compare with (3.3)].

When the volume of normalization, V, is made large, the distribution of the normal-mode-points in the k-space becomes dense. This also means that the density of the mode-points in the ω-domain becomes high. In the large-volume limit, we may convert the sum over $\{\vec{k}\}$ in eq.(11.4) into an ω-integral, and obtain

$$E_t(\beta) = E_{t,o} + \int_0^\infty d\omega\, \hbar\omega\, n_o(\hbar\omega)\, D_t(\omega) \qquad (11.6)$$

$$E_{t,o} \equiv \frac{1}{2}\int_0^\infty d\omega\, \hbar\omega\, D_t(\omega), \qquad (11.7)$$

where $D_t(\omega)$ is the density of states (modes) in angular frequency defined such that

the number of modes in the interval $(\omega, \omega + d\omega) = D_t(\omega)\, d\omega$.

(11.8)

The $E_{t,o}$ in (11.7) represents the sum of zero point energies, which is independent of temperature.

The density of states, $D_t(\omega)$, in the frequency domain, may be calculated as follows. First, we note that the angular frequency ω is related to the wave vector $\vec{k}$ by $\omega = c_t\, k$.

The constant-ω surface in the k-space therefore is the sphere of radius $k = \omega/c_t$ with center at the origin. Consider another concentric sphere of radius $k + dk = (\omega + d\omega) / c_t$. We know from (10.11) that each mode-point is located at the simple cubic lattice sites with unit spacing π / L_0 and only in the first octant $(k_x, k_y, k_z > 0)$. The number of mode-points within the spherical shell between the two spheres can be obtained by dividing one eighth of the k - volume of the shell,

$$\frac{1}{8} (4\pi k^2 dk) = \frac{1}{2} \pi \frac{\omega^2 d\omega}{c_t^3} ,$$

by the unit-cell k-volume, $(\pi/L_0)^3$. Multiplying the result by 2 in consideration of the polarization multiplicity, we then obtain

$$D_t(\omega) = \pi \frac{\omega^2 d\omega}{c_t^3} / \left(\frac{\pi}{L_0} \right)^3 = \frac{L_0^3}{\pi^2} \frac{\omega^2 d\omega}{c_t^3} = \frac{V}{\pi^2} \frac{\omega^2 d\omega}{c_t^3} . \quad (11.9)$$

We may treat the longitudinal waves in a similar manner. The average energy $E_\ell (\beta) \equiv \langle E_\ell [\{n_{\vec{k}}\}] \rangle$ can be written in the same form as (11.4) with the density of states

$$D_\ell (\omega) = \frac{V}{2\pi^2} \frac{\omega^2 d\omega}{c_\ell^3} . \quad (11.10)$$

Adding these contributions together we obtain

$$E \equiv E_t + E_\ell$$

$$= E_0 + \int_0^\infty d\omega\, \hbar\omega \quad n_0(\hbar\omega)\, D(\omega), \qquad \text{[use of (11.6)]} \tag{11.11}$$

$$E_0 \equiv \frac{1}{2} \int_0^\infty d\omega\, \hbar\omega\, D(\omega), \tag{11.12}$$

where

$$D(\omega) \equiv D_t(\omega) + D_\ell(\omega)$$

$$= \frac{V}{2\pi^2} \left(\frac{2}{c_t^3} + \frac{1}{c_\ell^3} \right) \omega^2 \tag{11.13}$$

denotes the density of states in frequency for the combined wave modes. It is possible to show that the density of states in the large-volume(bulk) limit is given by this formula regardless of the shape of the solid. The formula is also valid for any boundary condition. In fact, if the cube-shaped periodic boundary condition with the periodicity L_0 is assumed, the normal modes are characterized by the quantized k-vectors given in (1.14) and (1.15), and can be represented by the points forming a simple cubic lattice with the lattice constant $2\pi/L_0$ in the k-space (k_x, k_y, k_z = positive, negative or zero). This configuration also yields the same number of mode-points in the spherical shell between the two spheres, and therefore the same result for the density of states.

For a real crystal of N molecules, the number of degrees of freedom is 3 N. Therefore, there exist exactly 3 N normal modes. The elastic body, whose dynamical state is described by the vector field $\vec{u}$, has an infinite number of degrees of freedom, and therefore it has infinitely many modes. The model of a macroscopic elastic body is a reasonable representation of a real solid for low-frequency modes, that is, long-wavelength modes. [This property can be demonstrated explicitly as was done in §§3.10* and 3.11* for a linear chain of coupled harmonic oscillators.] At high frequencies, the continuum model does not provide normal modes expected of a real solid. Debye overcame this difficulty by assuming that (1) the "corrected" density of states, $D_D(\omega)$, has the same value as given by expression (11.13) up to the maximum frequency ω_D and vanishes thereafter :

$$D_D(\omega) = \begin{cases} \dfrac{V}{2\pi^2}\left[\dfrac{2}{c_t^3} + \dfrac{1}{c^3}\right]\omega^2, & \omega < \omega_D \\ 0 & \text{otherwise} \ ; \end{cases} \tag{11.14}$$

and (2) the limit frequency, called the Debye frequency, is chosen such that the number of modes for the truncated continuum model equals 3 N,(the number of normal modes for the crystal):

$$\int_0^\infty d\omega\, D_D(\omega) = \int_0^{\omega_D} d\omega\, D(\omega) = 3\ N. \tag{11.15}$$

Substitution from (11.14) yields that [Problem 11.1]

$$\omega_D = \left[\, 18\,\pi^2 n \,/\left(\frac{2}{c_t^3} + \frac{1}{c_\ell^3}\right)\right]^{1/3}, \tag{11.16}$$

where $n \equiv N/V$ represents the number density.

The density of states $D_D(\omega)$ grows quadratically near the origin, reaches the maximum at $\omega = \omega_D$, and vanishes thereafter. This behavior is schematically

shown in Figure 11.14

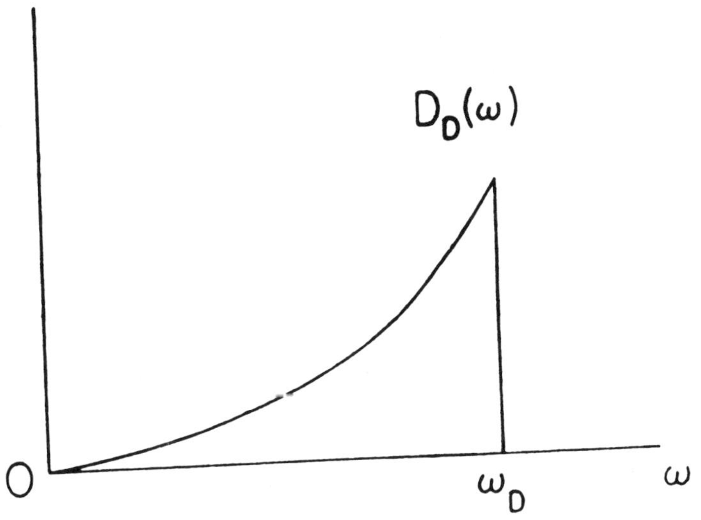

The density of modes in angular frequency for the Debye continuum model has a sharp cut-off at $\omega = \omega_D$ (Debye frequency).

We now compute the heat capacity C_V for the Debye model. Introducing (11.14) in (11.11), we obtain

$$E(T) - E_0 = \int_0^{\omega_D} d\omega\, \hbar\omega\, n_0(\hbar\omega) \frac{V}{2\pi^2} \left(\frac{2}{c_t^3} + \frac{1}{c_\ell^3} \right) \omega^2$$

$$= \frac{9N\hbar}{\omega_D^3} \int_0^{\omega_D} d\omega\, \omega^3 \,[\exp(\hbar\omega/k_BT) - 1\,]^{-1}. \tag{11.17}$$

[use of (11.5) and (11.6)]

Let us introduce the Debye temperature Θ_D defined by

$$\Theta_D \equiv \hbar\omega_D / k_B, \tag{11.18}$$

which is proportional to the cut-off frequency ω_D, and has the dimension of temperature. Using this definition, we can rewrite (11.17) as

$$E(T) - E_0$$

$$= 9N k_BT\, (T / \Theta_D)^3 \int_0^{x_D} dx \frac{x^3}{e^x - 1}, \tag{11.19}$$

where

$$x_D \equiv \frac{\hbar\,\omega_D}{k_B\,T} = \frac{\Theta_D}{T}, \quad x \equiv \frac{\hbar\,\omega}{k_B\,T}. \tag{11.20}$$

Consider first the high temperature region where $x \equiv \hbar\omega / k_B T \ll 1$. Then, $e^x - 1 \simeq x$, and

$$\int_0^{x_D} dx \frac{x^3}{e^x - 1} \simeq \int_0^{x_D} dx\, x^2 = \frac{x_D^3}{3} = \frac{1}{3}\left(\frac{\Theta_D}{T}\right)^3, \quad x_D \ll 1. \tag{11.21}$$

Using this approximation in (11.19), we obtain

$$E = E_o + 3 N k_B T, \qquad T >> \Theta_D. \tag{11.22}$$

Differentiating this with respect to T, we obtain $C_V = \frac{\partial E}{\partial T} = 3 N k_B$, which is in agreement with the classical value.

At very low temperatures, $x_D \equiv \hbar\omega_D / k_B T$ becomes very large. Then, the upper limit x_D of the x - integral may be replaced by ∞ :

$$\int_0^{\omega_D} dx \frac{x^3}{e^x - 1} \rightarrow \int_0^{\infty} dx \frac{x^3}{e^x - 1} = \frac{\pi^4}{15}. \tag{11.23}$$

[Note that this last integral appeared in (3.12).] We thus obtain from (11.19)

$$E = E_o + \frac{3}{5} \pi^4 N k_B T^4 \Theta_D^{-3}. \qquad (T << \Theta_D) \tag{11.24}$$

Differentiating this expression with respect to T, we get

$$\boxed{C_V = \frac{12\pi^4}{5} N k_B \left(\frac{T}{\Theta_D} \right)^3.} \tag{11.25}$$

According to this formula, the heat capacity at very low temperatures decreases with the cube of the absolute temperature. This is known as the Debye-T^3 law. The decrease is therefore slower than the Einstein heat capacity, given by (7.9), which decreases exponentially. In Figure 11.15 experimental heat capacity *versus* T^3 for argon (Ar) is shown. Agreements here are excellent.

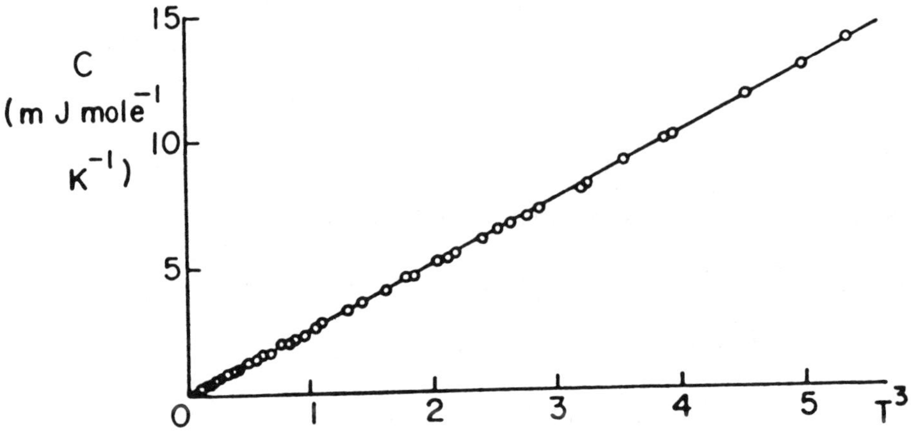

Fig. 11.15 The heat capacity data for solid Ar below 2 K are plotted against T^3. The straight-line confirms the Debye T^3-law in this temperature range. [reproduced from Introductory Statistical Mechanics by Turner and Betts [8]].

At intermediate temperatures, the integral may be evaluated by expressing it as an infinite series. By differentiating the result for E(T), we can obtain the heat capacity. A more accurate and practical way is as follows. Differentiating (11.17) with respect to T, we obtain

$$C_V = \frac{\partial E}{\partial T} = \frac{9N\hbar}{\omega_D^3} \int_0^{\omega_D} d\omega\, \omega^3 \frac{d}{dT} [e^{\hbar\omega/k_BT} - 1]^{-1}$$

$$= 9Nk_B \left(\frac{T}{\Theta_D} \right)^3 \int_0^{x_D} dx \frac{x^4 e^x}{(e^x - 1)^2} . \qquad (11.26)$$

[use of (11.18); $x_D \equiv \Theta_D/T$]

We may perform this integral numerically by computer . At any rate, we can express the molar heat of solid as a function of T/Θ_D only. The theoretical curve so obtained is shown in solid line Figure 11.16 . The points in the

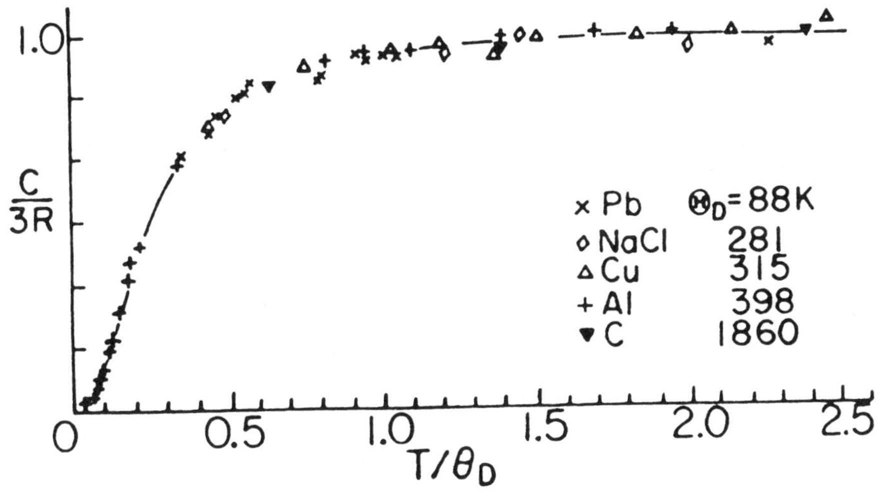

Fig. 11.16 The heat capacities of solids. The solid line represents the theoretical Debye curve obtained from (11.26). Experimental data for various solids are shown with optimum Debye temperatures. After Sears and Salinger [9].

figure are experimental data for various solids. The Debye temperatures adjusted at optimum fitting are also shown in the same figure. Roughly speaking, when T/Θ_D is greater than 1, or when the actual temperature exceeds the Debye temperature, the solid behaves classically and the molar heat is nearly equal to 3R in agreement with the Dulong-Petit's law. When the temperature is less than the Debye temperature, the quantum effect sets in and the heat capacity can be less than 3R. Thus for lead (a soft metal), with a Debye temperature of only 88 K, room temperature is well above the Debye temperature while diamond (a hard crystal), with a Debye temperature of 1860 K, is a quantum solid even at room temperature.

The Debye temperatures for several typical solids are given in Table 11.2.

Solids	Debye temperature Θ_D in K
NaCl	281
KCl	230
Pb	88
Ag	225
Zn	308
Diamond	1860

Table 11.2. Approximate values for Debye temperatures

The Debye theory of the heat capacity gives quite satisfactory results for almost all non-conducting crystals. Exceptions arise if the crystals have striking anisotropies like graphite (black carbon crystal) or if finite-size crystals in the form of powder are considered. For conducting materials, we must take account of the electronic contribution to the heat capacity, which was discussed in Chapter 10. As mentioned earlier, we can look at a solid as a collection of coupled harmonic oscillators, and develop a theory from this viewpoint. This will be done in the following section. This theory will shed further light on the fundamental assumptions of the Debye model. Also, it can be used to treat exceptional cases mentioned above.

Problem 11.1 Derive (11.16), using (11.15) and (11.14).

11.12* More about the Heat Capacity. Lattice Dynamics.

In the present section, we will develop a general theory of the heat capacity based on the lattice dynamics. In terms of the results of this theory, we then discuss the basic assumptions of the Debye model and a few related topics. The discussion requires knowledge of the theory of small oscillations as presented in §§3.9* - 3.12*, and may be omitted for the first reading.

Let us first review the dynamics of coupled oscillators. For definiteness, let us take a set of N particles of equal mass m attached to a light string of length $(N+1)\ell$ in equilibrium, stretched to a tension τ. This is the same system which we discussed in §3.9*.

The kinetic energy T and the potential energy V for the system were calculated earlier, and are given by

$$T \equiv \frac{1}{2} m (\dot{y}_1^2 + \dot{y}_2^2 + \dots + \dot{y}_N^2) \quad \text{[see (3.9.1)]} \tag{12.1}$$

$$V \equiv \frac{\tau}{2\ell} [y_1^2 + (y_2-y_1)^2 + \dots + (y_N-y_{N-1})^2 + y_N^2], \tag{12.2}$$

[see (3.9.5)]

where y_j denotes the transverse displacement of the particle j. We construct the Lagrangian $L = T - V$ as

$$L = \frac{1}{2} m \dot{y}_1^2 - \frac{\tau}{2\ell} y_1^2 + \sum_{j=2}^{N-1} [\frac{1}{2} m \dot{y}_j^2 - \frac{\tau}{2\ell}(y_{j+1} - y_j)^2]$$

$$+ \frac{1}{2} m \dot{y}_N^2 - \frac{\tau}{2\ell} y_N^2 . \tag{12.3}$$

Lagrange's equations of motion generate coupled linear homogeneous differential equations for y_j. The associated characteristic equation (determinant equation) yields N characteristic frequencies : ω_1, ω_2, ... ω_N. By suitably choosing linear combinations of y_j (normal-mode coordinates) :

$$Q_k \equiv \sum_{j=1}^{N} a_{kj}\, y_j, \quad k = 1,2,\ldots, N, \tag{12.4}$$

we can re-express the Lagrangian L in the form [see(3.10.16)]

$$L = \sum_{k=1}^{N} \frac{1}{2} \left(\dot{Q}_k^{\,2} - \omega_k^{\,2} Q_k^{\,2} \right). \tag{12.5}$$

We note that this Lagrangian is the sum of the simple - harmonic - oscillator Lagrangians over N normal modes. Also note that the number of the normal modes equals the number of the degrees of freedom for the system.

After introducing the canonical momenta :

$$P_k \equiv \frac{\partial L}{\partial \dot{Q}_k} = \dot{Q}_k \,, \qquad \text{[use of (12.5)]} \tag{12.6}$$

we can derive the Hamiltonian H for the system as [see (3.10.18)]

$$H = \sum_{k=1}^{N} \frac{1}{2} \left(P_k^{\,2} + \omega_k^{\,2} Q_k^{\,2} \right). \tag{12.7}$$

Let us now take a three-dimensional crystal composed of N molecules. The potential energy V depends on the configuration ($\vec{r}_1$, $\vec{r}_2$, ..., $\vec{r}_N$) of N molecules. This energy V can also be regarded as a function of the displacements of the molecules,

$$\vec{u}_j \equiv \vec{r}_j - \vec{r}_j^{(o)}, \tag{12.8}$$

measured from the equilibrium positions $\vec{r}_j^{(o)}$. Let us expand the potential $V = V(\vec{u}_1, \vec{u}_2, .., \vec{u}_N) \equiv V(u_{1x}, u_{1y}, u_{1z}, u_{2x}, \ldots)$ in terms of small displacements $\{u_{j\mu}\}$:

$$V = V_o + \sum_{j=1}^{N} \sum_{\mu=x,y,z} u_{j\mu} \left(\frac{\partial V}{\partial u_{j\mu}} \right)_o$$

$$+ \frac{1}{2} \sum_j \sum_\mu \sum_k \sum_\nu u_{j\mu} u_{k\nu} \left(\frac{\partial^2 V}{\partial u_{j\mu} \partial u_{k\nu}} \right)_o + \ldots, \tag{12.9}$$

where all partial derivatives are to be evaluated at $\vec{u}_1 = \vec{u}_2 = .. = \vec{u}_N = 0$. We may set the constant V_o equal to zero with no loss of generality. By assumption, the lattice is stable at the equilibruim configuration at which the potential V must have a minimum value. This means that all the first-order derivatives must vanish :

$$\frac{\partial V}{\partial u_{j\mu}} = 0. \tag{12.10}$$

For small oscillations, we may keep terms of second order in $u_{j\mu}$ only. [This is called the harmonic approximation.] We then have

$$V \simeq V' \equiv \frac{1}{2} \sum_j \sum_\mu \sum_k \sum_\nu A_{j\mu\, k\nu}\, u_{j\mu}\, u_{k\nu}\,, \tag{12.11}$$

where

$$A_{j\mu k\nu} \equiv \left(\frac{\partial^2 V}{\partial u_{j\mu}\, \partial u_{k\nu}} \right)_o \tag{12.12}$$

are constants. The kinetic energy of the system is given by

$$T \equiv \sum_{j=1}^{N} \frac{1}{2} m \dot{r}_j^{\,2} = \sum_{j=1}^{N} \frac{1}{2} m \dot{u}_j^{\,2} \equiv \sum_j \sum_\mu \frac{1}{2} m \dot{u}_{j\mu}^{\,2}\,. \tag{12.13}$$

We can now write down the Lagrangian $L \equiv T - V \cong T - V'$ as

$$L = \sum_j \sum_\mu \frac{1}{2} m \dot{u}_{j\mu}^{\,2} - \frac{1}{2} \sum_j \sum_\mu \sum_k \sum_\nu A_{j\mu k\nu}\, u_{j\mu}\, u_{k\nu}\,. \tag{12.14}$$

This Lagrangian L in the harmonic approximation is quadratic in $u_{j\mu}$ and their time derivatives $\dot{u}_{j\mu}$ just as the Lagrangian (12.3) for the coupled linear chain. According to a theory of the <u>principal-axis transformations</u> [6], we can in principle transform the Hamiltonian (total energy) $H = T + V'$ for such a system into the sum of the energies of the normal-modes of oscillations :

$$H = \sum_{k=1}^{3N} \frac{1}{2} \left(P_k^2 + \omega_k^2\, Q_k^2 \right), \tag{12.15}$$

where $\{Q_k, P_k\}$ are the normal coordinates and momenta, and ω_k are characteristic frequencies. Note that there are exactly 3N normal modes.

Let us now calculate the heat capacity, using (12.15). This Hamiltonian is quadratic in canonical variables (Q_k, P_k). Therefore in the classical treatment, the equipartition theorem holds: the average thermal energy for each mode is $k_B T$. Multiplying it by the number of modes, 3N, we obtain $3Nk_BT$ for the average energy of the system, $\langle H \rangle$. Differentiating this with respect to T, we obtain $3Nk_B$ for the heat capacity, which is in complete agreement with Dulong-Petit's law. It is interesting to observe that we obtained this result without knowing the actual distribution of normal mode frequencies. The fact that there are 3N normal modes, played an important role.

Let us now look at the quantum theory of the heat capacity based on the Hamiltonian (12.15). The energy eigenvalues of the quantum Hamiltonian of the form (12.15) are given by

$$E\ [\{ n_k \}] = \sum_k \left(\frac{1}{2} + n_k \right) \hbar\, \omega_k, \tag{12.16}$$

where n_k are non-negative integers :

$$n_k = 0,\ 1,\ 2,\ \ldots . \tag{12.17}$$

If we introduce the concept of phonons, we can interpret (12.16) as follows : the energy of the lattice vibrations is characterized by the set of the numbers of phonons, $\{ n_k \}$, in the normal modes $\{ k \}$.

By taking the canonical-ensemble average of eq.(12.16), we obtain

$$\langle E\ [\{n_k\}] \rangle = \sum_k \left(\frac{1}{2} + \langle n_k \rangle \right) \hbar\, \omega_k$$

$$= \sum_k \left[\frac{1}{2} + n_0(\hbar\omega_k) \right] \hbar\omega_k \equiv E(T), \tag{12.18}$$

where

$$n_0(\varepsilon) \equiv [\exp(\varepsilon/k_B T) - 1]^{-1} \tag{12.19}$$

denotes the Planck distribution function.

Let us make a small digression here, and imagine a case in which all frequencies ω_k have the same value ω_0. Then, formula (12.18) reduces to (7.8) in Einstein's theory.

The normal-modes frequencies $\{\omega_k\}$ depend on the normalization volume V, and will be densely populated for large V. In the large-volume limit, we may then convert the sum over the normal modes into a frequency integral, and obtain

$$E(T) = E_0 + \int_0^\infty d\omega\, \hbar\omega\, n_0(\hbar\omega)\, D(\omega) \tag{12.20}$$

$$E_0 \equiv \frac{1}{2} \int_0^\infty d\omega\, \hbar\omega\, D(\omega), \tag{12.21}$$

where $D(\omega)$ is the density of states (modes) in angular frequency defined such that

$$\text{the number of modes in the interval } (\omega, \omega + d\omega) = D(\omega)\, d\omega. \tag{12.22}$$

The E_0 in (12.21) represents the total zero-point energy, which does not depend on the temperature.

Differentiating E(T) with respect to T, we obtain for the heat capacity

$$C_V = \frac{\partial E}{\partial T} = \int_0^\infty d\omega\, \hbar\omega\, \frac{\partial n_0(\hbar\omega)}{\partial T}\, D(\omega). \qquad (12.23)$$

This general expression was obtained under the harmonic approximation only. This approximation should be reasonable at low temperatures, where only small-amplitude oscillations are excited.

To proceed further, we have to know the density of normal modes, $D(\omega)$. Let us recall that to find the set of characteristic frequencies $\{\omega_k\}$ requires solving an algebraic equation of 3N-th order, and we need the frequency distribution for large N. This is no simple matter. In fact, a branch of mathematical physics whose principal aim is to find the frequency distribution, is called lattice dynamics.

Figure 11.17 represents a result obtained by Walker [10] after the analysis of the X-ray scattering data for aluminum in terms of lattice dynamics. Some remarkable features of the curve are as follows.

(a) At low frequencies,

$$D(\omega) \propto \omega^2 . \qquad (12.24)$$

(b) There exists a maximum frequency ω_m such that

$$D(\omega) = 0 \text{ for } \omega \geqq \omega_m. \qquad (12.25)$$

(c) A few sharp peaks exist below ω_m.

The feature (a) is common to all crystals in three dimensions. The low frequency modes can be described adequately in terms of longitudinal and transverse elestic waves which were discussed in §§ 11.9-10. Note that this region can be represented very well by the Debye model calculation, indicated by

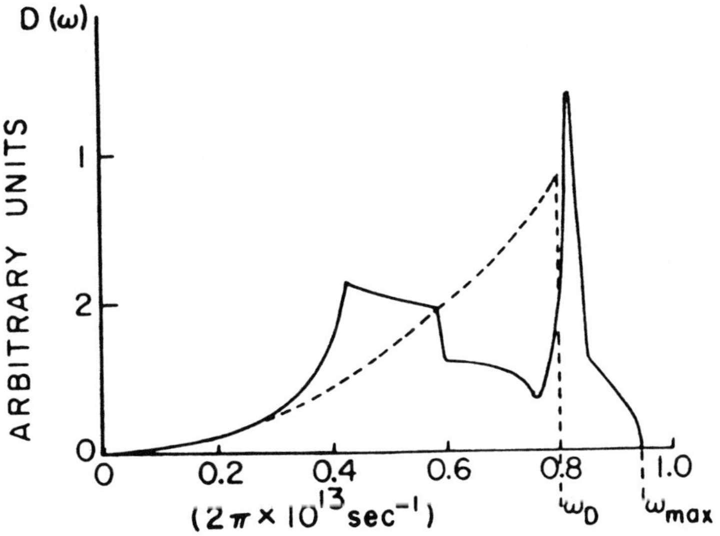

Fig. 11.17 The density of normal modes in the angular frequency for aluminum. The solid curve represents the data deduced from x-ray scattering measurements due to Walker [10]. The broken lines indicate the Debye frequency distribution (11.14) with Θ_D = 328 K.

the broken line. The feature (b) is connected with the basic lattice structure as mentioned in § 3.11. Briefly, the normal modes of extreme short wavelength (much shorter than the lattice constant) and extreme high frequency cannot exist for a real lattice. Existence of sharp peaks was first discovered by Van Hove [11] on topological grounds, and sharp peaks are often referred to as Van Hove singularities.

So far, we have tacitly assumed large perfect crystals. If the solids have significant anisotropies, defects or if the sizes of crystals are small the low-temperature thermal properties of such materials can be quite different from the standard Debye behavior.

References

[1] e.g. Lorrain P. and Corson, D.R., Electromagnetism Freeman, San Francisco (1978); other references are listed in Bibliography section.

[2] Moon, P., Scientific Basis of Illuminating Engineering, Dover, New York [1936], P.111.

[3] Kittel, C., Introduction to Solid State Physics, Wiley, New York, N.Y. 10016, Fifth Edition (1976)

[4] Einstein, A., Ann. Physik, 22, 186 (1907)

[5] Kibble, T.W.B., Classical Mechanics, McGraw-Hill, London (1966)

[6] Goldstein, H., Classical Mechanics, Addison-Wesley, Reading Mass., Second Edition (1980)

[7] Debye, P., Ann. Physik, 39, 789 (1912)

[8] Turner, R.E. and Betts, D.S., Introductory Statistical Mechanics,

[9] Sears, F.W. and Salinger, G.L., Thermodynamics, Kinetic Theory, and Statistical Thermodynamics, Addison-Wesley, Reading, Mass., Third Edition (1975)

[10] Walker, C.B., Phys.Rev. 103, 547 (1956)

[11] Van Hove, L., Phys.Rev. 89, 1189 (1953)

Review Questions

1. Write down mathematical expressions, figures, etc. and explain briefly.

(a) The wave equation for the electric field in vacuum.

(b) The plane-wave solution of a wave equation. The wave vector and angular frequency

(c) Plane-polarized wave. The polarization vector

(d) The energy of electromagnetic fields within a volume V

(e) A classical expression for the electromagnetic (radiation) field energy in terms of the normal mode coordinates

(f) A quantum mechanical expression for the radiation field energy

(g) A quantum mechanical expression for the energy of black-body radiation at temperature T

(h) Planck distribution function

(i) Stephan-Boltzmann law of radiation

(j) Photon flux through a small hole

(k) Radiation pressure

(l) Simple, face-centered, and body-centered, cubic lattices

(m) Einstein's theory of the heat capacity of a solid

(n) Einstein temperature for diamond

(o) Young modulus

(p) Shear modulus

(q) Poisson's ratio

(r) The wave equations for transverse and longitudinal elastic waves

(s) The Hamiltonian for elastic waves

(t) Debye's theory of the heat capacity of a solid

(u) Debye temperature for diamond

(v) Debye T^3-law

(w) The harmonic approximation in lattice dynamics

General Problems

1. Consider a homogeneous metallic wire of length ℓ and cross section πR^2 We bend the wire to form a planar arc of curvature κ . Show that the elastic energy stored is given by

$$U = \frac{1}{8} \pi Y R^4 \ell \kappa^2,$$

where Y is the Young modulus. Note that the curvature is defined as the inverse of the radius of contact circle.

2. Consider the same wire as that in Problem 1. We twist it to a small angle α with no change in shape. Show that the elastic energy stored at the final state is given by

$$U = \frac{1}{4} \pi S R^4 \ell^{-1} \alpha^2,$$

where S is the shear modulus.

3. Let us take a thin metallic spring of a helical shape of base radius a and pitch $2\pi b$ in equilibrium. By assumption, the spring has a configurational energy of the form

$$U = \frac{1}{8} \pi R^4 \int_0^{\ell} ds \left\{ Y(\kappa - \kappa_0)^2 + 2S (\tau - \tau_0)^2 \right\},$$

where $\kappa(s)$ and $\tau(s)$ represent respectively the curvature and torsion of the central line of the wire, which are functions of the arc length s from one end ; $\kappa_0 = a / (a^2 + b^2)$ and $\tau_0 = b / (a^2 + b^2)$ are those corresponding to the equilibrium shape; other symbols are defined in Problem 1 and 2. Let us suppose a uniform extension of the wire $2\pi\Delta b$ along the helical axis. Show that the increase in energy per unit length (of the central line) is given by

$$\Delta u = \frac{1}{8} \pi R^4 (a^2 + b^2)^{-2} \left[(b/a)^2 Y + 2S \right] \Delta b^2 + O(\Delta b^4).$$

For a normal spring, the base circumference $2\pi a$ is much greater than the pitch $2\pi b$. The above result therefore indicates that the change in energy mostly comes from the twisting deformation.

4. A graphite (carbon) crystal has a quasi-two-dimensional structure, and its experimental heat capacity indicates considerable deviations from the Debye T^3-law at very low temperatures. Let us consider a two-dimensional lattice. By applying the Debye continuum model, investigate the heat capacity of the lattice .

5. Experiments on a sample of graphite crystallites which form a common black carbon, indicate a linear-in-T heat capacity at very low temperatures. The magnitude of the heat capacity however cannot be explained entirely in terms of the electronic contribution. The rest must come from the lattice contribution. The bending oscillations of a thin plate has a dispersion relation of the form

$$\omega = a\, k^2,$$

where $\vec{k}$ represents the two-dimensional k-vector and $k^2 = |\vec{k}|^2$. Show that the excitation of such an oscillation leads to a heat capacity linearly proportional to the absolute temperature T at very low temperatures.

Chapter 12. SPIN AND MAGNETISM. PHASE TRANSITIONS. POLYMER FONFORMATION

This chapter consists of three main parts. In the first part, §§ 12.1-7, the quantum angular momentum, quantum spins, and paramagnetism due to isolated ions and degenerate electrons are discussed. In the second part, §§ 12.8-11, ferromagnetism and phase transition are treated in terms of an internal field model and an Ising Model. The conformation of various types of polymers in solution is discussed in the third part, §§ 12.12-13.

12.1° Angular Momentum in Quantum Mechanics

Consider a particle whose motion is described by the Cartesian coordinates and momenta, (x,y,z,p_x,p_y,p_z). Its angular momentum $\vec{j}$ about the origin is defined by

$$j_x \equiv yp_z - zp_y \; , \quad j_y \equiv zp_x - xp_z \; , \quad j_z \equiv xp_y - yp_x \; . \tag{1.1}$$

These components (operators) (j_x, j_y, j_z) are Hermitean by virtue of the fact that x, y, z, p_x, p_y and p_z are all Hermitean by postulate [Problem 1.1].

Let us evaluate the <u>commutators</u> of angular momentum components and dynamical variables x, p_x, ... etc. We obtain

$$[j_z, x] \equiv [xp_y - yp_x, x]$$

$$= x[p_y, x] + [x, x]\, p_y - y\,[p_y, x] - [y, x]\, p_x \qquad \text{[use of (8.14.5 and 6)]}$$

$$= 0 + 0 - y(-i\hbar) - 0 \qquad \text{[use of (8.14.1)]}$$

$$= i\hbar\, y$$

$$[j_z, y] \equiv [xp_y - yp_x, y] = x[p_y, y] = -i\hbar x$$

$$[j_z, p_x] \equiv [xp_y - yp_x, p_x] = [x, p_x]\, p_y = i\hbar p_y, \text{ etc.}$$

In summary, we have

$$[j_z, x] = i\hbar y, \quad [j_z, y] = -i\hbar x, \quad [j_z, z] = 0 \tag{1.2}$$

$$[j_z, p_x] = i\hbar p_y, \quad [j_z, p_y] = -i\hbar p_x, \quad [j_z, p_z] = 0 \tag{1.3}$$

with corresponding relations for j_x and j_y.

Using these results, we obtain

$$[j_y, j_z] = [zp_x - xp_z, j_z] = z[p_x, j_z] - [x, j_z]\, p_z$$

$$= -i\hbar\, zp_y + i\hbar\, yp_z = i\hbar j_x, \qquad \text{[use of (1.2) and (1.3)]}$$

$$[j_y, j_z] = i\hbar j_x, \quad [j_z, j_x] = i\hbar j_y, \quad [j_x, j_y] = i\hbar j_z. \tag{1.4}$$

Here we see that <u>the angular momentum components do not commute with each other</u>, unlike the position (or momentum) components. The sign in eqs. (1.2), (1.3), and (1.4) may be memorized with the rule that the + sign occurs when the three variables, consisting of the two in the brackets on the l.h.s. and the one on the r.h.s., are in the cyclic order (xyz), and the - sign occurs otherwise. These relations (1.2) - (1.4) have analogues in the classical theory with the correspondence: $(i\hbar)^{-1}\,[A, B] \leftrightarrow \{A, B\}$. [see Problem 3.6.3]

Eq. (1.4) may be put in the vector form:

$$\vec{j} \times \vec{j} = i\hbar\vec{j}. \tag{1.5}$$

This is truly a quantum expression since the classical vector product of any vector and itself should vanish.

We can express the three sets of equations, (1.2), (1.3) and (1.4), in a unified manner:

$$[j_z, A_x] = i\hbar A_y, \quad [j_z, A_y] = -i\hbar A_x, \quad [j_z, A_z] = 0, \tag{1.6}$$

where $\vec{A} = (A_x, A_y, A_z)$ represents $\vec{r}$, $\vec{p}$ or $\vec{j}$. In fact, it can be shown that the relations (1.6) hold for any vector $\vec{A}$ which can be constructed from $\vec{r}$ and $\vec{p}$ [see Problem 1.2].

Let B_x, B_y, B_z be a second set of components of a vector, which satisfy eqs. (1.6). We obtain

$$[j_z, A_xB_x + A_yB_y + A_zA_z]$$

$$= [j_z, A_x] B_x + A_x [j_z, B_x] + [j_z, A_y] B_y + A_y [j_z, B_y]$$

$$+ [j_z, A_z] B_z + A_z [j_z, B_z]$$

$$= i\hbar A_yB_x + i\hbar A_xB_y - i\hbar A_xB_y - i\hbar A_yB_x = 0 \qquad \text{[use of (1.6)]}$$

or

$$[j_z, \vec{A} \cdot \vec{B}] = 0 . \tag{1.7}$$

This means that <u>any scalar commutes with j_z</u>.

The results obtained for a single particle can simply be generalized for a system composed of several particles. Let their angular momenta be

$$\vec{j}_1 \equiv \vec{r}_1 \times \vec{p}_1, \quad \vec{j}_2 \equiv \vec{r}_2 \times \vec{p}_2, \ \ldots \tag{1.8}$$

The dynamical variables describing different particles commute by postulate. For example,

$$[x_1, p_{2x}] = [x_2, p_{1x}] = [y_1, p_{2y}] = 0. \tag{1.9}$$

From these, we obtain

$$\vec{j}_k \times \vec{j}_\ell = -\vec{j}_\ell \times \vec{j}_k, \quad \ell \neq k \tag{1.10}$$

We note that these relations resemble the properties of the classical mechanical vectors. On the other hand, for the same particle, we have, from (1.5),

$$\vec{j}_k \times \vec{j}_k = i\hbar\, \vec{j}_k, \quad k = 1,2,\ldots \tag{1.11}$$

The <u>total angular momentum</u> $\vec{J}$ is defined by

$$\vec{J} \equiv \sum_k \vec{j}_k. \tag{1.12}$$

Then,

$$\vec{J} \times \vec{J} = (\sum_k \vec{j}_k) \times (\sum_\ell \vec{j}_\ell)$$

$$= \sum_k \vec{j}_k \times \vec{j}_k + \sum_k \sum_\ell (\vec{j}_k \times \vec{j}_\ell + \vec{j}_\ell \times \vec{j}_k)$$

$$= i\hbar \sum_k \vec{j}_k + 0 \qquad \text{[use of (1.10) and (1.11)]}$$

$$= i\hbar\, \vec{J}$$

or

$$\vec{J} \times \vec{J} = i\hbar\, \vec{J}. \tag{1.13}$$

This result has the same form as eq.(1.5).

We can further show that relations similar to (1.6) and (1.7) hold for $\vec{J}$. That is,

$$[J_z, A_x] = i\hbar\, A_y, \quad [J_z, A_y] = -i\hbar\, A_x, \quad [J_z, A_z] = 0 \tag{1.14}$$

$$[J_z, \vec{A} \cdot \vec{B}] = 0, \tag{1.15}$$

where $\vec{A}$ and $\vec{B}$ are vectors constructed in terms of $\vec{r}_1, \vec{p}_1, \vec{r}_2, \vec{p}_2, \ldots$.

Problem 1.1 Show that j_x, j_y, j_z are Hermitean by virtue of the fact that x, y, z, p_x, p_y and p_z by postulate are Hermitean.

Problem 1.2 Show that if (A_x, A_y, A_z) and (B_x, B_y, B_z) are any two sets of vector components satisfying the relations (1.6), then the components of the vector $\vec{C} = \vec{A} \times \vec{B}$ also satisfy the relations (1.6).

12.2° Properties of Angular Momentum

We have seen in the preceding section that the three components j_x, j_y, j_z of angular momentum do not commute but do satisfy the following relations:

$$[j_x, j_y] = i\hbar j_z, \quad [j_y, j_z] = i\hbar j_x, \quad [j_z, j_x] = i\hbar j_y \ . \tag{2.1}$$

The square of angular momentum $\vec{j}$, that is,

$$j^2 \equiv j_x^{\ 2} + j_y^{\ 2} + j_z^{\ 2} \equiv \vec{j}\cdot\vec{j} \tag{2.2}$$

is a scalar. According to (1.7), it must then commute with j_x, j_y or j_z:

$$[j_x, j^2] = [j_y, j^2] = [j_z, j^2] = 0. \tag{2.3}$$

From the two sets of relations (2.1) and (2.3), we can derive important properties of the quantum mechanical angular momentum, which will be discussed in the present section.

Let us suppose that we have a fictitious system for which j_x, j_y and j_z are the only dynamical variables. We know from Problem 1.1 that the angular momentum constructed by the rule $\vec{r} \times \vec{p}$ is Hermitean. So is the square of the angular momentum j^2. Since $[j_z, j^2] = 0$, there exist <u>simultaneous eigenstates for j^2 and j_z</u> such that

$$j^2|\beta, m\rangle = \beta\hbar^2|\beta, m\rangle \tag{2.4}$$

$$j_z|\beta, m\rangle = m\hbar|\beta, m\rangle , \tag{2.5}$$

where $\beta\hbar^2$ and $m\hbar$ are the eigenvalues for j^2 and j_z, respectively. Note that β and m are dimensionless real numbers. Furthermore, β must be a non-negative number since j_x, j_y and j_z all have real eigenvalues.

We now wish to show that

(a) $\beta = j'(j'+1)$, (2.6)

where j' denotes a non-negative integer or half-integer, and

(b) for a given j', the possible values of m are

$$-j', -j'+1, -j+2, \ldots, j'-2, j'-1, j'. \tag{2.7}$$

For example, for $j' = 1/2$, $\beta = 1/2(1/2 + 1) = 3/4$, and the possible values of m are -1/2 and 1/2; for $j' = 1$, $\beta = 1(1+1) = 2$, and the possible values of m are -1, 0 and 1.

Let us introduce

$$j_+ \equiv j_x + ij_y, \quad j_- \equiv j_x - ij_y = (j_+)^\dagger. \tag{2.8}$$

Using the commutation relations (2.1), we obtain

$$[j_z, j_+] = [j_z, j_x+ij_y] = [j_z, j_x] + i[j_z, j_y]$$

$$= i\hbar j_y + i(-i\hbar j_x) \qquad \text{[use of (2.1)]}$$

$$= \hbar(j_x+ij_y) = \hbar j_+$$

or

$$[j_z, j_+] = \hbar j_+ . \tag{2.9}$$

Similarly, we obtain

$$[j_z, j_-] = -\hbar j_- . \tag{2.10}$$

Applying eq.(2.9) on the ket $|\beta, m\rangle$ and rearranging terms, we obtain

$$0 = \{[j_z, j_+] - \hbar j_+\}|\beta, m\rangle$$

$$= j_z j_+|\beta, m\rangle - j_+ j_z|\beta, m\rangle - \hbar j_+|\beta, m\rangle$$

$$= j_z(j_+|\beta, m\rangle) - (m+1)\hbar(j_+|\beta, m\rangle) \qquad \text{[use of (2.5)]}$$

or

$$j_z(j_+|\beta, m\rangle) = (m+1)\hbar(j_+|\beta, m\rangle) \ . \tag{2.11}$$

This means that if $j_+|\beta, m\rangle$ is not zero, then it is the eigenstate of j_z with the eigenvalue $(m+1)\hbar$. By repeating the same process, we can show that $(m+2)\hbar$ is another possible eigenvalue unless $(j_+)^2|\beta, m\rangle = 0$. In this manner, we can obtain a series of possible eigenvalues:

$$m\hbar,\ (m+1)\hbar,\ (m+2)\hbar,\ \ldots,\ (m+k)\hbar,$$

which ends if

$$(j_+)^k|\beta, m\rangle = 0 \ . \tag{2.12}$$

Similarly, using (2.5) and (2.10), we get

$$j_z(j_-|\beta, m\rangle) = (m-1)\hbar(j_-|\beta, m\rangle) \ . \tag{2.13}$$

This implies that if $j_-|\beta, m\rangle = 0$, then $(m-1)\hbar$ is another eigenvalue. Therefore, another series of possible eigenvalues is

$$m\hbar,\ (m-1)\hbar,\ (m-2)\hbar,\ \ldots,\ (m-k)\hbar, \tag{2.14}$$

which ends if $(j_-)^k|\beta, m\rangle = 0$. We thus find that the possible eigenvalues are separated by the unit step $\hbar$ in agreement with statement (b).

We see in (2.11) that the operator j_+, acting on the eigenket $|\beta, m\rangle$ of j_z, generates another eigenket with the raised m-value if $j_+|\beta, m\rangle \neq 0$. Therefore, j_+ is often called an m-raising or simply a raising operator. On the other hand, the operator j_-, acting on $|\beta, m\rangle$, generates another ket with the lowered m-value, as we can see in (2.13). Therefore, j_- is called a lowering operator. The raising and lowering operators are not Hermitean, but they are mutually Hermitean conjugates: $(j_+)^\dagger = j_-$. They do not correspond to any observable physical quantities but they are useful quantum dynamical operators.

Let us now take j_+j_-, which can be transformed as follows:

$$\begin{aligned} j_+j_- &\equiv (j_x+ij_y)(j_x-ij_y) \\ &= j_x^2 + j_y^2 - i[j_x, j_y] \\ &= j^2 - j_z^2 - i(i\hbar j_z) \qquad \text{[use of (2.1) and (2.2)]} \\ &= j^2 - j_z^2 + \hbar j_z \ . \end{aligned}$$

or

$$j_+j_- = j^2 - j_z^2 + \hbar j_z \ . \tag{2.15}$$

In a similar manner, we obtain

$$j_-j_+ = j^2 - j_z^2 - \hbar j_z \ . \tag{2.16}$$

Let us look at

$$\begin{aligned} \langle\beta, m|j_+j_-|\beta, m\rangle &= \langle\beta, m|(j^2-j_z^2+\hbar j_z)|\beta, m\rangle \qquad \text{[use of (2.15)]} \\ &= (\beta-m^2+m)\hbar^2\langle\beta, m|\beta, m\rangle \ . \\ &\qquad\qquad \text{[use of (2.4) and (2.5)]} \end{aligned} \tag{2.17}$$

The element $\langle\beta, m|j_+j_-|\beta, m\rangle$ on the l.h.s. is non-negative since it can be regarded as the squared length of the ket $j_-|\beta, m\rangle$, and vanishes if and only if $j_-|\beta, m\rangle = 0$. Since $\langle\beta, m|\beta, m\rangle > 0$ by assumption, we obtain

$$\beta - m^2 + m \geqq 0 \ . \qquad (2.18)$$

The eigenvalues $\hbar^2$ of the operator j^2 are non-negative [see the sentence preceding eq. (2.6)]. Introducing

$$\beta \equiv j'(j'+1) \geqq 0 \ , \qquad (2.19)$$

we obtain from (2.18)

$$j'(j'+1) - m^2 + m \geqq 0 \ . \qquad (2.20)$$

We plot

$$j'(j'+1) - m^2 + m = -(m - \tfrac{1}{2})^2 + j'(j'+1) + \tfrac{1}{4}$$

against m in Figure 12.1. From the figure we observe that the lowest permissible value for m satisfying the inequality (2.20) is equal to $-j'$.

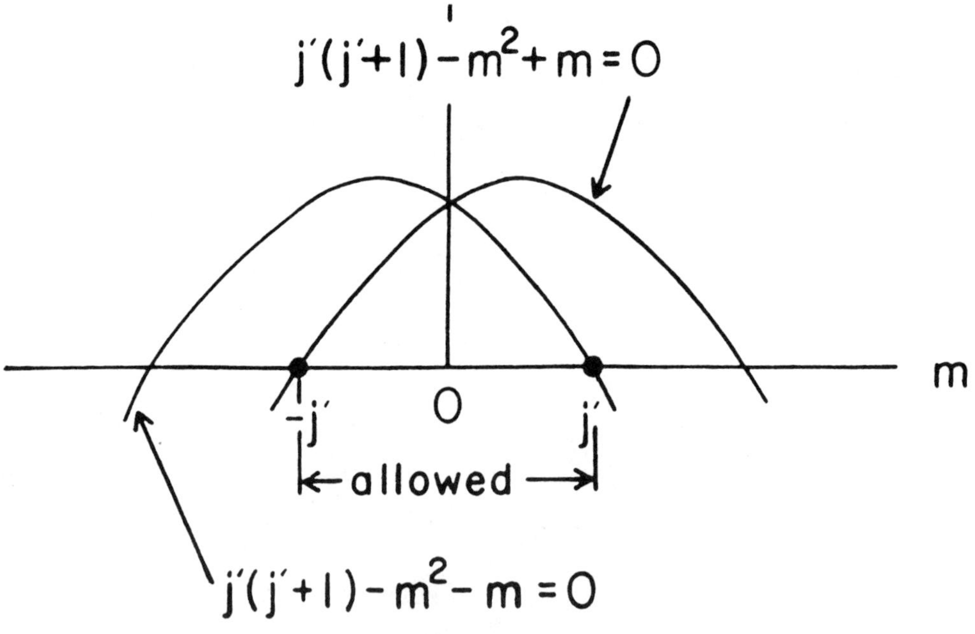

Fig. 12.1. The values of m are restricted to $-j', -j'+1, \ldots, j'-1, j'$. The The limits are imposed by the two inequalities $j'(j'+1) - m^2 + m \geqq 0$ and $j'(j'+1) - m^2 - m \geqq 0$.

Similarly using (2.5) and (2.16), we obtain

$$\langle\beta,\ m|j_-j_+|\beta,\ m\rangle = (\beta-m^2-m)\hbar^2\langle\beta,\ m|\beta,\ m\rangle \geqq 0\ , \tag{2.21}$$

and hence

$$\beta - m^2 - m \geqq 0\ , \tag{2.22}$$

or

$$j'(j'+1) - m^2 - m \geqq 0\ . \qquad [\text{use of } (2.6)] \tag{2.23}$$

Plotting $j'(j'+1) - m^2 - m$ against m in the same figure we see that the highest permissible value of m is j'. We know from the previous study that all possible values of m are separated by integral numbers. The three restrictions on m are satisfied only if j' is either an integer or a half-integer. Q.E.D.

We have seen that the possible eigenvalues for the square of the angular momentum, j^2, are $\beta\hbar^2 = j'(j'+1)\hbar^2$ where j' are either integers or half-integers. It is customary to call $j'\hbar$, rather than $(\beta)^{\frac{1}{2}}\hbar = [j'(j'+1)]^{\frac{1}{2}}\hbar$, the <u>magnitude of the angular momentum</u>. This is convenient because the possible values for j' are

$$0,\ \frac{1}{2},\ 1,\ \frac{3}{2},\ \ldots \tag{2.24}$$

while the possible values of $\beta^{\frac{1}{2}}$ are more complicated numbers.

Problem 2.1 Assume that $|j',\ m\rangle$, $|j',\ m-1\rangle$ and $|j',\ m+1\rangle$ are all normalized to unity. Calculate the following matrix elements:
(a) $\langle j',\ m+1\,|j_+|\ j',\ m\rangle$ (b) $\langle j',\ m-1|\ j_-\ |j',\ m\rangle$.

12.3° The Spin Angular Momentum

The angular momentum, arising from the motion of a particle and defined by the rule $\vec{r} \times \vec{p}$, will be called orbital angular momentum. In quantum mechanics, there exists, by postulate, another kind of angular momentum called spin angular momentum. The spin angular momentum of a particle should be regarded as arising from a certain internal motion of the particle, but has no analogue in classical mechanics. We will hereafter denote the orbital angular momentum by $\vec{\ell}$ and the spin angular momentum $\vec{s}$. The theory dealing with the spin will be developed in parallel to that dealing with the orbital angular momentum.

We postulate that the components of the spin accompanying the particle are represented by Hermitean operators (s_x, s_y, s_z) which satisfy the following commutation relations:

$$[s_y, s_z] = i\hbar\, s_x, \quad [s_z, s_x] = i\hbar s_y, \quad [s_x, s_y] = i\hbar\, s_z \,. \tag{3.1}$$

These relations can also be written by the vector equation:

$$\vec{s} \times \vec{s} = i\hbar\, \vec{s}. \tag{3.2}$$

Eqs.(3.1) and (3.2) are analogues of eqs.(1.4) and (1.5), the corresponding equations for orbital angular momentum. By postulate, each of the spin components s_x, s_y, s_z commute with all dynamical variables x, y, z, p_x, p_y, p_z describing the motion of the particle:

$$[s_x, x] = [s_x, y] = [s_x, z] = 0$$

$$[s_x, p_x] = [s_x, p_y] = [s_x, p_z] = 0 \,, \tag{3.3}$$

and similar relations for s_y and s_z. From these, we obtain

$$\vec{s} \times \vec{\ell} = - \vec{\ell} \times \vec{s}. \quad (3.4)$$

It is further postulated that the spin and orbital angular momentum, $\vec{s}$ and $\vec{\ell}$, can be added vectorically:

$$\vec{j} \equiv \vec{s} + \vec{\ell}. \quad (3.5)$$

This sum defines the <u>total angular momentum</u> $\vec{j}$.

We now look at

$$\vec{j} \times \vec{j} \equiv (\vec{s} + \vec{\ell}) \times (\vec{s} + \vec{\ell})$$

$$= \vec{s} \times \vec{s} + (\vec{s} \times \vec{\ell} + \vec{\ell} \times \vec{s}) + \vec{\ell} \times \vec{\ell}$$

$$= i\hbar\vec{s} + 0 + i\hbar\,\vec{\ell} \qquad \text{[use of (3.2), (3.4) and (1.6)]}$$

$$= i\hbar\,(\vec{s} + \vec{\ell}) = i\hbar\vec{j} \qquad \text{[use of (3.5)]}$$

or

$$\vec{j} \times \vec{j} = i\hbar\vec{j}. \quad (3.6)$$

This has the same form as (1.6). We can further show that

$$[j_z, A_x] = i\hbar\, A_y, \quad [j_z, A_y] = -i\hbar\, A_x, \quad [j_z, A_z] = 0, \quad (3.7)$$

where A_x, A_y, A_z are the three components of any vector which is constructed from $\vec{r}$, $\vec{p}$, and $\vec{s}$. [See Problem 3.1]

We can also show that

$$[j_z, \vec{A} \cdot \vec{B}] = 0, \quad (3.8)$$

where $\vec{A}$ and $\vec{B}$ are any two vectors which are constructed from $\vec{r}$, $\vec{p}$, and $\vec{s}$.

In summary, the properties obtained for the orbital angular momentum hold quite generally for the total angular momentum $\vec{j}$. This means that the properties of angular momentum concerning its eigenvalues are also valid for $\vec{j}$.

By direct calculation, we can show that

$$[s_i, s^2] \equiv [s_i, s_x^2 + s_y^2 + s_z^2] = 0, \quad i = x, y, \text{ or } z. \tag{3.9}$$

[This can also be regarded as a special case of relation (3.8).]
Because of this and eq.(3.1), all results of §12.2 can be applied to the spin angular momentum.

It has been found that every elementary particle has a spin whose magnitude is equal to either half-integral or integral multiples of Planck's constant $\hbar$. For example, electrons, positrons (electrons with positive charge), protons, neutrons, μ-mesons, all have a spin whose magnitude is $\frac{1}{2}\hbar$. Photons have a spin of magnitude $\hbar$. Gravitons, massless quanta corresponding to the gravitational wave, are believed to have spin of magnitude $2\hbar$. It has been pointed out earlier in §9.2 that those particles with half-integer spins obey the Fermi-Dirac statistics and are thus called fermions, and those with integer spins obey the Bose-Einstein statistics and are called bosons. If particles carry electric charge, they have intrinsic magnetic moments which are closely associated with their spins. This topic will be discussed in §12.5.

In contrast, the orbital angular momentum arises from the motion of the particle about a certain origin. Its magnitude may vary but must have an integral multiple of $\hbar$. The fact that its magnitude cannot be half-integers in units of $\hbar$, which is not excluded according to our study in §12.2, can be argued for on general physical grounds, but will not be discussed here. The interested reader should study this point in a standard graduate-level textbook on quantum mechanics [1] or read Dirac's book [2], pp. 144-148.

Problem 3.1 Prove eqs.(3.7).

Problem 3.1 Using the results of the preceding problem, prove eq.(3.8).

12.4° The Spin of the Electron

Electrons have a spin of magnitude $\frac{1}{2}\hbar$. This is found from various experimental data.

In dealing with an angular momentum whose magnitude is $\frac{1}{2}\hbar$, it is convenient to introduce the Pauli spin operator $\vec{\sigma}$ defined by

$$\vec{s} \equiv \frac{1}{2}\hbar\vec{\sigma} . \tag{4.1}$$

Note that this $\vec{\sigma}$ is a dimensionless quantum vector. After substituting its components into (3.1), we can obtain

$$[\sigma_y, \sigma_z] = 2i\sigma_x, \quad [\sigma_z, \sigma_x] = 2i\sigma_y, \quad [\sigma_x, \sigma_y] = 2i\sigma_z . \tag{4.2}$$

The eigenvalues of s_z are $\frac{1}{2}\hbar$ and $-\frac{1}{2}\hbar$ according to our study in §12.2. From (4.1), the eigenvalues of σ_z are then equal to 1 and -1. From this, we can conclude that σ_z^2 has only one eigenvalue 1. This means that σ_z^2 is the unit operator I, which, by definition, has the unique eigenvalue 1 [see §8.3]. By symmetry, σ_x^2 and σ_y^2 must also equal I. Therefore, we have

$$\sigma_x^2 = \sigma_y^2 = \sigma_z^2 = I . \tag{4.3}$$

The product of the unit operator I and any quantity generates the very same quantity, and can be looked upon as multiplication by one. [See Appendix D.] From now on, we will denote the unit operator by 1.

Let us take $\sigma_y{}^2\sigma_z - \sigma_z\sigma_y{}^2$, which vanishes since $\sigma_y{}^2 = 1$:

$$\sigma_y{}^2\sigma_z - \sigma_z\sigma_y{}^2 = 0 \ .$$

Rewriting this, we obtain

$$\sigma_y\sigma_y\sigma_z - \sigma_y\sigma_z\sigma_y + \sigma_y\sigma_z\sigma_y - \sigma_z\sigma_y\sigma_y$$

$$= \sigma_y[\sigma_y, \sigma_z] + [\sigma_y, \sigma_z]\sigma_y$$

$$= 2i\sigma_y\sigma_x + 2i\sigma_x\sigma_y \qquad \text{[use of (4.2)]}$$

$$= 2i(\sigma_y\sigma_x + \sigma_x\sigma_y) = 0$$

or

$$\sigma_y\sigma_x + \sigma_x\sigma_y = 0 \ .$$

Similar equations can be obtained permuting the indices (x,y,z). Thus we have

$$\sigma_y\sigma_z + \sigma_z\sigma_y = 0$$

$$\sigma_z\sigma_x + \sigma_x\sigma_z = 0$$

$$\sigma_x\sigma_y + \sigma_y\sigma_x = 0 \ . \qquad (4.4)$$

Two operators which commute except for a minus sign are said to <u>anticommute</u>. Thus, σ_y anticommutes with σ_z, and with σ_x. From (4.2 - 4.4) we also deduce that

$$\sigma_y\sigma_z = i\sigma_x, \quad \sigma_z\sigma_x = i\sigma_y, \quad \sigma_x\sigma_y = i\sigma_z \qquad (4.5)$$

$$\sigma_x\sigma_y\sigma_z = i \ . \qquad (4.6)$$

Eqs.(4.3 - 4.6) are the fundamental properties satisfied by the operator σ describing a spin of magnitude $\frac{1}{2}\hbar$.

Let us now find a matrix representation for σ_x, σ_y and σ_z. Since these three operators do not commute, we cannot diagonalize them simultaneously. By convention, let us choose that the z-component operator σ_z is diagonal. Since the eigenvalues of σ_z are +1 and -1, the matrix for σ_z can be represented by

$$\sigma_z = \begin{pmatrix} 1 & 0 \\ 0 & -1 \end{pmatrix} . \tag{4.7}$$

Using this and eqs.(4.3 - 4.6) and also the fact that all σ_x, σ_y and σ_z are Hermitean operators, we can obtain matrix representations for σ_x and σ_y. [See Problem 4.1.] The representations are not unique, but the following representation,

$$\sigma_x = \begin{pmatrix} 0 & 1 \\ 1 & 0 \end{pmatrix}, \quad \sigma_y = \begin{pmatrix} 0 & -i \\ i & 0 \end{pmatrix}, \quad \sigma_z = \begin{pmatrix} 1 & 0 \\ 0 & -1 \end{pmatrix}, \tag{4.8}$$

is one of the simplest and most commonly used. These matrices are called Pauli's spin matrices. [They were introduced earlier in Problem 8.3.1.]

For a complete description of the motion of an electron, we need the spin variables as well as Cartesian coordinates x,y,z and momenta p_x, p_y, p_z. The spin variables commute with these coordinates and momenta. Therefore, we can choose (x,y,z,σ_z) as a complete set of commuting observables. The corresponding eigenvalues (x',y',z',σ_z') can be used to characterize the quantum state for the electron. Then, any quantum-state vector $|\psi\rangle$ can be expanded in terms of the eigenkets $|x',y',z',\sigma_z'\rangle$ as follows:

$$|\psi\rangle = \int\int\int dx'\,dy'\,dz' \sum_{\sigma_z' = \pm 1} |x',y',z',\sigma_z'\rangle\langle x',y',z',\sigma_z'|\psi\rangle$$

$$\equiv \int d^3r' \sum_{\sigma_z'} |\vec{r}', \sigma_z'\rangle\langle\vec{r}', \sigma_z'|\psi\rangle . \tag{4.9}$$

This is a generalization of the relation (8.6.7): $|\psi\rangle = \int dx'|x'\rangle\langle x'|\psi\rangle$, which is valid without consideration of the spin and in one dimension.

We note that the spin quantum number σ_z' can take on the values ± 1. In the wave-mechanical description, the quantum states are usually characterized by the two sets of wave functions:

$$\psi_+(x,y,z) \equiv \langle x,y,z,\sigma_z' = 1|\psi\rangle$$
$$\psi_-(x,y,z) \equiv \langle x,y,z,\sigma_z' = -1|\psi\rangle . \tag{4.10}$$

Alternatively, we may choose p_x, p_y, p_z, σ_z as a complete set of commuting observables. In terms of their eigenvalues, an arbitrary state vector $|\psi\rangle$ can be expanded as follows:

$$|\psi\rangle = \sum_{p_x'}\sum_{p_y'}\sum_{p_z'}\sum_{\sigma_z'} |p_x',p_y',p_z',\sigma_z',\rangle\langle p_x',p_y',p_z',\sigma_z'|\psi\rangle$$
$$\equiv \sum_{\vec{p}'}\sum_{\sigma_z'} |\vec{p}',\sigma_z'\rangle\langle\vec{p}',\sigma_z'|\psi\rangle . \tag{4.11}$$

Problem 4.1 Assume that σ_x is of the form $\begin{pmatrix} a_1 & a_2 \\ a_3 & a_4 \end{pmatrix}$. Using the fact that σ_x is Hermitean, and the relation $\sigma_z\sigma_x = -\sigma_x\sigma_z$, show that $a_1 = a_4 = 0$, $a_2 = a_3^*$, $|a_2| = 1$. Therefore, σ_x is of the form $\begin{pmatrix} 0 & e^{i\alpha} \\ e^{-i\alpha} & 0 \end{pmatrix}$ where α is a real number. Similarly, one finds that σ_y is of this form. By choosing $\sigma_x = \begin{pmatrix} 0 & 1 \\ 1 & 0 \end{pmatrix}$, and using $\sigma_y = i\sigma_x \sigma_y = i\sigma_x\sigma_z$, show that $\sigma_y = \begin{pmatrix} 0 & -i \\ i & 0 \end{pmatrix}$.

12.5 The Magneto-gyric Ratio

Let us consider a classical electron describing a circle in the x-y plane as shown in Figure 12.2. The angular momentum $\vec{j} \equiv \vec{r} \times \vec{p}$ points in the positive z - axis and its magnitude is given by

$$m\ r\ v. \tag{5.1}$$

According to the electromagnetic theory [3] a current loop generates a magnetic moment $\vec{\mu}$ (vector) whose magnitude equals the current times the area of the loop and whose direction is specified by the right-hand screw rule. The magnitude of the moment generated by the electron motion therefore, is given by

$$(\text{current}) \times (\text{area}) = \left(\frac{ev}{2\pi r}\right)(\pi r^2) = \tfrac{1}{2}\, e\, v\, r, \tag{5.2}$$

and the direction is along the negative z - axis. We observe here that <u>magnetic moment</u> $\vec{\mu}$ is <u>proportional to the angular momentum $\vec{j}$</u>. We may express this relation by

$$\boxed{\vec{\mu} = \alpha \vec{j}\,.} \tag{5.3}$$

This relation, in fact, holds not only for this circular motion but in general. The proportionality factor

$$\alpha = \frac{-e}{2m} \tag{5.4}$$

is called the <u>magneto-gyric</u> or <u>magneto-mechanical ratio</u>. We note that the ratio is inversely proportional to the mass m and proportional to the charge -e.

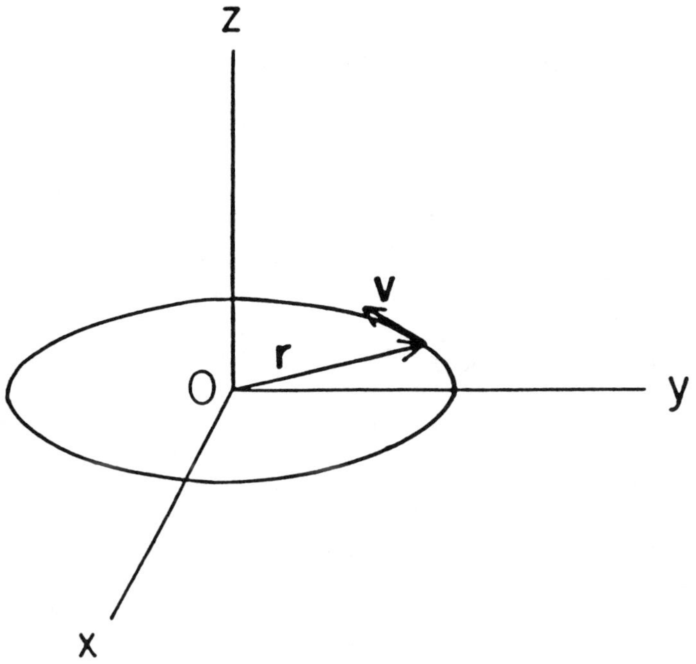

Fig. 12.2 An electron in a circular motion generate a magnetic moment $\vec{\mu}$ proportional to its angular momentum $\vec{j}$.

Let us assume that a magnetic field $\vec{B}$ is applied along the positive z-axis. The potential energy V of a magnetic dipole with moment $\vec{\mu}$ is given by

$$\boxed{V = -\mu B \cos\vartheta = -\mu_z B,} \tag{5.5}$$

where ϑ is the angle between the vectors $\vec{\mu}$ and $\vec{B}$.

We may expect that a general relation such as (5.3) holds also in quantum mechanics. In the preceding sections we saw that the angular momentum (eigenvalues) is quantized in the units of $\hbar$. Let us now take simple cases.

A. Free Electron

The electron has a spin $\vec{s}$ whose z-component can assume either $\frac{1}{2}\hbar$ or $-\frac{1}{2}\hbar$. Let us write

$$s_z' = \frac{1}{2}\hbar\,\sigma_z' \equiv \frac{1}{2}\hbar\,\sigma, \tag{5.6}$$

where

$$\sigma \equiv \sigma_z' = \pm 1. \tag{5.7}$$

In analogy with eq.(5.3) we may assume that $\mu_z \propto s_z' \propto \sigma$. We will write this relation in the form :

$$\boxed{\mu_z = \frac{1}{2} g\,\mu_B\,\sigma,} \tag{5.8}$$

where

$$\mu_B \equiv \frac{e\hbar}{2mc} = 0.927 \times 10^{-20} \quad \text{ergs gauss}^{-1}, \tag{5.9}$$

called the Bohr magneton , has the dimensions of magnetic moment. The constant g in (5.8) is a numerical factor of order 1, and is called the g factor. If the magnetic moment of the electron is accounted for by the "spinning" of the charge around a certain axis , the g - factor should be exactly one. Comparison with the experiments however shows that this factor is close to 2. This so-called spin anomaly is an important indication of the quantum nature of the spin.

In the presence of a magnetic field $\vec{B}$, the electron whose spin is directed along $\vec{B}$, that is, the electron with the "up-spin" will have a lower energy than the electron whose spin is directed against $\vec{B}$, that is, the electron with the "down-spin". The difference is, according to (5.5) and (5.8),

$$\Delta\varepsilon = \frac{1}{2} g \mu_B(+1) - \frac{1}{2} g \mu_B (-1) = g \mu_B B. \qquad (5.10)$$

For B = 7000 Gauss and g = 2, we obtain the numerical estimate : $\Delta\varepsilon \simeq 1^o k_B$.

If an electromagnetic wave with the frequency ν satisfying $h\nu = \Delta\varepsilon$ is applied, the electron may absorb a photon of the energy $h\nu$, and jump up to the upper energy level. See Figure 12.3. This phenomenon is known as the <u>electron spin resonance</u>. The frequency corresponding to $\Delta\varepsilon = 1^o k_B$ is

$$\nu = 2.02 \times 10^{10} \text{ cycles sec}^{-1}.$$

This frequency falls in the microwave region of the electromagnetic radiation spectrum.

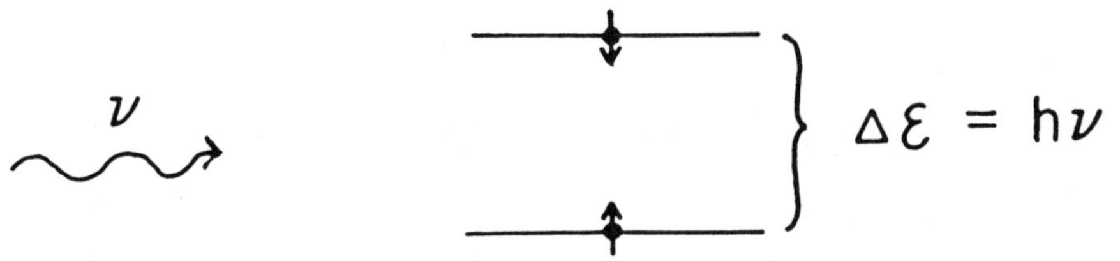

Fig. 12.3. An electron with the up-spin may absorb a photon of the energy $h\nu$ and jump to the upper energy level by flipping its spin provided that $\Delta\varepsilon = h\nu$. (Electron spin resonance).

B. Free Proton

The proton carries a positive charge, whose magnitude is equal to that of the electron charge, and a spin of magnitude $\frac{1}{2}\hbar$. We can therefore expect that it has a magnetic moment proportional to the spin quantum number s'_z. According to eq.(5.4) the classical magnetogyric ratio is inversely proportional to the mass. Since the proton has about 1840 times heavier mass than the electron, the magnetic moment will be smaller by the same factor.

The natural unit of magnetic moment for the proton may be defined by

$$\mu'_B \equiv \frac{e\hbar}{2m_p c} = \frac{e\hbar}{2m_e c} \left[\frac{m_e}{m_p} \right]$$

$$= \mu_B \left[\frac{m_e}{m_p} \right] = 5.05 \times 10^{-24} \text{ ergs gauss}^{-1}, \qquad (5.11)$$

where m_p is the mass of a proton. This unit is sometimes called the <u>nuclear magneton</u>. In analogy with the case of electron, we may assume that

$$\mu_z = \frac{1}{2} g_p \mu_B' \sigma . \qquad (5.12)$$

The <u>proton g - factor,</u> g_p, defined here and compared with experiments, is

$$g_p = 2.78 . \qquad (5.13)$$

This value again indicates the quantum nature of the proton spin.

Under a magnetic field, the energy levels of a proton will be split into two, allowing a possibility of <u>nuclear magnetic resonance</u> (N M R). The energy difference is smaller by three orders of magnitude compared with a similar condition for an electron.

C. Free Neutron

The neutron carries a spin of magnitude $\frac{1}{2}\hbar$ but no charge. In the classical picture, no neutral particle should carry a magnetic moment. In reality, the neutron is a quantum particle and has a nuclear magnetic moment characterized by eq.(5.12) with the <u>neutron g-factor</u>

$$g_n = -1.913. \tag{5.14}$$

How can a neutral particle like neutron have a magnetic moment? This is a good question. Study of this question however must involve difficult quantum mechanical calculations [4].

D. Atomic Nuclei

An atomic nucleus is composed of a number of nucleons. For example, the α-particle, the nucleus of He^4, is composed of two protons and two neutrons. When a nucleus contains even numbers of protons and neutrons, the stable nucleus (in the lowest-energy state) is most likely to have zero net spin and no magnetic moment. On the other hand, the He^3 nucleus has two protons and one neutron; it has a net spin of magnitude $\frac{1}{2}\hbar$ and consequently a finite magnetic moment. As in this example, if a nucleus consists of an odd number of nucleons, it must have a spin of non-zero magnitude and a finite magnetic moment.

E. Atoms and Ions.

An atom contains a nucleus and a certain number of electrons orbiting around the nucleus. For example, the He atom has two electrons in the 1s states, which will be denoted $(1s)^2$ [see [1] for atomic orbital notations such as $(1s)^2$]; the Ne atom has ten electrons in the state, $(1s)^2 (2s)^2 (2p)^6$. If the atomic orbitals are filled to the fullest as in the cases of He or Ne in the ground state, the net spin of the electrons is zero; in more detail, the 1s orbitals are filled by a pair of electrons with up and down spins; so are 2s orbitals. Such atoms do not possess orbital angular momentum.

Ions formed from elements of the so-called transition groups in the periodic table usually have large magnetic moments. These groups are located at regions of the periodic table where inner electron shells grow from one inert configuration to a larger one. Examples are rare earth ions (Ce^{3+}, Gd^{3+}, Yb^{3+}, ..) and Iron-group ions (Fe^{3+}, Co^{++}, Ni^{++}, ..). Ions outside the transition groups normally have zero resultant angular momentum (whether spin or orbital in nature) and zero magnetic moment. Readers interested in learning about these features are encouraged to look at the introductory section of the excellent book, Paramagnetic Resonance by Pake [5].

12.6 Paramagnetism of Isolated Atoms. Curie's Law.

Let us consider an atom which has a spin of magnitude $\frac{1}{2}\hbar$ due to one unpaired electron and a magnetic moment $\vec{\mu}$. If this atom is placed in a magnetic field $\vec{B}$ directed along the positive z-axis, its spin points either up along the positive z-axis or down along the negative z-axis. If the spin points up : $\sigma = 1$, the potential energy

$$V(\sigma) = -\mu_z B = -\tfrac{1}{2} g \mu_B \sigma B \qquad \text{[use of (5.7)]} \qquad (6.1)$$

is equal to $-\mu_B B$ where we chose $g = 2$. If the spin points down : $\sigma = -1$, the potential energy is then, $\mu_B B$. If a canonical distribution is assumed, the difference in energy makes the probability of finding an up-spin atom greater, and generates a finite average magnetic moment $\langle\mu_z\rangle$.

The canonical-ensemble average of the magnetic moment μ_z can be calculated as follows :

$$\langle\mu_z\rangle \equiv \frac{\sum_\sigma \mu_z e^{-\beta V(\sigma)}}{\sum_\sigma e^{-\beta V(\sigma)}} = \frac{\sum_{\sigma=1,-1} \mu_B \sigma e^{\beta \mu_B B \sigma}}{\sum_\sigma e^{\beta \mu_B B \sigma}}$$

$$= \mu_B \frac{e^{\beta \mu_B B} - e^{-\beta \mu_B B}}{e^{\beta \mu_B B} + e^{-\beta \mu_B B}}$$

or

$$\langle\mu_z\rangle = \mu_B \tanh(\mu_B B / k_B T), \qquad (6.2)$$

where

$$\tanh x \equiv \text{hyperbolic tangent } x \equiv \frac{e^x - e^{-x}}{e^x + e^{-x}} . \qquad (6.3)$$

In Figure 12.4, the reduced average moment $\langle \mu_z \rangle / \mu_B = \tanh x$ versus $x \equiv B \mu_B / k_B T$ is drawn. At very high temperatures where $x \equiv B \mu_B / k_B T \ll 1$, the reduced average moment approaches zero, indicating that the spin is equally likely to point up or down, and the net moment tends to zero. At lower temperatures the atom "prefers" occupying the lower energy state with the up spin. The average moment $\langle \mu_z \rangle$ grows monotonically and reaches the maximum value μ_B as the temperature T is lowered to 0 K.

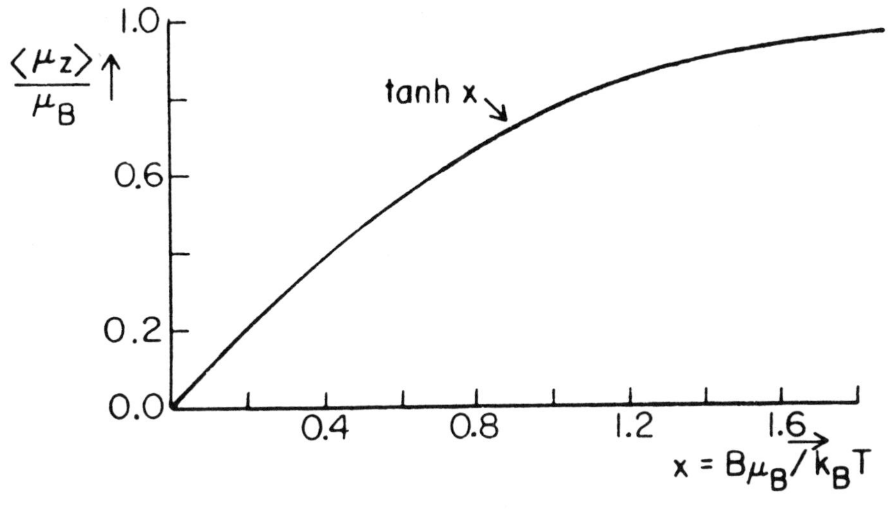

Fig. 12.4 The reduced average moment $\langle \mu_z \rangle / \mu_B$ against the reduced magnetic field $x \equiv B \mu_B / k_B T$.

Let us now consider a set of isolated atoms in a volume V. The magnetization I, that is, the magnetic moment per unit volume, is given by

$$I = \frac{\langle N \mu_z \rangle}{V} = n \langle \mu_z \rangle, \tag{6.4}$$

where n represents the number density of magnetic atoms. Introducing (6.2) in (6.4), we obtain

$$I = n \mu_B \tanh (\mu_B B / k_B T). \tag{6.5}$$

Since

$$\tanh x \simeq x \qquad \text{for} \quad x \ll 1, \tag{6.6}$$

the magnetization I is proportional to B for small B. Therefore, in this domain, we have the linear relation:

$$I = \frac{n \mu_B^2}{k_B T} B, \qquad (\text{small } B). \tag{6.7}$$

The magnetic susceptibility χ is defined through

$$I = \chi B \qquad \text{for small } B. \tag{6.8}$$

Comparing the last two expressions, we obtain

$$\chi = \frac{n \mu_B^2}{k_B T} . \tag{6.9}$$

We note that the susceptibility χ is positive, that is, the system is paramagnetic. We further note that the susceptibility χ is inversely proportional to the temperature T. This is known as Curie's law.

We have so far considered the case in which the magnetic atom has a spin of magnitude $\frac{1}{2}\hbar$. Some magnetic ions [see §12.5] are known to have angular momenta of higher magnitudes. Let us take the case of an angular momentum of the magnitude $j\hbar$. From (2.7) and (5.3), the magnetic moment along the field (along the positive z-axis), μ_z, can be written in the form

$$\mu_z = g\,\mu_B\, m, \tag{6.10}$$

where m can take on any element of the set

$$(-j,\ -j+1,\ \ldots,\ j-1,\ j). \qquad \text{[see (2.7)]}$$

The potential energy corresponding to the state m is

$$V = -\mu_z B = -g\,\mu_B\, B\, m. \tag{6.11}$$

The average magnetic moment $\langle\mu_z\rangle$ is then given by

$$\langle\mu_z\rangle = \frac{\sum_m \mu_z\, e^{-\beta V}}{\sum_m e^{-\beta V}} = \frac{\sum_{m=-j}^{j} m g\mu_B \exp[\beta g\mu_B\, B\, m]}{\sum_m \exp\,[\beta g\mu_B\, B\, m]}$$

$$= \frac{\partial}{\partial\,(\beta B)}\,\ln\left[\sum_m \exp\,[\beta g\mu_B\, B\, m]\right]. \tag{6.12}$$

Since

$$\sum_{m=-j}^{j} \exp\,(mx) = e^{-jx} + e^{-(j-1)x} + \ldots + e^{jx}$$

$$= e^{jx}\,[1 + e^{-x} + \ldots + e^{-2jx}]$$

$$= e^{jx}\,\frac{1 - e^{-(2j+1)x}}{1 - e^{-x}} = \frac{e^{-\frac{1}{2}x}\left[e^{(j+\frac{1}{2})x} - e^{-(j+\frac{1}{2})x}\right]}{e^{-\frac{1}{2}x}\left[e^{\frac{1}{2}x} - e^{-\frac{1}{2}x}\right]}$$

$$= \frac{\sinh(j + \frac{1}{2})x}{\sinh \frac{1}{2} x}\,, \tag{6.13}$$

we obtain

$$\ln \left\{ \sum_m \exp\,[\beta g \mu_B\, B\, m] \right\}$$

$$= \ln \sinh\,[(j + \tfrac{1}{2})\,\beta g \mu_B\, B] - \ln \sinh\,[\tfrac{1}{2}\beta g \mu_B\, B]. \tag{6.14}$$

Differentiating this with respect to βB, we obtain [Problem 6.1]

$$\langle \mu_z \rangle = g\, \mu_B\, j\, B_j(\beta g\, \mu_B j B)$$

or

$$\boxed{\langle \mu_z \rangle = g\, \mu_B\, j\, B_j(g\, \mu_B j B/k_B\, T)\,,} \tag{6.15}$$

where

$$B_j(x) \equiv \left(1 + \frac{1}{2j}\right) \coth\left[\left(1 + \frac{1}{2j}\right) x\right] - \frac{1}{2j} \coth\left[\frac{1}{2j}\, x\right] \tag{6.16}$$

is called the Brillouin function.

For $j = 1/2$, $B_{1/2}(x) = \tanh x$, and (6.15) reduces to (6.2) ($g = 2$). The behavior of B_j for a few j is shown in Figure 12.5. All $B_j(x)$ start off lineraly near the origin and approach the maximum value 1 as x tends to ∞.

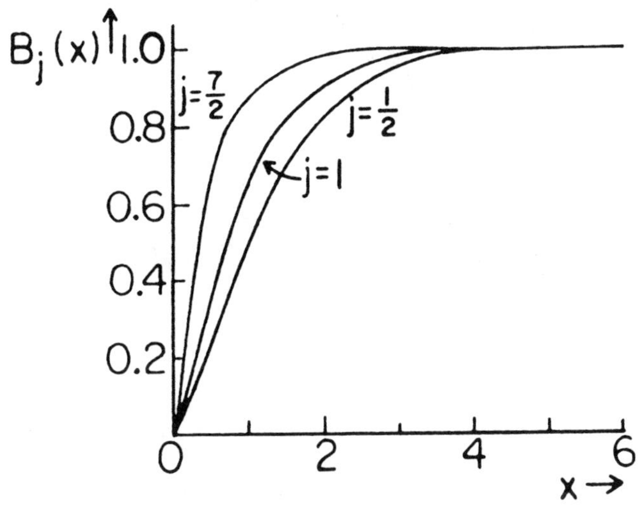

Fig. 12.5 The Brillouin functions $B_j(x)$ start off linearly and approach unity as x tends to ∞ .

The magnetization I is given by

$$I = n \langle \mu_z \rangle = g \mu_B nj B_j(x) \quad [x = g \mu_B jB/k_B T.] \tag{6.17}$$

In Figure 12.6 , the mean magnetic moments of a few different ions are shown against B / T (gauss deg^{-1}). We note that the agreements between theory and experiment are excellent, indicating, among other things, the quantization of the angular momentum.

The susceptibility χ can be obtained by studying the behavior of $B_j(x)$ for small x, and is given by [problem 6.2].

$$\chi = n \frac{g^2 \mu_B^2 j(j+1)}{3k_B T} . \tag{6.18}$$

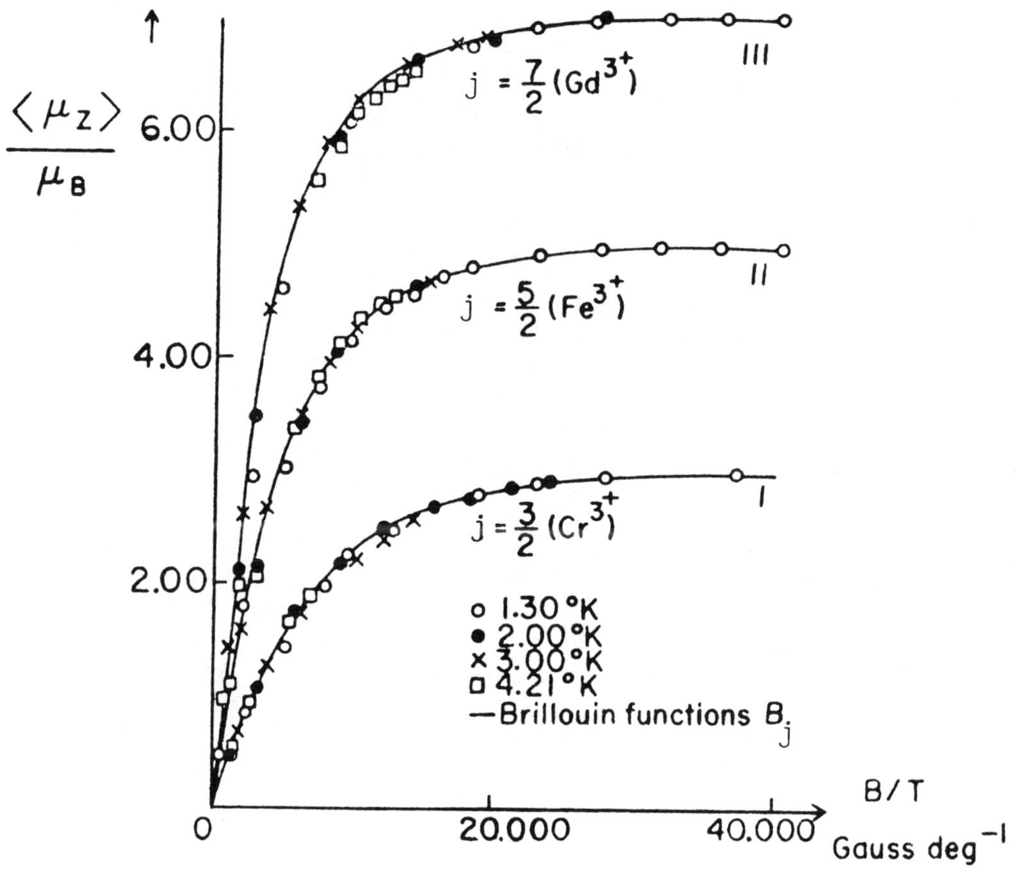

Fig. 12.6 The mean magnetic moment $\langle \mu_z \rangle / \mu_B$ reduced in units of the Bohr magneton μ_B against the ratio B/T. The solid curves represent Brillouin functions, and the expermental points are due to Henry [6].

This expression shows that Curie's law : $\chi \propto T^{-1}$, is valid for any value of j.

Problem 6.1. Derive (6.15) from (6.12), using (6.14).

Problem 6.2. Derive (6.18) from (6.16) and (6.17).

12.7 Paramagnetism of Degenerate Electrons. (Pauli Paramagnetism)

Let us consider an electron moving in free space. The quantum states for the electron can be characterized by momentum $\vec{p}$ and spin σ (= ±1). If a weak constant magnetic field $\vec{B}$ is applied along the positive z-axis, the energy ε associated with the quantum state $(\vec{p},\sigma)$ is given by

$$\varepsilon = \frac{p^2}{2m} - \frac{1}{2} g \mu_B \sigma B \equiv \varepsilon(\vec{p},\sigma) , \qquad (7.1)$$

where the second term arises from the electromagnetic interaction, see (5.5) and (5.8). Since g=2 for the electron spin, we may simplify (7.1) to

$$\varepsilon = \frac{p^2}{2m} - \mu_B B \sigma . \qquad (7.2)$$

This expression shows that the electron with up spin has a lower energy than the electron with down spin. In other words, the spin degeneracy is removed in the presence of a magnetic field.

Let us now consider a collection of free electrons in equilibrium. At the absolute zero, the states with the lowest energies will be occupied by the electrons, the Fermi energy ε_F providing the upper limit. This situation is schematically shown in Figure 12.7, where the densities of states, $D_+(\varepsilon)$ and $D_-(\varepsilon)$, for electrons with up and down spins, are drawn against the energy ε. The density of states for free electrons was discussed in §10.4. In the absence of the field, both $D_+(\varepsilon)$ and $D_-(\varepsilon)$ are the same , and are given by one half of expression (10.4.9):

$$D_o(\varepsilon) \equiv V \frac{m^{3/2}}{\sqrt{2}\,\pi^2 \hbar^3} \varepsilon^{1/2} \qquad (7.3)$$

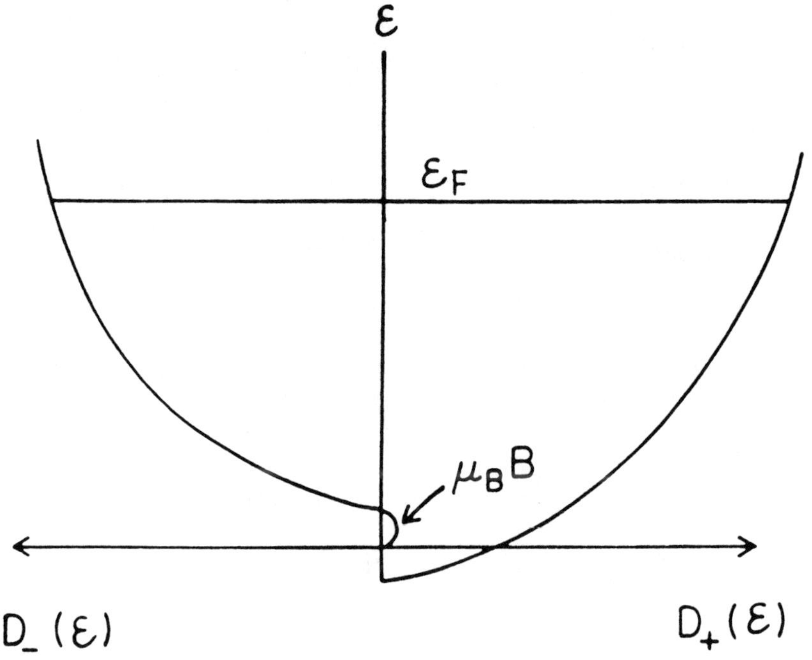

Fig. 12.7 The density of states, D_+ and D_-, for free electrons with up and down spins, are drawn against the energy ε which is directed upwards.

Because of the magnetic energy, $-\mu_B B$, the curve for the density of states $D_+(\varepsilon)$ for electrons with up spins will be displaced downward by $\mu_B B$ compared with that for zero field, and will be given by

$$D_+(\varepsilon) = V \frac{m^{3/2}}{\sqrt{2}\,\pi^2 \hbar^3} (\varepsilon + \mu_B B)^{1/2} = D_o(\varepsilon + \mu B), \quad \varepsilon \geq -\mu_B B. \tag{7.4a}$$

Similarly the curve for the density of states, $D_-(\varepsilon)$, for electrons with down spins is displaced upward by $\mu_B B$:

$$D_-(\varepsilon) = V \frac{m^{3/2}}{\sqrt{2}\,\pi^2 \hbar^3} (\varepsilon - \mu_B B)^{1/2} = D_o(\varepsilon - \mu_B B), \quad \varepsilon \gtreqqless \mu_B B. \tag{7.4b}$$

From Figure 12.7 , the numbers $N_\pm$ of the electrons with up and down spins are given by

$$N_+ = \int_{-\mu_B B}^{\varepsilon_F} d\varepsilon\, D_+(\varepsilon) = \int_0^{\varepsilon_F + \mu_B B} dx\, D_o(x) \qquad [\, x = \varepsilon + \mu_B B \,]$$

$$N_- = \int_{\mu_B B}^{\varepsilon_F} d\varepsilon\, D_-(\varepsilon) = \int_0^{\varepsilon_F - \mu_B B} dx\, D_o(x). \qquad [\, x = \varepsilon - \mu_B B \,] \tag{7.5}$$

The difference $N_+ - N_-$ generates a finite magnetic moment. Each electron with up spin contributes μ_B and each electron with down spin $-\mu_B$. Therefore, the total magnetic moment is $N_+\mu_B - N_-\mu_B$. Dividing this by volume V, we obtain, for the magnetization,

$$I = \frac{\mu_B}{V} [\, N_+ - N_- \,].$$

$$= \frac{\mu_B}{V} \left[\int_0^{\varepsilon_F + \mu_B B} dx \quad D_o(x) - \int_0^{\varepsilon_F - \mu_B B} dx \quad D_o(x) \right]$$

$$\cong \frac{2\,\mu_B^2 B}{V} \quad D_o(\varepsilon_F), \qquad [\text{use of } (7.5)] \tag{7.6}$$

where we retained the term proportional to B only. Using (7.3), we can re-express this as follows:

$$I = \frac{\sqrt{2}\, \mu_B^2\, m^{3/2}}{\pi^2\, \hbar^3}\, B\, \varepsilon_F^{1/2} > 0. \tag{7.7}$$

The last expresseion shows that the magnetization is positive, and is proportional to the external field B. That is, the system is paramagnetic. The susceptibility χ defined through the relation $I = \chi B$, is given by

$$\chi = \frac{\sqrt{2}\, \mu_B^2\, m^{3/2}}{\pi^2\, \hbar^3}\, \varepsilon_F^{1/2}. \tag{7.8}$$

By using the relation (10.4.11):

$$n = \frac{2}{3}\, \frac{\sqrt{2}\, m^{3/2}}{\pi^2\, \hbar^3}\, \varepsilon_F^{3/2},$$

we can represents (7.8) in a simple manner as follows:

$$\chi = \frac{3}{2}\, \frac{\mu_B^2\, n}{\varepsilon_F}. \tag{7.9}$$

This result was first obtained by Pauli, and is often referred to as the Pauli paramagnetism.

We note that Pauli paramagnetism is weaker than the paramagnetism of isolated atoms [see (6.9)] approximately by the factor $k_B T/\varepsilon_F$ (if this factor is small).

12.8 Ferromagnetism. Weiss' Internal Field Model

Let us take a piece of iron (Fe) such as a common nail. In the normal (untreated) state it does not have a magnetic moment. If we place it between magnetic pole-blocks, as indicated in Figure 12.8, we can bring it into a magnetized state in which it has a finite magnetic moment parallel to the magnetic field. The magnetization I, that is, the magnetic moment per unit volume, can be increased by applying higher magnetic fields. The magnetization I saturates for a field of moderate strength, a few hundred Gausses. If we reduce the magnetic field, the magnetization decreases but does so in an irreversible

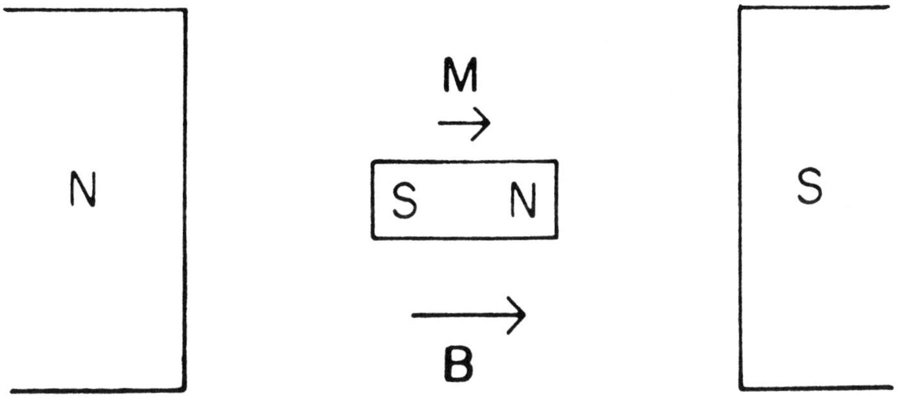

Fig. 12.8. A piece of iron is magnetized when it is placed between magnetic pole-blocks.

manner; the magnetization is therefore not a unique function of the field but depends on past history. In Figure 12.9 the magnetization as a function of an applied field for a typical ferromagnet is shown. We note that the

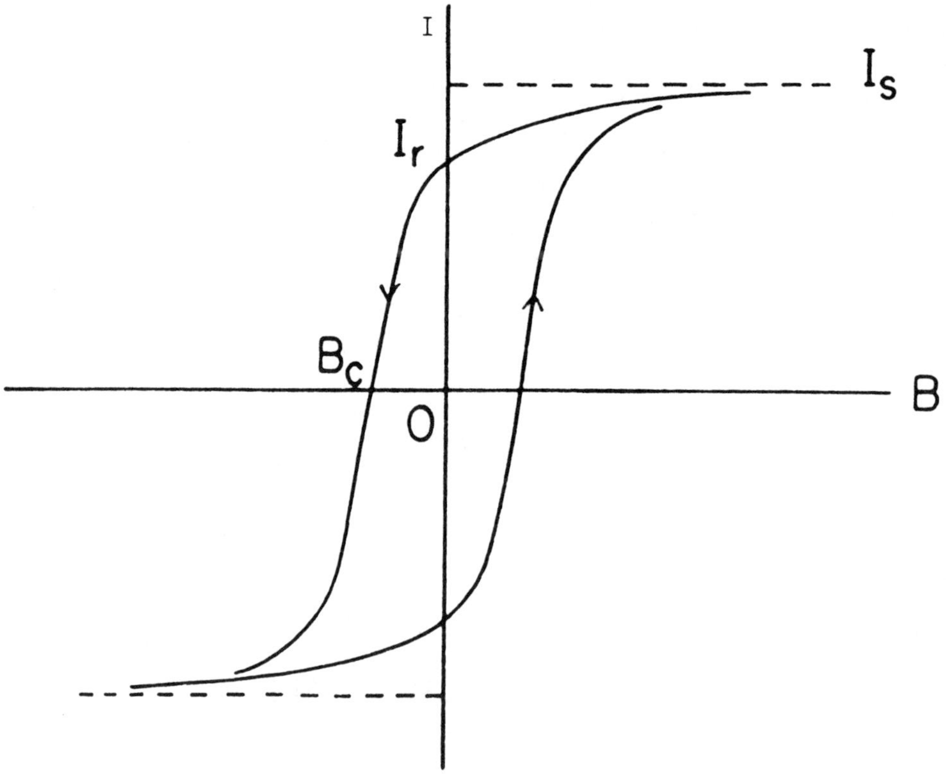

Fig. 12.9 A magnetization curve for a ferromagnet. The curves are irreversible.

magnetization at zero field, called the remnant magnetization, depends on the direction in which the specimen was last saturated. Also note that in order to bring the magnetization to zero, a finite opposing field, called the coercive field B_c, is necessary. The magnetization saturates to a value I_s at high fields.

The physical understanding of the ferromagnetic behavior had long been lacking. The magnetic hysteresis indicated in Figure 12.9, is a dissipative phenomenon ; in fact, the area enclosed by the magnetization curves, corresponds to the amount of energy converted into heat in one hysteresis cycle. The phenomenon therefore involves some sort of friction.

In 1907 P. Weiss made a fundamental step in the basic understanding of the phenomenon based on the magnetic domain hypothesis. According to his theory

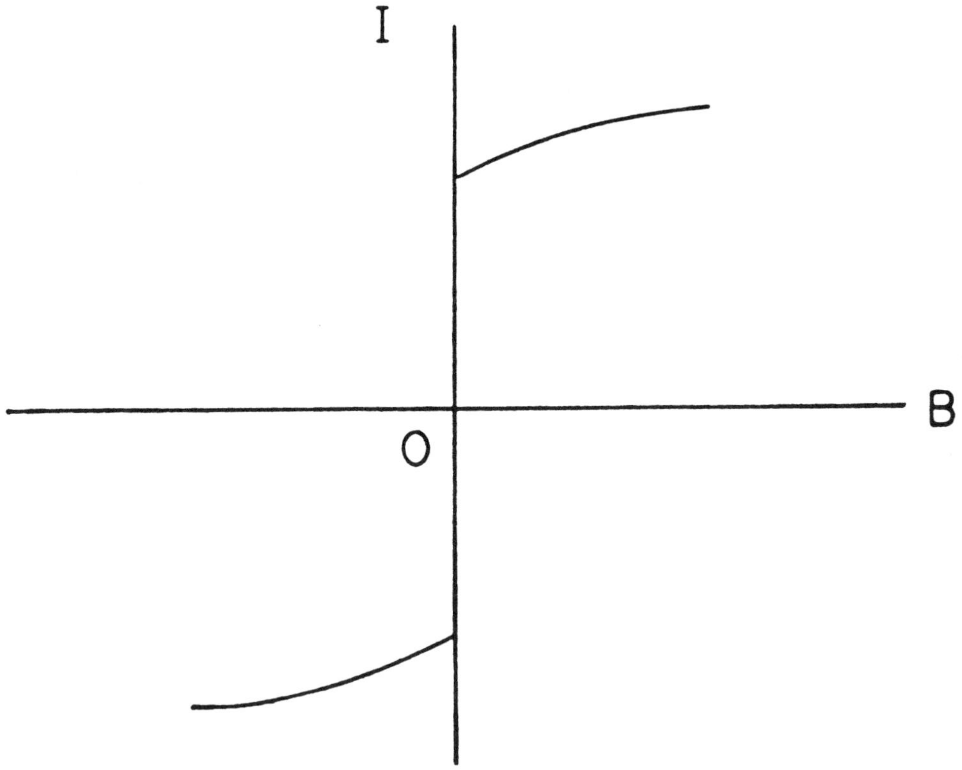

Fig. 12.10 A magnetization curve for a single ferromagnetic domain

a single domain of a microscopic size was supposed to have a magnetization curve of the type shown in Figure 12.10, where the magnetization appears as a definite function of temperature and field. Thermodynamics and statistical mechanics may then apply to individual domains. Obvious similarity exists between this and the magnetization curve of Figure 12.4. They both show saturation at high fields. The difference is a finite spontaneous magnetization at zero filed. According to Weiss, those aspects of Figure 12.10 not reproduced in Figure 12.9, should be explained by the assumption of domains, which are spontaneously magnetized but are not free to follow external fields in a reversible manner. The existence of such magnetic domains are verified experimentally in the 1950s. We will not discuss this aspect here. Interested readers should refer to the literature such as Ferromagnetic Domain Theory by Kittel and Galt, [7]. The thermodynamics of a ferromagnet, represented by Figure 12.10, will now be discussed in the remainder of the present section.

The original hypothesis of Weiss is that the internal field should be proportional to the magnetization. This appears to be reasonable because such effect is known in electromagnetism. Let us modify the r.h.s. of eq.(6.5) by making the substitution :

$$B \rightarrow B + \alpha I , \tag{8.1}$$

where α is a proportionality factor. We then get

$$I = n \mu \tanh [\mu \beta (B + \alpha I)]. \tag{8.2}$$

Here we dropped the suffix B on μ for the sake of simplicity.

We will postulate that eq.(8.2) represents the equation of state for a ferromagnet, which interrelates among the thermodynamic variables(I, B, T).

Let us first consider the spontaneous magnetization. Setting B equal to zero, we obtain from (8.2)

$$I = n\mu \tanh(\alpha \mu I/k_B T). \tag{8.3}$$

This equation can be solved graphically as follows. We plot tanh $(\alpha \mu I/k_B T) \equiv \tanh x$ and $I/n\mu = (k_B T/\alpha n \mu^2)x$ against x as indicated in Figure 12.11. For small T, the two curves

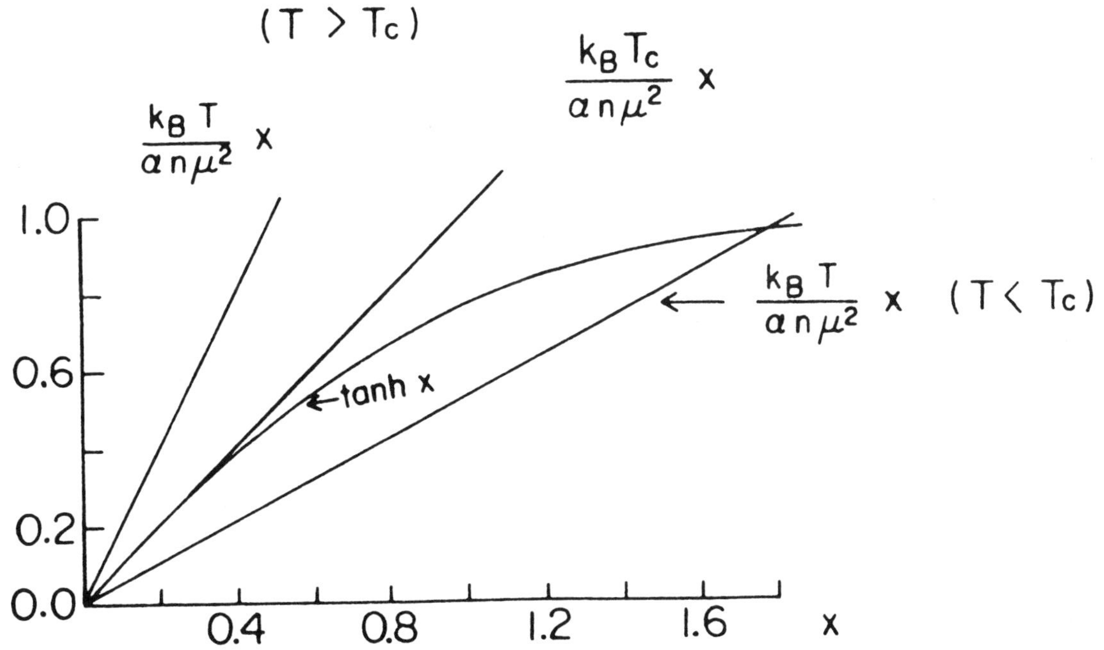

Fig.12.11 The graphical solutions of eq.(8.3).

intersect twice. For large T, the two curves intersect only once at the origin. There will be a temperature T_c at which the curves tanh x and $(k_B T/\alpha n \mu^2)x$ have the same slope at the origin. Since $\tanh x \cong x$ for $|x| \ll 1$, the slope of tanh x is unity. Equating this with $k_B T_c/\alpha n \mu^2$, we obtain

$$k_B T_c = \alpha n \mu^2. \tag{8.4}$$

In high temperature region where $T > T_c$, the only solution is I = 0, that is, there is no spontaneous magnetization. In low temperature region where $T < T_c$, a finite magnetization is possible, and the piece is ferromagnetic. The temperature T_c which separate these two regions, is called the Curie temperature.

In Figure 12.12 the spontaneous magnetization-versus-temperature curve predicted by (8.3) is shown together with experimental results for iron, cobalt, and nickel ; we see that the agreement is fairly good.

Let us look at the behavior of the magnetization near the critical temperature in more detail. Expanding tanh x in powers of x and retaining the first two nonvanishing terms, we have

$$\tanh x \simeq x - \frac{1}{3} x^3, \quad |x| \ll 1, \tag{8.5}$$

which is a good approximation for small x. Applying this formula to eq.(8.3), we obtain

$$\frac{I}{n\mu} = \frac{\alpha \mu I}{k_B T} - \frac{1}{3} \left(\frac{\alpha \mu I}{k_B T} \right)^3. \tag{8.6}$$

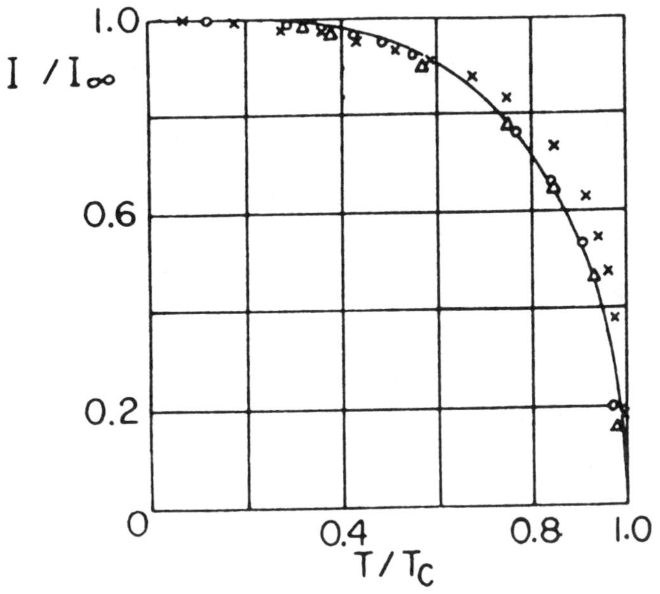

Fig. 12.12 Reduced spontaneous magnetization, I/I_∞ , versus reduced temperature T/T_C; the full line corresponds to the theoretical curve predicted from eq.(8.4) and experimental points are for iron (x), nickel (o) and (Δ), [after R. Becker [8]]

Below the critical temperature T_c, the magnetization I is finite. We can therefore divide eq.(8.6) by I , and obtain

$$\frac{1}{n\mu} = \frac{\alpha\mu}{k_B T} - \frac{1}{3}\left(\frac{\alpha\mu}{k_B T}\right)^3 I^2 . \tag{8.7}$$

We now look at the region, in which the temperature difference

$$\Delta T \equiv T_c - T \quad (>0) \tag{8.8}$$

is very small compared with the critical temperature T_c :

$$\Delta T \ll T_c \quad \text{or} \quad \frac{T_c - T}{T_c} \ll 1. \tag{8.9}$$

Solving eq.(8.7) for I^2 and using the relations (8.4) and (8.9), we obtain

$$I^2 = 3\left(\frac{k_B T}{\alpha\mu}\right)^3 \left[\frac{\alpha\mu}{k_B T} - \frac{1}{n\mu}\right]$$

$$= 3\left(\frac{k_B T}{\alpha\mu}\right)^3 \frac{\alpha n \mu^2 - k_B T}{n\mu k_B T}$$

$$= 3\left(\frac{k_B T}{\alpha\mu}\right)^3 \frac{1}{n\mu} \frac{T_c - T}{T} \qquad \text{[use of (8.4)]}$$

$$= 3\left(\frac{\alpha n \mu^2}{\alpha\mu}\right)^3 \frac{1}{n\mu} \frac{T_c - T}{T_c} \qquad \text{[use of (8.4); } T \simeq T_c\text{]}$$

$$= 3\, n^2 \mu^2 \frac{\Delta T}{T_c}. \qquad [\Delta T \equiv T_c - T] \tag{8.10}$$

Taking the square-root of this equation we obtain

$$I = \sqrt{3}\, n\mu \left(\frac{\Delta T}{T_c}\right)^{\frac{1}{2}}, \quad T_c - T \ll T_c . \tag{8.11}$$

We note that the magnetization I rises like $(\Delta T/T_c)^{\frac{1}{2}}$ with the power $\frac{1}{2}$ when ΔT is raised.

We will come back to this power law in later discussions. Above the Curie temperature T_c, the spontaneous magnetization vanishes. If a small magnetic field B is applied, the magnetization I becomes finite. If we use (8.2) and expand $\tanh[\mu(B+\alpha I)/k_B T]$ for small B and I, we obtain

$$I \simeq n\mu^2 \frac{1}{k_B T}(B+\alpha I) \quad \text{for small B and I.} \tag{8.12}$$

Solving this with respect to I, we get

$$I = \frac{n\mu^2 B}{k_B(T-T_c)} . \qquad \text{[use of (8.4)]} \tag{8.13}$$

This shows that the magnetization is proportional to B, and the susceptibility χ is given by

$$\boxed{\chi = \frac{n\mu^2}{k_B(T-T_c)} > 0.} \tag{8.14}$$

Since this χ is positive, the material is <u>paramagnetic above the critical temperature</u>. The susceptibility χ is inversely proportional to the temperature difference $T - T_c$. This behavior called the <u>Curie-Weiss law</u> is observed generally for ferromagnetic materials. Figure 12.13 represents a plot of the inverse of the susceptibility (χ^{-1}) against the absolute temperature (T) for a gadolinium (Gd, rare earth element) metal. The intercept of the straight line with the temperature axis gives $T_c' = 310$ K while the metal becomes ferromagnetic below 289 K. Thus, there exists a small but non-negligible discrepancy between theory

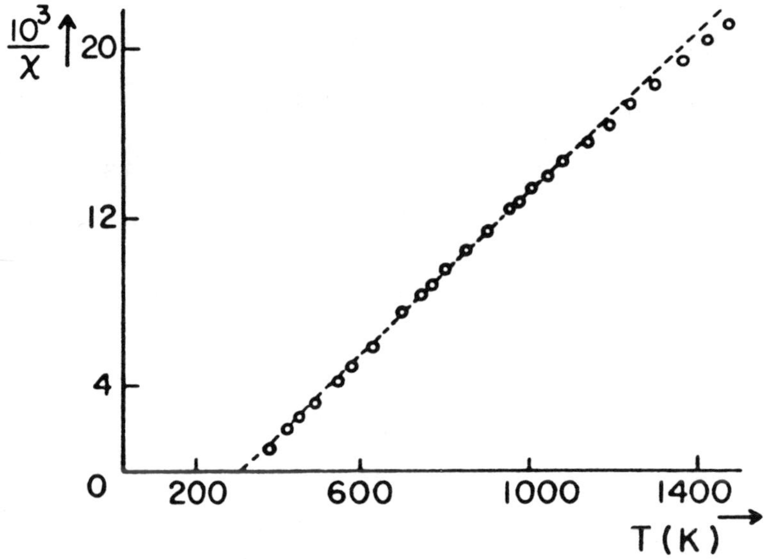

Fig. 12.13 The inverse of the susceptibility data for gadolinium metal above the Curie temperature is plotted against the absolute temperature T. The linear fit confirms the Curie-Weiss law (8.14). Experimental data are due to Arajs and Colvin [9].

and experiment.

In summary, Weiss' internal-field model appears to explain the general behavior of a ferromagnet very well. In the model, a phenomenological parameter α was introduced. What is the molecular basis for this parameter α ? This question will be studied in the following few sections.

12.9* The Ising Model. Solution of the One-Dimensional Model.

In 1925 Uhlenbeck and Goudsmit put forward the hypothesis that the electron has a spin whose direction is quantized such that it orients either parallel or anti-parallel to the applied magnetic field. In the same year Ising published an important model for a ferromagnet [10]. If an interaction between spins is introduced such that the parallel spin configuration has a lower energy than the anti-parallel configuration, then at sufficiently low temperatures the spins, placed on the lattice sites, would all be aligned, and such a model might provide a microscopic description of ferromagnetism.

The Hamiltonian for the Ising model can be written in the form

$$H = -J\sum_{(j,k)} \sigma_j\sigma_k - \mu B\sum_j \sigma_j \ , \qquad (9.1)$$

where J (>0 for a ferromagnet) represents the interaction energy between spins, called the <u>exchange energy</u>. The variables σ_j take either of two values ± 1:

$$\sigma_j = \begin{cases} +\ 1 \text{ for up spin} \\ -\ 1 \text{ for down spin} \end{cases} ; \qquad (9.2)$$

and the summation (j,k) extends over all <u>nearest neighbor pairs</u>. A criterion to test the model is whether it gives rise to the characterisitic pattern associated with ferromagnetism, and in particular whether a Curie temperature T_c can be found such that a finite spontaneous magnetism exists for a temperature below that temperature T_c.

Ising solved the problem exactly for the one-dimensional lattice, and found that the solution is analytic without any singularity. He inadvertently concluded that the model was not adequate for a ferromagnet. As it turns out, the phenomenon of phase transition depends on the dimensionality of the system. A number of later studies have shown that the

Ising Hamiltonian in two- and three-dimensions, do give rise to a phase transition.

In the present section, we will solve and discuss the one-dimensional Ising problem.

For mathematical convenience, see below, we add the term

$$-J\,\sigma_0\sigma_1 \tag{9.3}$$

to the one-dimensional form of the Hamiltonian H in (9.1) so that we have the new Hamiltonian

$$H' \equiv -\sum_{j=1}^{N} (J\,\sigma_{j-1} + \mu B)\sigma_j . \tag{9.4}$$

The prime which indicates the above modification will be dropped hereafter.

The partition function $Z(\beta,N)$ for the system is given by

$$Z(\beta,N) \equiv \mathrm{Tr}\{\exp(-\beta H)\} \equiv \sum_{\text{states}} \exp(-\beta H). \tag{9.5}$$

Since each state is specified by the set of σ's, we can write it as

$$Z(\beta,N) = \sum_{\sigma_1=\pm 1}\sum_{\sigma_2=\pm 1}\cdots\sum_{\sigma_N=\pm 1} e^{[-\beta H(\sigma_0,\sigma_1,\ldots,\sigma_N)]}. \tag{9.5a}$$

Clearly, there are 2^N distinct states. If we now assume that up and down spins correspond to right and up steps respectively, each state for the system can be represented by a path of the walker on the square lattice in Fig. 2.9, which was used earlier for the discussions of correlated walks. For example, the straight-line path from the origin to the site (N,0) corresponds to the state for which $\sigma_1 = \sigma_2 = \ldots = \sigma_N = 1$.

We now assert that the sum over states $\{\sigma_j\}$ is equivalent to the following sum over paths:

$$Z = \sum_{\text{paths}} \exp(-\beta H) , \tag{9.6}$$

where the statistical weights arising from the canonical distribution can be obtained by prescribing the step probabilities (p_1' , q_1', p_2', q_2') as follows:

$$\begin{bmatrix} p_1' & q_1' \\ p_2' & q_2' \end{bmatrix} = \begin{bmatrix} \exp[\beta(J+\mu B)] & \exp[\beta(-J-\mu B)] \\ \exp[\beta(-J+\mu B)] & \exp[\beta(J-\mu B)] \end{bmatrix} . \tag{9.7}$$

The assertion may be ascertained in an elementary manner by working it out explicitly for small N [Problem 9.1].

In order to complete this correspondence it was necessary to introduce the term (9.3) with σ_0 representing the state of the ghost spin 0, which by assumption carries no magnetic energy. [This addition will not change the macroscopic behavior of the system with a large number of spins as we will see later]. Let us assume that

$$\sigma_0 = 1 \tag{9.8}$$

corresponding to the choice that the zeroth spin points upward, and that the walker arrived at the origin with the right step.

Clearly the correlated walks defined here are almost identical to those described in §2.6. The only difference is that the step probabilities (p_1', q_1', p_2', q_2') are unnormalized as can be seen from (9.7). Except for the normalization, the dynamics of the walker including the recurrence relations (2.6.3), is the same, and can be solved by the same techniques.

The partition function Z given in the form (9.6) can simply be expressed by

$$Z_N = \sum_{X=0}^{N} Q(X,N) = \varphi(\xi=1,N), \tag{9.9}$$

(which is not equal to unity since the step probabilities are not normalized.)

The grand partition function defined by [see (7.12.1)]

$$\Xi \equiv \sum_{N=0}^{\infty} \nu^N Z_N$$

$$\equiv \sum_{N=0}^{\infty} \nu^N \sum_{X=0}^{N} Q(X,N) \qquad (9.10)$$

can be expressed in terms of the generating function ψ which is defined through (2.6.14). This Ξ can be written down from (2.6.21) as follows :

$$\Xi(\nu) = \psi(\xi=1,\nu) = \frac{1+\nu(q_1'-q_2')}{(1-\nu q_2')(1-\nu p_1') - \nu^2 p_2' q_1'} . \qquad (9.11)$$

Hereafter, primes on p's and q's will be dropped.

In the field-free case (B=0), we obtain from (9.7)

$$p_1, q_2 \rightarrow \exp(\beta J) \equiv x$$

$$p_2, q_1 \rightarrow \exp(-\beta J) = x^{-1}. \qquad (9.12)$$

Then, eq.(9.11) is reduced to [Problem 9.2]

$$[\Xi(\nu)]_{B=0} = [1-\nu(x+x^{-1})]^{-1}$$

$$= [1-2\nu\cosh(\beta J)]^{-1} \qquad [\text{ use of (9.12) }]$$

$$= \sum_{k=0}^{\infty} \nu^k [2\cosh(\beta J)]^k. \qquad (9.13)$$

Comparing this expression with (9.10), we obtain

$$Z_N = [2 \cosh (\beta J)]^N. \tag{9.14}$$

The free energy per spin in the bulk limit is defined by

$$f \equiv \lim_{N\to\infty} \frac{F}{N}$$

$$\equiv \lim_{N\to\infty} -\frac{1}{N} k_B T \ln Z_N. \tag{9.15}$$

Introducing (9.14) here, we obtain

$$f = - k_B T \ln [2 \cosh(J/k_B T)]. \tag{9.16}$$

Since this is an analytic function of T, we can expect all thermodynamic functions to be analytic.

In particular, the average energy per spin, $\bar{e}$, is given by

$$\bar{e} \equiv \lim_{N\to\infty} \frac{1}{N} [- \frac{\partial}{\partial\beta} \ln Z_N] = \frac{\partial}{\partial\beta} [\beta f] \quad \text{[use of (9.15)]}$$

$$= - \frac{\partial}{\partial\beta} \ln [2 \cosh (J\beta)] \quad \text{[use of (9.16)]}$$

$$= - J \tanh (\beta J) \equiv - J \tanh (J/k_B T). \tag{9.17}$$

Differentiating this with respect to T, we obtain the heat capacity per spin :

$$c \equiv \frac{\partial\bar{e}}{\partial T} = \frac{J^2}{k_B T^2} \operatorname{sech}^2 (J/k_B T). \tag{9.18}$$

The behavior of the heat capacity versus temperature is shown in Figure 12.14. Note that the heat capacity is a smooth function of the temperature T and exhibits a maximum at a temperature of the order of J/k_B.

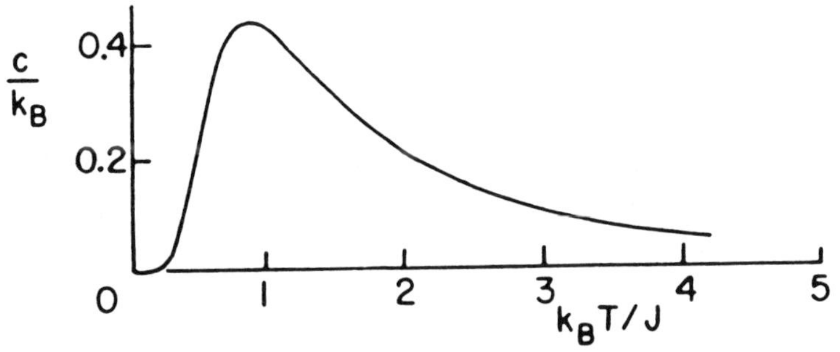

Fig. 12.14 The heat capacity per spin for a linear Ising model exhibits a smooth maximum at a temperature of the order of J/k_B.

We now go back to the case of a finite magnetic field. We take expression (9.11) and expand it in powers of ν : [Problem 9.3]

$$\Xi = [1 + \nu(q_1-q_2)] / [(p_1 q_2 - p_2 q_1)(\nu-\nu_1)(\nu-\nu_2)]$$

$$= \frac{1 + \nu(q_1 - q_2)}{(p_1q_2 - p_2q_1)} \frac{1}{\nu_2 - \nu_1} \left(\frac{1}{\nu_1 - \nu} - \frac{1}{\nu_2 - \nu} \right)$$

$$= \frac{1 + \nu(q_1-q_2)}{(p_1q_2-p_2q_1)\ (\nu_2-\nu_1)\nu_1} \sum_{N=0}^{\infty} \left(\frac{\nu}{\nu_1}\right)^N [1-\left(\frac{\nu_1}{\nu_2}\right)^{N+1}], \qquad (9.19)$$

where ν_1 and ν_2 are the two roots of the equation:

$$(1 - \nu q_2)\ (1 - \nu p_1) - \nu^2\ p_2q_1$$

$$= (p_1q_2 - p_2q_1)\ (\nu - \nu_1)\ (\nu-\nu_2) = 0. \qquad (9.20)$$

Written explicitly, they are given by [Problem 9.3]

$$\frac{1}{2(p_1q_2 - p_2q_1)} [p_1+q_2 \pm \sqrt{(p_1 + q_2)^2 - 4(p_1q_2 - p_2q_1)^2}\ ]$$

$$= \frac{e^j}{2(e^{2j} - e^{-2j})} [e^b + e^{-b} \pm \sqrt{(e^b - e^{-b})^2 + 4\ e^{-4j}}\ ],$$

[use of(9.7)] (9.21)

where

$$j \equiv J / k_B T, \qquad b \equiv \mu B / k_B T. \qquad (9.22)$$

We observe that both roots are positive. Let us denote the minimum root by ν_1, and the other root by ν_2 :

$$\nu_1 \equiv \frac{e^j}{2(e^{2j}-e^{-2j})} [e^b + e^{-b} - \sqrt{(e^b - e^{-b})^2 + 4 e^{-4j}}] < \nu_2. \tag{9.23}$$

From the expansion(9.19), the partition function Z_N for large N is obtained as follows :

$$Z_N \simeq \frac{1}{(\nu_2-\nu_1)(p_1q_2-p_2q_1)\nu_1^N} \left(\frac{1}{\nu_1} + (q_1 - q_2) \right). \tag{9.24}$$

Using this result, we obtain

$$f \equiv \lim_{N\to\infty} [-\frac{1}{N} k_B T \ln Z_N]$$

$$= - k_B T \lim_{N\to\infty} \frac{1}{N} [\ln \nu_1^{-N} + (\text{non-contributing terms in the limit})]$$

or

$$\boxed{f = k_B T \ln \nu_1 .} \tag{9.25}$$

It is interesting to note that, in the bulk-limit, the free energy is determined by the minimum root ν_1 alone.

If we go back and examine the derivation of expression (9.11) or (2.6.21) in §2.6, we can see that the denominator $(1 -\nu q_2)(1 -\nu p_1) - \nu^2 p_2 q_1$ arises from the form of the recurrence equations (2.6.11). The initial condition (2. 6.12), which corresponds to the end condition (9.8), determines the form of the numerator. We therefore see that <u>the bulk-limit behavior does not depend on the end condition</u> employed in the model. This means that the extra term (9.3) added for convenience does not affect the bulk-limit behavior.

The average magnetic moment of the system is defined by

$$\langle \mu \sum_k \sigma_k \rangle \equiv \frac{Tr\{\mu \sum_k \sigma_k \; e^{-\beta H}\}}{Tr\{e^{-\beta H}\}}. \tag{9.26}$$

Using the explicit form (9.4) of the Hamiltonian H, we can express this average moment in terms of the partition function Z as follows:

$$\langle \mu \sum_k \sigma_k \rangle = \frac{\partial}{\beta \partial B} \ln Tr\{ \exp[\beta \sum_k (J\sigma_{k-1} + \mu B)\sigma_k]\}$$

$$= \frac{\partial}{\beta \partial B} \ln Z_N. \qquad \text{[use of (9.5)]} \tag{9.27}$$

By using (9.23), (9.25) and (9.27), we can calculate the magnetic moment per spin, μm, in the following manner:

$$\mu m \equiv \lim_{N\to\infty} \frac{1}{N} \langle \mu \sum_k \sigma_k \rangle = \frac{\partial}{\beta \partial B}(-\beta f) \qquad \text{[use of (9.15) and (9.27)]}$$

$$= -\frac{\partial}{\partial B} \ln \nu_1 \qquad \text{[use of (9.25)]}$$

$$= \mu \frac{(e^b - e^{-b})}{\sqrt{(e^b - e^{-b})^2 + 4e^{-4j}}}. \qquad \text{[use of (9.23)]} \tag{9.28}$$

In zero field limit ($b \equiv \mu B/k_B T = 0$), the spontaneous magnetization vanishes for all temperature :

$$[\mu m]_{b=0} = 0. \tag{9.29}$$

Expanding m (B) in powers of B, we obtain, from (9.28),

$$m = \frac{\mu}{k_B T} e^{2J/k_B T} B + O(B^2). \tag{9.30}$$

We then obtain the susceptibility χ per spin as follows :

$$\chi = \frac{\mu^2}{k_B T} e^{2J/k_B T}. \tag{9.31}$$

It is interesting to note that the temperature dependence of the susceptibility χ is a little different from the Curie's law due to the exponential factor $e^{2J/k_B T}$.

Problem 9.1 We wish to show that the partition function $Z(\beta,N)$ defined by (9.5) can be obtained from the sum over paths, as defined by (9.6) and (9.7).

(a) Write out explicitly the partition function Z from both for the cases of N = 1, 2 and 3.

(b) Establish the equivalence for a general N.

Problem 9.2 Derive (9.13) from (9.11) by assuming the limits (9.12).

Problem 9.3 Verify (9.19) and (9.21).

12.10 Bragg-Williams' Approximation

In the present section we will present an approximate calculation of the thermodynamic properties of an Ising model. The results of this theory can clarify greatly the microscopic foundation of the internal field model discussed in §12.8.

For definiteness, let us consider a system of spins located on a square lattice. We denote up spins by + and down spins by −. A typical configuration is shown in Figure 12.15. The numbers of the up spins and down spins will be

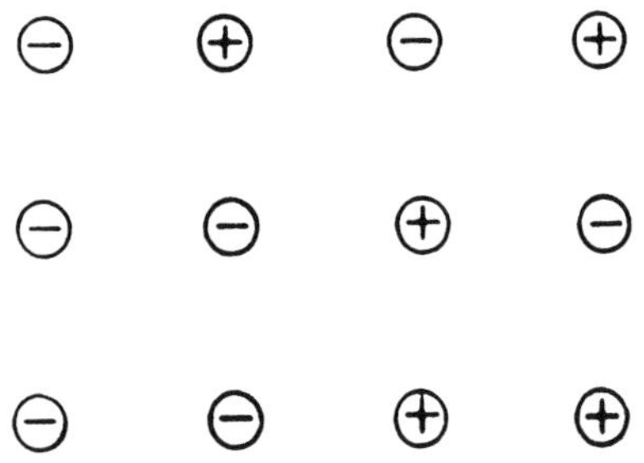

Fig. 12.15 A typical configuration of the spins on the square lattice.

denoted by N_+ and N_-, respectively. The sum of these numbers equals the total number N of lattice points :

$$N_+ + N_- = N. \tag{10.1}$$

The ratios

$$c_+ \equiv \frac{N_+}{N} \text{ and } c_- \equiv \frac{N_-}{N} \tag{10.2}$$

can be regarded as the average concentrations of up spins and down spins, respectively. From (10.1), they satisfy the relation

$$c_+ + c_- = 1. \tag{10.3}$$

Let us fix the numbers of up and down spins. Within this restriction, there exist in general many possible configurations. We wish to find an approximate energy for these configurations. Take an arbitrary site. This site may carry an up spin or down spin. The probabilities of the first and second cases are given by c_+ and c_- . Let us assume the first case, in which the site carries an up spin. We now consider one of its nearest-neighbor sites, which may carry an up spin or a down spin. This will occur with the probabilities c_+ or c_- if we neglect the correlation. In the former case, the spin pair (+, +) will generate the interaction energy -J. In the latter case, the spin pair (+,-) will generate J. Therefore, the pair of spins on the neighboring sites is likely to contribute

$$c_+ (-J\, c_+ + J\, c_-) = -J\, c_+ (c_+ - c_-) \tag{10.4}$$

to the average energy. In the second case, we have a down spin at the chosen site. By similar arguments, the pair of spins on the neighboring sites is likely to contribute

$$c_- (J c_+ - J c_-) = J c_- (c_+ - c_-). \tag{10.5}$$

Adding (10.4) and (10.5), we have $-J (c_+ - c_-)^2$. The same arguments should hold for any nearest neighbor. Therefore, the chosen site and its nearest neighbors will contribute $-zJ (c_+ - c_-)^2$, where z denotes the number of the nearest neighbors. Clearly for the square lattice exhibited in Figure 12.15, $z = 4$. Since the reference site could be any of N lattice sites, we might multiply this energy by N. But by doing so, we count the contribution of each pair twice. Therefore, we should multiply it by $\frac{1}{2}$ N and obtain

$$-\frac{1}{2} NzJ (c_+ - c_-)^2 = E. \tag{10.6}$$

This should represent the approximate energy of the system with a fixed set of concentrations (c_+, c_-). We note that when $c_+ = 1$ and $c_- = 0$, the energy is negative and lowest. The energy E increases as the concentration c_+ is reduced from 1 to 1/2; it reaches the maximum value 0 when $c_+ = c_- = 1/2$.

The number of possible configurations depends on c_+, and hence on c_-. This number grows as c_+ is reduced from unity. Conseqently the entropy S of the system should increase as c_+ is reduced. We now postulate that the entropy S is given by the entropy of mixing :

$$S = -Nk_B [c_+ \ln c_+ + c_- \ln c_-], \tag{10.7}$$

which we discussed earlier in § 7.9. [See eq.(7.9.13)]

The free energy F of the system is defined by

$$F \equiv E - TS. \tag{10.8}$$

This energy F must be at minimum in thermal equilibrium : (See § 6.7)

$$\delta F = 0. \tag{10.9}$$

By using this extremum condition, we can determine the concentration pair (c_+, c_-) as functions of temperature as we will see presently.

The total magnetic moment M can be expressed in terms of the numbers of up and down spins as follows:

$$M = N_+ \mu - N_- \mu.$$

$$= N \mu (c_+ - c_-). \quad \text{[use of (10.2)]} \tag{10.10}$$

We represent the average magnetic moment per site, M/N, by $m \mu$:

$$M / N \equiv m \mu . \tag{10.11}$$

From the last two equations , we obtain

$$m = c_+ - c_-$$

$$= 2 c_+ - 1, \quad \text{[use of (10.3)]} \tag{10.12}$$

which shows that the number m is a unique function of the up-spin concentration c_+.

Let us now express the free energy F in terms of m. Using (10.3), (10.6), (10.7) and (10.8), we obtain [Problem 10.1]

$$F \equiv E - TS$$

$$= -\frac{1}{2}\ zJN\ m^2 + Nk_BT\{\ \frac{1}{2}(1+m)\ \ell n[\frac{1}{2}(1+m)]$$

$$+\frac{1}{2}(1-m)\ \ell n[\frac{1}{2}(1-m)]\}. \tag{10.13}$$

The extremum condition (10.9) can be represented by

$$\frac{dF}{dm} = 0\ . \tag{10.14}$$

Introducing F from (10.13), we then obtain [Problem 10.1]

$$-zJNm + Nk_BT\{\frac{1}{2}\ \ell n[\frac{1}{2}(1+m)] - \frac{1}{2}\ell n[\frac{1}{2}(1-m)] + \frac{1}{2} - \frac{1}{2}\}$$

$$= \ -zJNm + \tfrac{1}{2}Nk_BT\ \ell n\ [(1+m)\ /\ (1-m)] = 0. \tag{10.15}$$

Solving this for (1 - m) / (1 + m), we have

$$\frac{1-m}{1+m} = \exp\ [\ \frac{-2z\ Jm}{k_BT}\ ]. \tag{10.16}$$

It can be shown [Problem 10.2] that

$$\frac{1-y}{1+y} = e^{-2x} \tag{10.17}$$

is equivalent to

$$y = \tanh x. \tag{10.18}$$

Using this equivalence, we can rewrite equation (10.16) as

$$m = \tanh\left[\frac{zJm}{k_B T}\right]. \tag{10.19}$$

This equation has the same form as eq.(8.3). Comparison between the two shows that

$$\alpha = \frac{zJ}{n\mu^2}. \tag{10.20}$$

Thus, the parameter α which was introduced in the internal field model, is expressed in microscopic terms. In particular, note that the parameter α, which brings out an alignment of spins at low temperatures, is proportional to the exchange energy J.

Using (8.4) and (10.20), we obtain

$$k_B T_c = \alpha n \mu^2 = \frac{zJ}{n\mu^2} n\mu^2 = zJ, \tag{10.21}$$

which relates the Curie temperature T_c with the exchange energy J in a simple manner.

The statistical mechanical calculations carried out in the present section are approximate ones but they can shed light on the important general feature of the phase transition. At low temperatures the spins tend to align themselves to gain exchange energy. At high temperature, the spins orient themselves at random. The transition from the ferromagnetic to random configuration occurs abruptly at the critical temperature T_c.

We can also look at the phase transition in terms of the Helmholtz free energy as given in eq. (10.8) : $F \equiv E - TS$. At very low temperatures, where T is small, the second term -TS is a minor addition to the first term E. In the low temperature limit, the energy E alone will determine the equilibrium state of the system. In contrast, at higher temperatures the term -TS becomes more important and favors the random orientation of spins which makes the entropy S large. At temperatures near the Curie temperature T_c, both terms E and -TS for the free energy F are comparable and important. To predict the precise behavior of the phase transition, more careful calculations of the free energy [or the partition function] must be done [see further discussions in §12.11].

Problem 10.1 Derive (10.13), and (10.15).

Problem 10.2 Show the equivalence between (10.17) and (10.18).

12.11* More About the Ising Model

Let us take an L x L' lattice shown in Figure 12.16a. The lattice points will be numbered from the left-hand corner as indicated. Let us now imagine a helical lattice of a circumference of length L, which is shown in b. We assume that the spin at kL + 1, k $\geqq$ 1, interacts not only with the spin at kL + 2, (k-1)L + 1, and (k+1)L + 1 but also with the spin at kL.

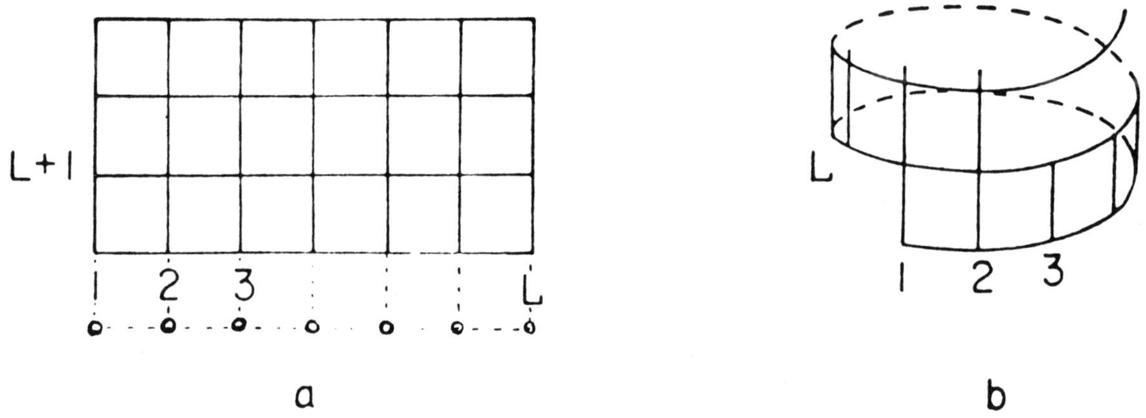

Fig. 12.16 A square lattice with ghost spin sites o A helical lattice

The Hamiltonian H' of N(=LL') spin system is then given by

$$H' = -J \sum_{M=2}^{N} \sigma_{M-1}\sigma_M - J' \sum_{M=L+1}^{N} \sigma_{M-L}\,\sigma_M - \mu B \sum_{M=1}^{N} \sigma_M, \qquad (11.1)$$

where J and J' represent the exchange energies along horizontal and vertical directions, respectively. This Hamiltonian H' differs from the Hamiltonian of the N spins on the rectangular lattice in Figure 12.16a, but the difference suffered by the boundary spins is expected to become negligibly small in the bulk limit in which both L and L' tend to infinity.

For later convenience, we add the energy

$$-J\,\sigma_0\sigma_1 \;-\; J' \sum_{M=1}^{L} \sigma_{M-L}\,\sigma_M \tag{11.2}$$

so that we have a new Hamiltonian

$$H \;\equiv\; H' - J\,\sigma_0\,\sigma_1 - J' \sum_{M=1}^{L} \sigma_{M-L}\,\sigma_M$$

$$= -\sum_{M=1}^{N} (J'\,\sigma_{M-L} + J\,\sigma_{M-1} + \mu B)\sigma_M . \tag{11.3}$$

The energy in (11.2) may be regarded as the energy of the interaction with fictitious spins (ghost spins) located below the first row of the lattice, see Figure 12.16a ; these ghost spins, by assumption, carry no energy of the interaction with the external field.

The partition function for the system:

$$Z(\beta,N) \;\equiv\; \sum_{\text{states}} \exp\,(-\beta H) \tag{11.4}$$

can again be re-expressed as the sum over paths on the auxiliary lattice in Figure 2.9:

$$Z \;=\; \sum_{\text{paths}} \exp(-\beta H) \tag{11.5}$$

just as in the one-dimensional case.

The statistical weights arising from the canonical distribution law, exp $(-\beta H)$, are obtained from (11.3); the M-th spin should carry the weight

$$\exp[(j'\sigma_{M-L} + j\,\sigma_{M-1} + b)\sigma_M] \equiv g(\sigma_{M-L}, \sigma_{M-1}, \sigma_M), \tag{11.6}$$

where

$$j' \equiv \beta J', \quad j \equiv \beta J, \quad b \equiv \beta\mu B. \tag{11.7}$$

When the path of the walker on the auxiliary lattice is given, the weight can be obtained by looking at the (M-L)-th and M-th steps. If the M-th step is the right step corresponding to $\sigma_M = 1$, the weights will be characterized by

$$\begin{pmatrix} a_{11} & a_{12} \\ a_{21} & a_{22} \end{pmatrix} \equiv \begin{pmatrix} g(1,1,1) & g(1,-1,1) \\ g(-1,1,1) & g(-1,-1,1) \end{pmatrix} = \begin{pmatrix} e^{j'+j+b} & e^{j'-j+b} \\ e^{j-j'+b} & e^{-j-j'+b} \end{pmatrix} ; \tag{11.8}$$

these weights depend on the past steps of the walker, more explicitly on the last and the L-th last steps. If the M-th step is upward corresponding to $\sigma_M = -1$, the weights are given by

$$\begin{pmatrix} b_{11} & b_{22} \\ b_{21} & b_{22} \end{pmatrix} \equiv \begin{pmatrix} g(1,1,-1) & g(1,-1,-1) \\ g(-1,1,-1) & g(-1,-1,-1) \end{pmatrix} = \begin{pmatrix} e^{-j'-j-b} & e^{-j'+j-b} \\ e^{j'-j-b} & e^{j'+j-b} \end{pmatrix} . \tag{11.8a}$$

Note that all the weights, a's and b's, are non-negative. Except for the difference in the statistical weight, the partition function Z can be formulated in the same manner as for the one-dimensional system which was discussed in §12.9. In particular, we obtain

$$Z = \sum_{X=0}^{N} P(X,N), \tag{11.9}$$

where P(X,N) is the sum of the products of accumulated weights over all paths from the origin to the site (X,N).

For the one-dimensional case, it is possible to evaluate expression (11.9) exactly as we saw in §12.9. In the present case, if we introduce the probabilities of arrival with the last L steps specified,

$$P_{j_{N-L}\cdots j_{N-1}\, j_N}(X,N) \equiv P_\alpha, \tag{11.10}$$

we can show that P_α's obey a closed set of linear equations. The number of unknowns, however, is 2^L, and the equations are difficult to solve for large L. In the present section we will discuss an approximate solution.

Let us rewrite the Hamiltonian H in (11.3) as follows :

$$H = -J \sum_{M=1}^{N} \sigma_{M-1}\, \sigma_M - \mu \sum_{M=1}^{N} B_M\, \sigma_M , \tag{11.11}$$

where

$$B_M \equiv B + \frac{J'}{\mu}\, \sigma_{M-L} . \tag{11.12}$$

The quantity B_M can be regarded as the effective inhomogeneous field acting on the spin M. Note that it depends on the state of the spin M-L, σ_{M-L}.

Let us now assume, with a spirit of the Bragg-Williams treatment, that the spins along the vertical have no correlation and the spin M-L points up with probability p and down with probability q. The pair (p,q) are normalized such that

$$p + q = 1. \tag{11.13}$$

With this assumption, the weights can be obtained simply adding the contributions corresponding to the two possible cases. For example, if the M-th and (M-1)-th spins are both up (right steps), then the step probability is given by $p\, g(1,1,1) + q\, g(-1,1,1) = p\, e^{j'+j+b} + q\, e^{-j'+j+b} \equiv p_1$, where (11.8) is used. Notice that we could obtain this step probability by looking at spins along the horizontal just as for the case of a linear chain. Applying similar arguments, we may represent the weights (step probabilities) by

$$\begin{bmatrix} p_1 & q_1 \\ p_2 & q_2 \end{bmatrix} = \begin{bmatrix} e^{j+b}(p\, e^{j'} + q\, e^{-j'}) & e^{-j-b}(p\, e^{-j'} + q\, e^{j'}) \\ e^{-j+b}(p\, e^{j'} + q\, e^{-j'}) & e^{j-b}(p\, e^{-j'} + q\, e^{j'}) \end{bmatrix} . \tag{11.14}$$

We can now write down the grand partition function Ξ from (9.11) as follows :

$$\Xi(\nu) = \frac{1 + \nu(q_1 - q_2)}{(1 - \nu q_2)(1 - \nu p_1) - \nu^2 p_2 q_1} . \tag{11.15}$$

This result contains the parameter (p,q), which may be determined from the <u>self-consistency condition</u> such that

$$p = \frac{\text{average number of up spins}}{\text{total number of spins}} \equiv \frac{\langle N_+ \rangle}{N}$$

$$q = \frac{\text{average number of down spins}}{\text{total number of spins}} \equiv \frac{\langle N_- \rangle}{N} . \qquad (11.16)$$

The free energy per spin in the bulk limit, f, is defined by

$$f \equiv \lim_{L\to\infty} \lim_{L'\to\infty} - [\frac{1}{\beta LL'}] \ell n \, Z_N. \qquad (11.17)$$

Following the same procedures as described in §12.9, we then obtain

$$f = \beta^{-1} \ell n \, \nu_1, \qquad (11.18)$$

where ν_1 is the minimum root of the equation

$$\nu^2(p_1 q_2 - p_2 q_1) - \nu(p_1 + q_2) + 1 = 0, \qquad (11.19)$$

the ℓ. h. s. of which is the denominator appearing in eq.(11.15).

Explicitly, the minimum root ν_1 is given by [Problem 11.2]

$$\nu_1 = \{ A - \sqrt{A^2 - B} \} / B, \qquad (11.20)$$

where

$$A \equiv e^j (\cosh j' \text{ cpsh } b) + m \sinh j' \sinh b \qquad (11.21)$$

$$B \equiv 2 \sinh (2j) (\cosh^2 j' - m^2 \sinh^2 j') \qquad (11.22)$$

$$m = p-q. \qquad [\text{use of } (10.12)] \qquad (11.23)$$

The average magnetization per spin can be expressed as follows:

$$\lim_{N\to\infty} \frac{1}{N} \langle \mu \sum_M \sigma_M \rangle = \mu \frac{\langle N_+ \rangle - \langle N_- \rangle}{N}$$

$$= \mu (p - q) = \mu m. \quad \text{[use of (11.16) and (11.23)]} \tag{11.24}$$

On the other hand, the average magnetization $m \mu$ is connected with the root ν_1 by the relation (9.28) :

$$m = - \frac{\partial}{\mu \partial B} \ln \nu_1 = - \frac{\partial}{\partial b} \ln \nu_1 . \tag{11.25}$$

Using eqs. (11.20) to (11.25), we then obtain [Problem 11.2]

$$m = e^j [\cosh j' \sinh b + m \sinh j' \cosh b] \times$$

$$\{e^{2j} [\cosh j' \cosh b + m \sinh j' \sinh b]^2$$

$$- 2 \sinh (2j) [\cosh^2 j' - m^2 \sinh^2 j']\}^{-\frac{1}{2}} . \tag{11.26}$$

Let us now study the case of no applied field (b=0). In this case, the last equation is reduced to

$$m = 0 \tag{11.27}$$

or

$$e^{2j} = 2 \sinh (2j) (\cosh^2 j' - m^2 \sinh^2 j'). \tag{11.28}$$

For the sake of simplicity, let us consider the case of isotropic interaction ($j'=j$). Solving eq.(11.28), we obtain [Problem 11.3]

$$m = \coth(j)\left[1 - \frac{e^{2j}}{4\sinh(j)\cosh^3(j)}\right]^{\frac{1}{2}}. \tag{11.29}$$

The quantity inside the square brackets should vanish at the critical temperature T_c; the value of T_c so obtained, is given by [Problem 11.3]

$$k_B T_c = \frac{2}{\ln(\sqrt{2}+1)} J = 2.269\, J, \tag{11.30}$$

Below the critical temperature, the spontaneous magnetization m is finite. The reduced magnetization computed numerically from (11.29) is shown in a dotted line in Fig.12.17. The full line corresponds to the exact solution

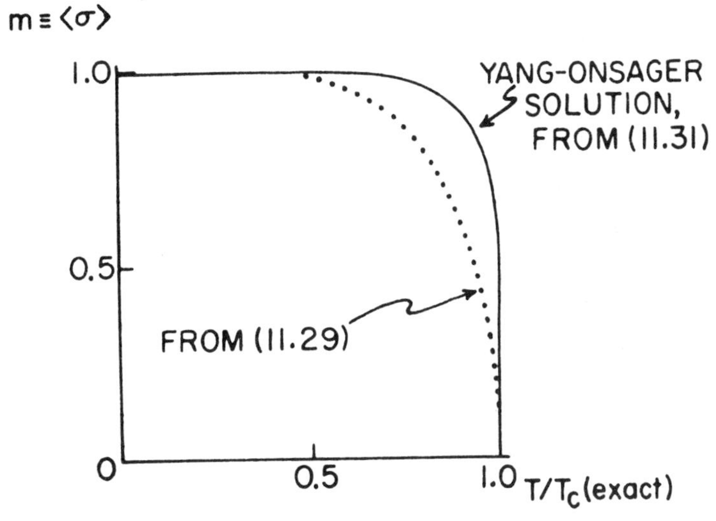

Fig. 12.17 Spontaneous magnetization against reduced temperature T/T_c. The exact Yang-Onsager solution (11.31) has a sharper rise characteristic of the critical power 1/8 compared with the approximation solution (11.29) with the critical power 1/2.

obtained by Onsager and Yang [11], which is represented by

$$m = \left[1 - \frac{1}{\sinh^4(2j)} \right]^{1/8}, \quad T < T_c, \tag{11.31}$$

which indicates that the magnetization curve falls off more steeply near the critical point T_c.

The internal energy and heat capacity per spin can be calculated from

$$u \equiv \lim_{N\to\infty} \left[-\frac{1}{N} \frac{\partial}{\partial\beta} \ln Z \right] \tag{11.32}$$

$$= \frac{\partial}{\partial\beta} \ln (\beta f) = \frac{\partial}{\partial\beta} \ln \nu_1 \quad \text{[use of (9.25)]}$$

$$c = \frac{\partial u}{\partial T} . \tag{11.33}$$

Using (11.20) - (11.23) and performing differentiations, we obtain [Problem 11.4]

$$u / J = \begin{cases} \dfrac{e^{-3j}}{\sinh j \sinh(2j)} \left[\cosh(4j) + 2\, e^{-j} \cosh j \right] - \dfrac{e^{j}}{\sinh j} & T < T_c \\ - 2 \tanh j & T > T_c \end{cases} \tag{11.34}$$

$$
c/k_B = \begin{cases} j^2 \dfrac{e^{-3j}}{\sinh j \, \sinh(2j)} \{ [3 + \coth j + 2\coth(2j)][\cosh(4j) + 2e^{-j}\cosh j\,] \\ \qquad - 4\sinh(4j) - 2e^{-j}\cosh j + 2e^{-j}\sinh j \} \\ \quad + j^2 \dfrac{e^{j}}{\sinh j}(1 - \coth j) & \text{if} \quad T < T_c \\ \dfrac{2j^2}{\cosh^2 j} & \text{if} \quad T > T_c . \end{cases} \tag{11.35}
$$

The heat capacity versus temperature computed from (11.35) is shown in dotted lines in Fig. 12.18. For comparison, the exact solution due to Onsager [12], is drawn in solid lines, which exhibits an extremely sharp logarithmic singularity at the critical point.

Onsager's work on the two-dimensional Ising model is regarded as a clasic masterpiece in modern mathematical physics. The mathematical techniques employed far exceed the level of the present text. Interested reader are recommended to look at an excellent introduction to the subject by Huang in his book on Statistical Mechanics [13] and the references cited there.

The approximate calculations presented in this section generate the critical temperature exactly. But the detailed behavior near the critical temperature is quite different. If we apply the Bragg-Williams approximation to the square lattice, we obtain the result shown in broken line, which has the critical temperature higher than the exact value, and which has vanishing heat capacity on the high temperature side. We may characterize the present model as something between the Braggs-Williams approximation and the exact solution.

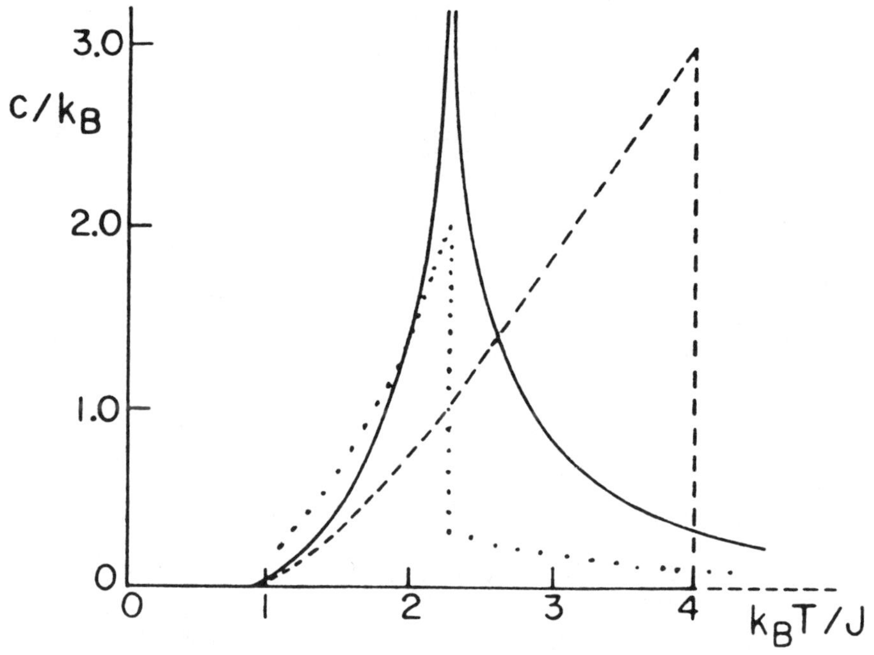

Fig.12.18 The heat capacity computed from (11.35) for the two-dimensional Ising model is shown in dotted lines and the exact solution due to Onsager in full lines. The Bragg-Williams approximation is represented by broken lines.

If we make the exchange energy J' along the vertical direction vanishingly small, we can reduce the present model to that of the linear chain, which undergoes no phase transition as we saw in § 12.9*. In this limiting process, the critical temperature T_c approaches zero. This behavior can correctly be described from eq.(11.26) [Problem 11.5]. This is significant since the Braggs-Williams approximation predicts a phase transition with $T_c = z\ J/k_B$ regardless of the dimension of the system.

The bahavior of the thermodynamic system near the critical temperature is singular. To correctly describe the "critical" behavior analytically is quite difficult. Recently a large number of theoretical and experimental works have been carried out on the critical behavior for many different materials [17] . The main findings of these works include

(a) The dimensionality

The phase transition depends on the dimension of the system. As we have already seen, the one-dimensional Ising model does not undergo a phase transition. This feature turns out to arise not from the particular form of the Hamiltonian but from the dimension of the system.

(b) The correlation at large distances .

As the system approaches the critical point, it exhibits correlation and fluctuations at large (macroscopic) distances. In fact, the fluid at the critical point with its large density fluctuations, scatters light to a great degree and looks milky-white. [This is called the critical opalescence.]

(c) The universality

The heat capacity, spontaneous magnetization and other thermodynamic properties diverge to infinity or go to zero following certain power laws as the system reaches the critical point. They do so independently of any specific nature of interaction. Onsager-Yang's calculations for the two-dimensional Ising model indicate that the spontaneous magnetization rises like $(T_c - T)^{1/8}$ at the immediate vicinity of the critical point. [See (11.31)]. The universality means that this power law holds for *any* two-dimensional ferromagnet. Furthermore, the critical exponents [like 1/8 for the spontaneous magnetization for a two-dimensional ferromagnet] for different thermodynamic properties satisfy a simple algebraic relation.

More about these and other fascinating recent discoveries for the critical behavior of the phase transition can be found in the growing references [17] cited earlier. In spite of all theoretician's efforts, the analytical solution of the three-dimensional Ising model has not been found up to the present time.

12.12.* Conformation of Polymers in Dilute Solution

A polymer molecule such as polyethylene is a giant molecule of chemical composition (CH_2 - CHR $)_n$ with a large number n (~ 10^5) of repetitive units, where R represents a radical ; for example R = H for polyethylene and R = CH_3 for polypropylene. Polymers in dilute solution may take various physical shapes, which are technically called <u>conformations</u>. Disregarding the <u>molecular configuration</u>, which refers to the chemical composition and the structural arrangement of the molecule, we may look at polymers as something like soft noodles in hot water as indicated in Figure 12.19.

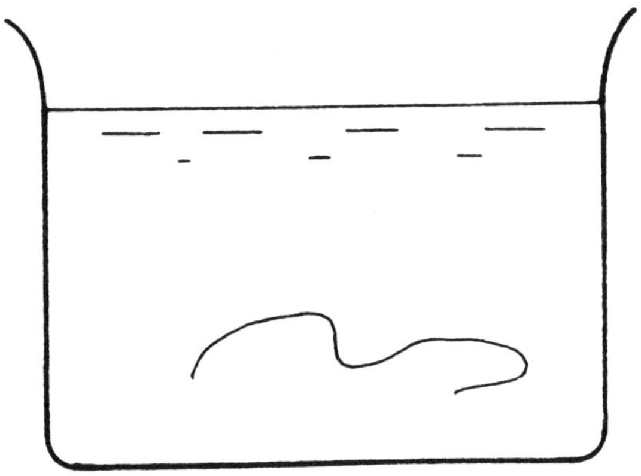

Fig. 12.19 The conformation of a simple polymer of large molecular weight in solution may be pictured as a soft noodle in hot water.

It is known that successive monomer units of a polymer cannot be bent at large angles, say, 180° without breaking the backbone chain. Such restriction on conformation, is called <u>excluded volume effect</u>.

In the following we will present a theoretical model of polymer conformation, which takes account of the short-range correlation of the backbone in a simple manner.

The basic model is as follows : A model polymer of a large number of segments is laid down on a simple cubic lattice with segment-segment correlation such that the polymer may be stretched out in the same direction as that of the previous segment with probability α or bent at right angles with probability γ but may not be reversed. The probabilities are normalized such that

$$\alpha + 4\gamma = 1, \tag{12.1}$$

where the numerical factor 4 corresponds to the number of possible turns at right angles.

The connection of the present model with real polymers may be obtained as follows : Let us take a polymer of simple structure and divide it into N segments of equal mass such that any neighboring segments can be bent at right angle but cannot be pointed in antiparallel directions. For example, the polyethylene has a simplest structure (CH_2 CH_2 $)_n$; and the size of the segment which satisfies the above requirement can be very small, say n = 2. The polypropylene (CH_2 C H C H_3) has a larger side element C H_3 and should be assigned a larger value, say n = 3, for the segment unit. The mean direction of each segment of a polymer of general shape will now be classified in one of the six general directions, i.e., x^+, x^-, y^-, z^+, and z^- directions.

The sequence of those directional vectors so obtained will then be mapped in one to one on the simple cubic lattice with the lattice constant a_o. If we choose a_o to be the average length of the segment, the mapped configuration will not be far from the original configuration of the polymer. When the probabilities (α, γ) which should characterize the segment-segment interaction and effects of solvents, and which in general depend on the temperature, are given, we may then discuss the polymer conformation in an approximate but fairly realistic manner.

We first note that our model can simply be interpreted in terms of the correlated walks in three dimensions, briefly discussed in §2.9. We can compute the mean square end-to-end distance of our model polymer by extending techniques used for the one-dimensional correlated walks in §§2.6- 2.8.

Before presenting the result, we make a slight generalization. We allow the correlated walker to reverse with probability β so that the step probabilities are normalized by

$$\alpha + 4\gamma + \beta = 1. \tag{12.2}$$

For this generalized model, the mean square end-to-end distance of N segments is found [15] to be

$$\boxed{\frac{\langle r^2 \rangle}{a_o^2} = \frac{1+\delta}{1-\delta} N - \frac{2\delta}{(1-\delta)^2}(1-\delta^N),} \tag{12.3}$$

where

$$\delta \equiv \alpha - \beta , \tag{12.4}$$

represents the degree of correlation, see below.
It is interesting to observe that expression (12.3) has the same functional form as (2.7.18), the corresponding expression for the linear correlated walks. In fact, expression (12.3) holds for any orthogonal lattice of arbitrary dimension.

Now to obtain the mean square end-to-end distance of our model polymer, we merely set β equal to zero. We therefore have

$$\boxed{\frac{\langle r^2 \rangle}{a_0^{\,2}} = \frac{1+\alpha}{1-\alpha} N - \frac{2\alpha}{(1-\alpha)^2} (1 - \alpha^N).} \tag{12.5}$$

Let us interpret the polymer conformation in terms of this expression. Earlier, we discussed a way of choosing the unit of segment for particular polymers. This choice basically depends on the molecular configuration. Once the segment unit is fixed, the parameters (α, γ) should be chosen to represent the bending property of the polymer. This parametrization in general depends on the environment such as the solvent and also on the temperature. If the segment-segment interaction is represented by a single activation-energy-like parameter U, we may assume the relation

$$\gamma = \alpha \exp (- U/k_B T) \tag{12.6}$$

on the basis of the Boltzmann factor arguments. Using the normalization condition $\alpha + 4\gamma = 1$, we obtain [Problem 12.1]

$$\alpha = (1 + 4 e^{-U/k_BT})^{-1}$$

$$\gamma = e^{-U/k_BT} (1 + 4 e^{-U/k_BT})^{-1}. \tag{12.7}$$

In the low temperature limit ($T \to 0$), we obtain, from (12.7),

$$\alpha \to 1, \qquad \gamma \to 0. \tag{12.8}$$

The polymer should be stretched out like a solid bar (if not perturbed by outside means). In fact, from (12.5), we obtain [see (2.7.24)]

$$\langle r^2 \rangle \to N^2 a_0^2 \qquad \text{as} \quad \alpha \to 1. \tag{12.9}$$

In the high temperature limit : $T \to \infty$, both α and γ approach 1/5 :

$$\alpha \to \frac{1}{5}, \quad \gamma \to \frac{1}{5} \qquad \text{as } T \to \infty. \tag{12.10}$$

In this case, the first term on the r.h.s. of eq.(12.5) dominates and approaches the following limit :

$$\langle r^2 \rangle_{T=\infty} = \frac{3}{2} N a_0^2 = \frac{3}{2} (N a_0) a_0 \quad \text{for large N.} \tag{12.11}$$

Observe that the mean square distance $\langle r^2 \rangle$ is proportional to the number of segments, N. The polymer in this limit can be pictured as a state of random coil. In finer detail, since Na_0 represents the total curved length of the polymer, expression (12.11) yields a greater average linear size ($\sqrt{\langle r^2 \rangle}$) for a molecule with a greater linear segment size (a_0). This effect may be understood by imagining a soft noodle in water : a thinner noodle coils more than a thicker one.

The long polymer thus changes from bar-like to coil-like conformation, accompanied by a large difference in the end-to-end distance. The transition in shape should occur at some temperature T_0 characterized by

$$k_B T_0 \sim U, \tag{12.10}$$

where the exponential function $\exp(-U/k_B T)$ change sharply. This transition for large N, can occur within a small range of temperature.

The interpretation of the bending parameters (α, γ) in terms of an excitation energy U should be valid if the solvent effect plays a minor role. If the solvent should obstruct the movement of the polymer substantially, a higher value for the effective activation energy may be assumed.

In the present model, the interaction between nearest-neighbor segments only is considered. Some important biological polymers such as polypeptides (proteins) and D N A in solution exhibit attraction between segments farther apart, and are known to form helical conformations in certain conditions. We will discuss the case of a helix-coil transition separately in the following section.

Problem 12.1 Verify (12.7).

12.13* Helix - Coil Transition of Polypeptides in Solution

In the mid-fifties Doty and his associates [16] and others [17] demonstrated experimentally that a single polypeptide in solution undergoes a helix-coil transition. Many important theories [17] followed this facinating discovery. In 1959 Zimm and Bragg (Z-B) published a classic paper [18] on the helix-coil transition in terms of the solutions of a modified Ising chain. In the last section, we discussed the conformation of a simple polymer based on the correlated walk model. Extending this model with inclusion of the hydrogen-bonding, we can construct a model of a polymer (polypeptide) capable of exhibiting a helix-coil transition [19]. This generalized model will be discussed in the present section.

Let us consider a polypeptide, $H\text{-}(NH\text{-}CHR\text{-}CO)_n$ - OH, where n is the number of amino acid residues (NH-CHR-CO). Let us choose each residue as a unit segment. From the nature of the residue, any neighboring segments may be bent at right angles but may not be pointed in anti-parallel directions. The global conformation of the polymer in solution will now be represented by a model-polymer stretched over on a simple cubic lattice with the following rules : A segment follows the direction of the preceding segment with probability α or turns at right angles (four possibilities) with probability γ but does not reverse. The probabilities are normalized such that $\alpha + 4\gamma = 1$. The lattice constant a_0 will be chosen to equal the average linear size of the unit amide residue. So far, the model is the same as that described in the last section.

Let us now consider a helical state of the polypeptide. In the Pauling-Corey alpha-helix structure [20] which is schematically shown in Figure 12.20,

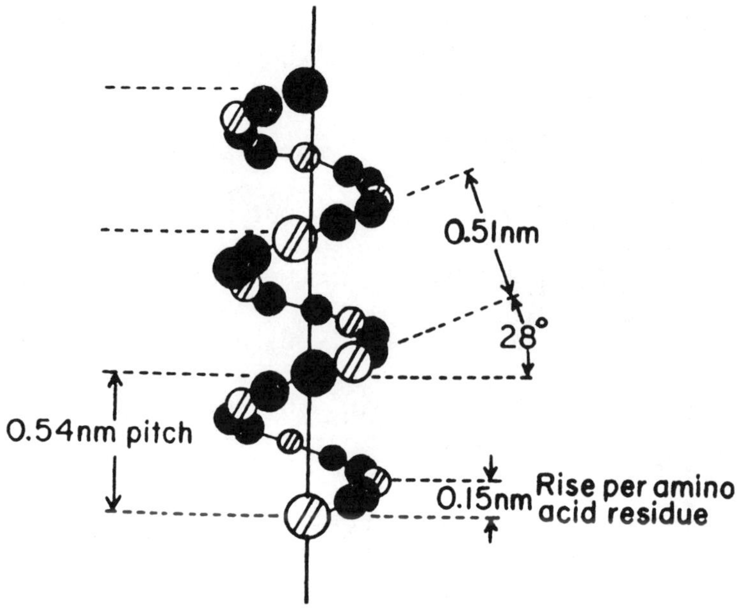

Fig. 12.20 (a) The Pauling-Corey alpha-helix structure [20]

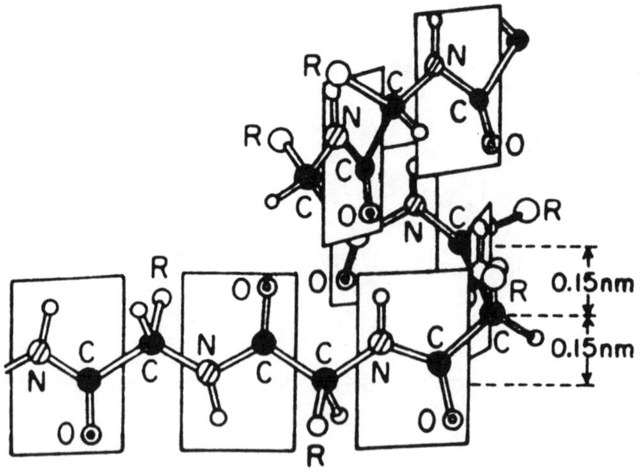

Fig. 12.20 (b) Formation of a right-handed α helix. The planes of the rigid peptide bonds are parallel to the long helical axis.

each amide (CONH) is hydrogen-bonded to the carbonyl oxygen of the third following amide group, and approximately four residues make up a helix unit [21]. Such a helix can be represented by the closed square formed by four segments as shown in Figure 12.21. We now postulate that the hydrogen bonding is attained only when a segment completes a new square with the preceding three segments. Thus for example, in Fig. 12.21 the fifth to eight segments from the end shown attain bonding.

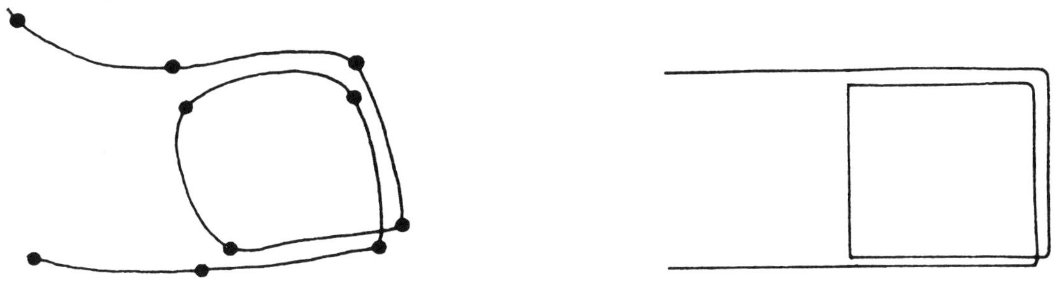

(a) (b)

Fig. 12.21 A polymer backbone forming helices(a) is represented by a model polymer stretched over on the orthogonal lattice (b). By assumption, the hydrogen bonding is attained when a segment completes a new square with the three preceding segments.

In this model, the complete helical state (ground state) corresponds to the polymer being wound around a single square repeatedly. This representation therefore neglects the obvious three-dimensional form of the polypeptide, and is a very crude approximation. Our model however contains two extreme states : random coil and helical states (phases). The transition between the two in our model can be demonstrated in a clear manner. Some important physical features in such transition can be learned from such model calculations.

Let us first consider a further simplified model. We take a model-polymer with several repeated squares (representing a polymer partially in helical states). We divide it in units of four segments. Counting from one end, each unit may or may not form a square. The state of jth unit will be represented by

$$\sigma_j = \begin{cases} 1 & \text{for closed square (loop)} \\ -1 & \text{for open figure.} \end{cases} \tag{13.1}$$

The statistical weight of a given set $\{\sigma_j\}$ of the chain of units is now assumed to be the product of the following factors :

(a) the <u>loop</u> (<u>square</u>) <u>formation factor</u> p if $\sigma_j = 1$ (close loop) or the factor $q = 1 - p$ if $\sigma_j = -1$ (open figure) ;

(b) the <u>Boltzmann factor</u> $\exp(\beta K)$ if a loop follows another loop and unity otherwise ; the energy K represents the bond energy between the units.

The loop formation factor p is prescribed to account for the fact that the segments in the unit must proceed in a well-defined orientation to form a square. Roughly speaking, since only one of four possible turns is "correct" to generate a square, this factor p may be chosen as

$$p = \left(\frac{\gamma}{4}\right)^3 ; \tag{13.2}$$

this is a small number compared with unity. This means that the statistical factor for an open figure is much greater than for a closed square. The physical meaning of the loop formation factor will further be discussed later in the section.

The total bond energy for the model-polymer can be represented by the Hamiltonian of the form

$$H = -K \sum_j \psi\,(\sigma_{j-1}, \sigma_j), \tag{13.3}$$

where

$$\psi\,(\sigma_1, \sigma_2) \equiv \frac{1}{4}\,(1 + \sigma_1)\,(1 + \sigma_2) = \begin{cases} 1 \text{ if } \sigma_1 = \sigma_2 = 1 \\ 0 \text{ otherwise.} \end{cases} \tag{13.3}$$

Note that this Hamiltonian is similar to the Hamiltonian for an Ising chain with the nearest neighbor interaction as represented by eq.(9.4). Only the statistical degeneracies of the states σ_j are different as stated in the rule (a).

The above assumption about the statistical factors represents a very crude approximation. In particular, a partially formed square followed by a complete closed square can get no benefit of bonding in this model. A theory improved upon this point will be discussed later.

The thermodynamic properties for this simplified model can be obtained simply in analogy with the Ising chain with a nearest neighbor interaction. If the M-th and (M -1) -th units are both closed loops (open figures), then the statistical weight is given by $p\ e^{\beta K}$ (q). Applying similar arguments to other possibilities, we can represent the set of weights (step probabilities) in the form of a matrix :

$$T = \begin{pmatrix} q & p \\ q & p\ e^{\beta K} \end{pmatrix}. \qquad \text{[corresponding to (9.7)]} \tag{13.4}$$

If we introduce

$$t \equiv p/q \tag{13.5}$$

$$s \equiv \frac{p}{q} e^{\beta K} = t\ e^{\beta K}, \tag{13.6}$$

the matrix T can be written as

$$T = q \begin{pmatrix} 1 & t \\ 1 & s \end{pmatrix}. \tag{13.7}$$

The grand partition function Ξ for the system can now be written down from (9.11) with the substitution ($p'_1 = q$, $q'_1 = qt$, $p'_2 = q$, $q'_2 = qs$) :

$$\Xi = \frac{1 + \nu q(t - s)}{(1 - \nu qs)(1 - \nu q) - \nu^2 q^2 t} . \qquad (13.8)$$

From this, we can obtain the partition function Z_n for large n as

$$Z_n \equiv \mathrm{Tr}\{ e^{-\beta H} \} \simeq \frac{1}{\nu_1^{\,n}} , \qquad \text{[see (9.24)]} \qquad (13.9)$$

where ν_1 is the minimum root of the equation :

$$(1 - \nu qs)(1 - \nu q) - \nu^2 q^2 t = 0,$$

[ℓ.h.s. = denominator of (13.8)]

and is given explicitly by

$$\nu_1 = \frac{1}{2} \{ 1 + s + [(1 - s)^2 + 4t]^{\frac{1}{2}} \}^{-1}. \qquad (13.10)$$

The free energy per unit, f, in the limit of large n, is given by

$$f = k_B T \ln \nu_1 . \qquad \text{[see (9.25)]} \qquad (13.11)$$

In the low temperature limit ($T \to 0$), we obtain from the last few equations,

$$s \equiv t\, e^{\beta K} \to \infty , \qquad \nu_1 \to s^{-1}$$

$$Z_n \to s^n , \qquad f \to -K \quad \text{as } T \to 0. \tag{13.12}$$

The system then approaches the full helix conformation with the lowest energy $-Kn$. In the high temperature limit ($T \to \infty$), where the system approaches a state of random coil, we obtain

$$s \to t, \qquad \nu_1 \to (1 + t)^{-1}$$

$$Z_n \to (1 + t)^n , \qquad f \to -k_B T \ln (1 + t) \quad \text{as } T \to \infty . \tag{13.13}$$

From (13.12), we can observe that the parameter s defined by (13.6) represents the helix contribution per unit to the partition function.

The ratio of the average number of helical loops to the total number of units, n, can be identified as the canonical ensemble average of $\sum_j \psi(\sigma_{j-1}, \sigma_j)/n$. This ratio $\langle\psi\rangle \equiv \langle \Sigma\, \psi \rangle/n$, called the average degree of helicity, can be calculated from

$$\langle\psi\rangle \equiv n^{-1}\, \mathrm{Tr}\{ \Sigma\, \psi\, e^{-\beta H} \} / Z_n$$

$$= n^{-1}\, \mathrm{Tr}\, \{\Sigma\, \psi\, e^{\beta K\, \Sigma\, \psi}\} \,/\, \mathrm{Tr}\{e^{-\beta H}\}$$

[use of (13.3)]

$$= \frac{1}{n} \frac{\partial}{\partial(\beta K)} \ln \mathrm{Tr} \{e^{-\beta H}\}$$

$$= \frac{\partial}{\partial(\beta K)} \ln \left(\frac{1}{\nu_1} \right) \qquad \text{[use of (13.9); large n]}$$

$$= s \nu_1 \frac{1 - \nu_1}{2 - s\nu_1 - \nu_1}. \qquad \text{[use of (13.10)]} \qquad (13.14)$$

The limit bahavior of this quantity $\langle\psi\rangle$ at low and high temperatures is given by

$$\langle \psi \rangle \to \begin{cases} 1 & \text{as } T \to 0 \\ \dfrac{t^2}{(1+t)^2} & \text{as } T \to \infty, \end{cases} \qquad (13.15)$$

which can be checked easily by using (13.12) and (13.13). For very small t, the abrupt change in $\langle \psi \rangle$ occurs near $s = 1$. In Figure 12.22, a few curves of $\langle \psi \rangle$ against $\ln s$ for selected values of t are shown.

The average degree of helicity $\langle \psi \rangle$ for a real polymer can be measured experimentally. Yang and Doty [16] measured the optical rotation of poly--benzyl-L-glutamate as a function of temperature in the transition region. By assuming that the fractional change in optical rotation is proportional to the average degree of helicity, the data for this polymer of molecular weight 350,000 ($n \sim 1500$) in solution are indicated by small circles ∘ in Figure 12.22, where the temperature scale is shown along the $\ln s$ scale.

Observe that the theoretical curve with $t = 2 \times 10^{-4}$ fits nicely with

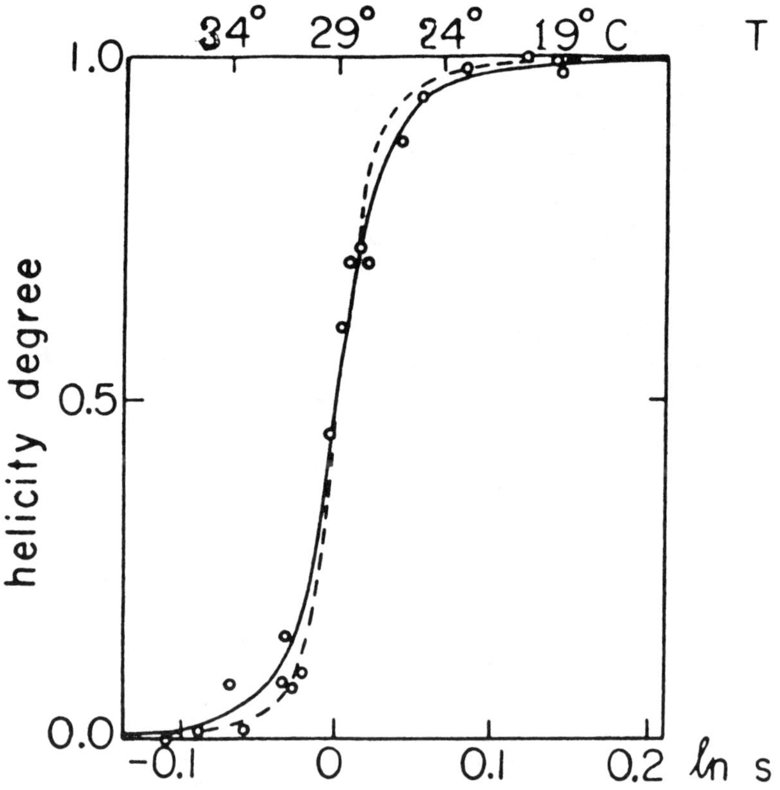

Fig. 12.22 The average degree of helicity $<\psi>$ as given by (13.14) is shown against ℓn s with the two t-values : —— $t = 2 \times 10^{-4}$, - - $t = 1 \times 10^{-4}$. The dots stand for experimental data of optical rotation by Doty and Yang [16]; the temperature scale (T) connected with ℓn s as given by (13.17) is shown with the choice of $t = 2 \times 10^{-4}$.

the experimental data. This matching was obtained with the assumptions that (a) the optical rotation changes halfway at the temperature T_0 while the average degree of helicity equals 0.5 at s = 1 (ℓn s = 0), and (b) the temperature scale is proportional to the ℓn s scale. The assumption (a) means that

$$\ln s\,(T = T_o) = \ln [t\, e^{K/k_B T_o}] = 0$$

or

$$t\, e^{K/k_B T_o} = 1\,, \quad \text{or} \quad -\ln t = \frac{K}{k_B T_o}\,. \qquad (13.16)$$

The assumption (b) is reasonable near the <u>transition temperature</u> since

$$\ln s \equiv \ln (t\, e^{K/k_B T}) \qquad \text{[use of (13.6)]}$$

$$= \ln (t\, e^{K/k_B T} / t\, e^{K/k_B T_o}) \qquad \text{[use of (13.16)]}$$

$$= \frac{K}{k_B} \left(\frac{1}{T} - \frac{1}{T_o} \right)$$

$$\simeq \frac{K}{k_B T_o^2} (T_o - T) \qquad [\,|T_o - T| \ll T_o]$$

$$= [-\ln t] \frac{(T_o - T)}{T_o} \qquad \text{[use of (13.16)]}$$

or

$$\ln s = \frac{T_o - T}{T_o} (-\ln t). \qquad (13.17)$$

With the choice of the numerical values

$$T_o = 29°C = 302.1\ K$$

$$t = 2 \times 10^{-4}, \tag{13.18}$$

we obtained a good fit between theory and experiment.

The theory described here is roughly equivalent to that explicitly worked out by Zimm and Bragg in their paper [18] although the presentation is quite different. Note especially that the sharpness of the transition is generated by the smallness of the parameter t. This parameter $t \equiv p / q$ represents the ratio of the phase-space volume for the helical conformation to that for any other conformation. In other words, the parameter t represents the degree of the difficulty for forming a helix. [It can also be related to the "correct" turn probability as indicated in (13.2).] Further note that the bond energy K and the transition temperature T_o are connected by the factor $-\ln t$ as indicated in (13.16). Since t is very small [see (13.18)], this means that the bond energy K is one order of magnitude higher than $k_B T_o$.

The model calculations presented are very crude, as noted earlier. We can improve the model in a substantial manner with slight mathematical complications.

In a refined model, we take each segment representing a single amide residue as one unit, and will postulate the following rules :

(1) If a segment follows the preceding segment in the correct direction, the statistical factor p' is given; otherwise the factor $q' \equiv 1 - p'$ is assigned.

(2) The Boltzmann factor $\exp(\beta K')$ is given if a segment makes a third correct turn following the two correct turns; otherwise unity is assigned.

The statistical factor p' represents the probability that the amide residue takes a suitable conformation for a right-hand or left-hand helix. This factor will be small compared with unity because the unsuitable conformation corresponds to much greater phase-space volume. A succession of three correct turns clearly generates a closed square. Therefore, the rule (2) for the attainment of the hydrogen bonding has the same content as the original postulate laid out earlier. In the vicinity of helix-coil transition, the tendency of forming helices to gain the bonding competes head-on against the accompanying decrease in the conformational entropy. The contribution of the conformational entropy to the partition function is given in terms of the probabilities (p', q') and that of the hydrogen bonding in terms of the Boltzmann factor $e^{\beta K'}$. The factor p' obviously is connected with the turn probability γ in the model introduced earlier and its connection will be discussed later. As far as the general feature of the helix-coil transition is concerned, the above two rules should describe the states of the model polymer in a semi-quantitative manner.

The partition function Z_n for a model polymer of n segments can be calculated in terms of a modified Ising chain with first to third nearest neighbor interactions [19]. In the limit of large n, the free energy per segment, f, and the average degree of helicity, $\langle \psi \rangle$, can be expressed in the forms of (13.11) and (13.14), respectively with ν_1 representing the minimum root of the equation : $1 - (1 + s')\nu + (s' - t')\nu^2 + t'(s' - t')\nu^3 = 0$, where $t' \equiv p'/q$ and $s' \equiv t' \exp(\beta K')$ (K' : bond energy). The further calculations show that the average degree of helicity $\langle \psi \rangle$ can be fitted well with the experimental data of Yang and Doty by the choice of

$$t' \equiv p'/q' = 0.01. \tag{13.19}$$

Thus, we see that both approximate calculations yield good fits with the same experimental data. In the second model, the "correct" turn probability is given by

$$p' = \frac{p'}{q'} q' \equiv t'(1 - p') \cong t' = 0.01. \tag{13.20}$$

For the first model, we may apply (13.2) and obtain

$$\frac{1}{4} \gamma = t^{1/3} = 0.059 \tag{13.21}$$

for the correct-turn probability. This value is considerably greater than 0.01. In summary, the qualitative behavior of the degree of helicity can be fitted without tight control as long as two parameters (t,s) characterizing statistical degeneracy and bond energy are introduced. The value 0.01 obtained for the refined model is quite consistent with the semi-empirical calculations of Scheraga _et al_ [17,22]. This means that the phase-space volume suitable for the helical conformation per amide group is about one percent of the total phase-space volume at the transition temperature.

Finally, we note that much the same model calculations can be applied to the double-helix-coil transition for DNA [19,21].

References

[1] e.g. Alonso, M. and Finn, E.J., Fundamental University Physics. III Quantum and Statistical Physics, Addison-Wesley, Reading, Mass. (1968)

[2] Dirac, P.A.M., Principles of Quantum Mechanics, Oxford University Press, London, 4th Edition (1958)

[3] E.g. Lorrain, P. and Corson, E.R., Electromagnetism, Freeman, San Fransicso (1978)

[4] Sakurai, J.J., Advanced Quantum Mechanics, Addison-Wesley, Reading Mass., (1967)

[5] Pake, G.E., Paramagnetic Resonance, Benjamin, New York (1962)

[6] Henry, W.E., Phys. Rev. 88, 561 (1952)

[7] Kittel,C. and Galt, J.K., Solid State Physics, vol.3, Ferromagnetic Domain Theory, Academic Press, New York (1956)

[8] Becker, R., Theorie der Wärme, Springer, Berlin (1955)

[9] Arajs, S. and Colvin, R.V., J. Appl. Phys. 32 (suppl.), 336 (1961)

[10] Ising, E., Z. Physik 31, 253 (1925)

[11] Yang, C.N., Phys. Rev. 85, 809 (1952)

[12] Onsager, L., Phys. Rev. 65, 117 (1944)

[13] Huang, K., Statistical Mechanics, Wiley, New York (1963)

[14] For extensive reviews and references, see Domb, C. and Green,M.S., (Editors), Phase Transitions and Critical Phenomena, multi-volume series, Academic Press, London, 1972-continued.

[15] Fujita, S., Okamura, Y. and Chen, J.T., J. Chem. Phys.72, 3993 (1980)

[16] Doty, P., and Yang, J. T., J. Am.Chem. Soc. 78, 498 (1956)

[17] Many significant experiments and theories of the helix coil transitions up to 1970 were collected and summarized in, Poland, D. and Scheraga,

H. A., Theory of Helix-coil Transitions in Biopolymers, Academic Press, New York (1970)

[18] Zimm, B.H. and Bragg, J.K., J. Chem. Phys. 31, 526 (1959)

[19] Fujita, S., Blaisten-Barojas, E., Torres, M. and Godoy, S.V., J. Chem. Phys. 75, 3097 (1981)

[20] Pauling, L. and Corey, R.B., Proc. Int. Wool Text. Res. Conf., B, 249 (1955)

[21] More about many fascinating facts and properties of biopolymers can be learned from, e.g., Lehninger, A.L., Biochemistry, Worth, New York (1975)

[22] e.g. Nemethy, G. and Scheraga, H.A., Biopolymers 3, 155 (1965); Scott, R.A. and Scheraga, H.A., J. Chem.Phys. 45, 2091 (1966)

Review Questions

1. Write down mathematical expressions, figures, etc., and explain briefly.

(a) Definition of the orbital angular momentum for a single particle

(b) Commutation relations among the components of the orbital angular momentum, (j_x, j_y, j_z)

(c) Commutation relations among the components of the spin angular momentum, (s_x, s_y, s_z)

(d) Quantization rules for the angular momentum

(e) Magnetogyric ratio

(f) Potential energy for a magnetic moment in a magnetic field

(g) g - factor

(h) Electron spin resonance

(i) Curie's law for paramagnetic materials

(j) Paramagnetism of degenerate electrons

(k)* Curie - Weiss law for a ferromagnet

(l)* Internal field model of Weiss

(m)* Ising Hamiltonian

(n)* Spontaneous magnetization

(o)* Polymer conformation. Random coil state

(p)* Excluded volume effect

(q)* Helix - coil transition for a polypeptide

General Problems

1. Consider a system of a free electrons moving in two dimensions. Assume that the density of electrons is high enough so that the system is statistically degenerate. Because of the magnetic moments associated with their spins, the electrons are expected to show a weak paramagnetism.

(a) Calculate the magnetic susceptibility due to the spins of the system at absolute zero temperature.

(b) If the electrons obey the Boltzmann statistics, what result would you expect? Compare it with case (a).

2. Imagine a hypothetical atom carrying a classical magnetic moment $\vec{\mu}$ which may point in any spatial direction.

(a) Write down an expression for the magnetic energy when the atom is placed under a magnetic field $\vec{B}$.

(b) Calculate the average magnetic moment by assuming a canonical ensemble.

(c) Find the magnetic susceptibility for a system of isolated atoms. Does it obey Curies law ?

3.* Let us consider the Ising model for a linear chain of spins characterized by the Hamiltonian

$$H = -\sum_{j=1}^{N} (J \sigma_{j-1} + \mu B) \sigma_j ,$$

where σ_j may take on -1, 0 or 1. Compute the grand partition function and investigate the thermodynamic properties including average energy, heat capacity, average magnetization for this system. Confirm that there is no phase transition.

Chapter 13. TRANSPORT PHENOMENA

This chapter deals with the non-equilibrium properties, in particular the electrical conductivity and diffusion coefficients of selected systems. The general theory of distribution functions developed in Chapter 4 can be applied here. In §§13.2 - 4, we set up and solve the Boltzmann equation for an electron-impurity system, and derive an expression for the electrical conductivity. This result is used to discuss the mobility of conduction electrons in semi-conductors in §13.4.

The method by means of a Boltzmann equation is a powerful microscopic theory for treating transport phenomena in a quantitative manner. This method however has a few drawbacks, including the difficulty of setting up a starting Boltzmann equation or its generalization for some particular systems. This is especially true for systems of charged particles subjected to a strong magnetic field. A modern theory of transport phenomena avoids the difficulty of setting up a kinetic equaiton initially, and proceeds in a different route. An introduction to this newer theory, called the correlation function method, and its simple applications will be discussed in the middle part of the present chapter, §§13.5 - 11. In the last part, §§13.12 - 14, a semi-microscopic approach to the diffusion problem is presented on the basis of correlated walks. This method is quite useful when microscopic methods formulated in terms of the Hamiltonian do not yield simple results as in the case of the atomic diffusion in a crystal.

13.1 Ohm's Law. The Electrical Conductivity. Matthiessen's Rule.

Let us consider a system of free electrons moving in a potential field of impurities which act as scatterers. The impurities are, by assumption, distributed uniformly in space.

Under the action of an electric field $\vec{E}$ pointed along the positive x-axis, a classical electron will move in accordance with the equation of motion:

$$m \frac{dv_x}{dt} = -e\,E \tag{1.1}$$

[in the absence of the impurity potential]. This gives rise to a uniform acceleration and therefore a linear change in the velocity along the direction of the field:

$$v_x = -\frac{e}{m} E\,t + v_x^o\,, \tag{1.2}$$

where v_x^o is the x-component of the initial velocity. For free electrons the velocity increases indefinitely and leads to infinite conductivity.

In the presence of the impurities, this uniform acceleration will be interrupted by scattering. When the electron hits a scatterer (an impurity), the velocity will suffer an abrupt change in direction and grow again following eq.(1.2) until the electron hits another scatterer. Let us denote the <u>average time between successive scatterings</u> by τ_f, which is also called the <u>mean free time</u>. The order of magnitude of the average velocity $\langle v_x \rangle$ is then given by

$$\langle v_x \rangle = -\frac{e}{m} E\,\tau_f. \tag{1.3}$$

In arriving at this expression, we assumed that the electron loses the memory of its preceding motion every time it hits the scatterer, and the average of the velocities just after the collisions vanishes:

$$\langle v_x^o \rangle = 0. \tag{1.4}$$

The charge current density (average current per unit volume), j_x, is given by

$$j_x = (\text{charge}) \times (\text{number density}) \times (\text{velocity}) = -en \langle v_x \rangle$$

$$= \frac{e^2}{m} n \tau_f E , \qquad \text{[use of (1.3)]} \tag{1.5}$$

where n is the number density of electrons.

According to <u>Ohm's Law</u>, the current density j_x is proportional to the applied field E when this field is small:

$$j_x = \sigma E. \qquad \text{(small E)} \tag{1.6}$$

The proportionality factor σ is called the <u>electrical conductivity</u>. It represents the facility with which the current is generated in response to the electric field. Comparing the last two equations, we obtain

$$\sigma = \frac{e^2}{m} n \tau_f. \tag{1.7}$$

This formula is very useful in the qualitative discussion of the electrical transport phenomenon. The inverse-law mass-dependence means that the ion contribution to the electric transport will be smaller by at least three orders of magnitude than the electron contribution. Notice that the conductivity is higher if the number density is greater and/or if the mean free time is greater.

The inverse of the mean free time τ_f:

$$\Gamma \equiv 1/\tau_f \tag{1.8}$$

is called the rate of collision or the collision rate. Roughly speaking, this Γ represents the mean frequency with which the electron is scattered by impurities. Earlier in §4.19, we saw [cf.(4.10.1)] that the collision rate Γ is given by

$$\Gamma = n_I v A, \tag{1.9}$$

where n_I and A are respectively the density of scatterers and the scattering cross section; we denote the cross section by A instead of σ as was done in §4.8; the symbol σ stands for the conductivity in this section.

If there is more than one kind of scatterer, the rate of collision may be computed by the addition law:

$$\begin{aligned}\Gamma &= n_1 v A_1 + n_2 v A_2 + \cdots \\ &= \Gamma_1 + \Gamma_2 + \cdots .\end{aligned} \tag{1.10}$$

This is often called Matthiessen's rule. The total rate of collision is the sum of collision rates computed separately for each kind of scatterer.

Historically and also in practice, the analysis of of resistance data for a conductor is done as follows: If the electrons are scattered by impurities and again by phonons, the total resistance will be written as the sum of the resistances due to each separate cause of scattering:

$$R_{total} = R_{impurity} + R_{phonon} . \tag{1.11}$$

This is the original statement of Matthiessen's rule. In further detail, the electron-phonon scattering depends on temperature because of the change in the phonon population, while the effect of the electron-impurity scattering is temperature-independent [see §13.4]. By separating the resistance in two parts, one temperature-dependent and the other temperature-independent, we may apply Matthiessen's rule. Since the resistance R is inversely proportional to the conductivity σ, eqs.(1.7) and (1.10) together imply eq.(1.11).

Problem 1.1 Free electrons are confined within a long rectangular planar strip shown below. Assume that each electron is <u>diffusely scattered</u> at the boundary so that it may move in all allowed directions without preference after the scattering. Find the mean free path along the long strip. Calculate the conductivity σ.

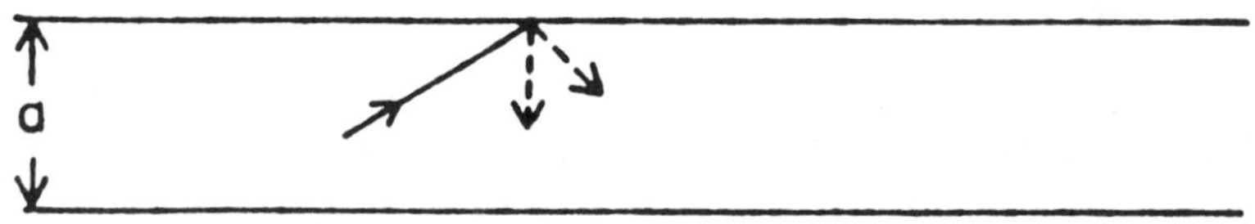

Problem 1.2 Do the same as in Problem 1.1 for the case in which electrons are confined within a long circular cylinder.

13.2 The Boltzmann Equation for an Electron-Impurity System

In the method of a Boltzmann equation the qualitative arguments in the simple kinetic theory will be formulated in more precise terms. In some limiting cases, the description of the transport phenomena by this method is exact. In cases of practical interest this method is an approximation, but is widely used.

Let us consider the same electron-impurity system. We introduce a momentum distribution function $\varphi(\vec{p},t)$ defined such that $\varphi(\vec{p},t)\, d^3p$ gives the relative probability of finding an electron in the element d^3p at time t. [see §4.7] This funciton will be normalized such that

$$\frac{1}{(2\pi\hbar)^3}\int d^3p\ \varphi(\vec{p},t) = \frac{N}{V} \equiv n. \tag{2.1}$$

The <u>electric current density</u> $\vec{j}$ is given in terms of φ as follows:

$$\vec{j} = \frac{-e}{(2\pi\hbar)^3 m}\int d^3p\ \vec{p}\ \varphi(\vec{p},t). \tag{2.2}$$

The function φ can be obtained by solving the Boltzmann equation, which may be set up in the following manner.

The change in the distribution function φ will be caused by the force acting on the electrons in the element d^3p <u>and</u> by collision. We may write this change in the form:

$$\frac{\partial\varphi}{\partial t} = \left(\frac{d\varphi}{dt}\right)_{\text{force}} + \left(\frac{d\varphi}{dt}\right)_{\text{collision}}. \tag{2.3}$$

The <u>force term</u> $(d\varphi/dt)_{force}$, caused by the force $-e\vec{E}$ acting on the electrons can be expressed by [see (4.9.12)]

$$\left(\frac{d\varphi}{dt}\right)_{force} = -e\vec{E} \cdot \frac{\partial\varphi}{\partial\vec{p}} \ . \tag{2.4}$$

If the density of impurities, n_I, is low, and the interaction between electron and impurity has a short range, the electron will be scattered by one impurity at a time. We may then write the <u>collision term</u> in the following form:

$$\left(\frac{d\varphi}{dt}\right)_{collision} = \int d\Omega \, \frac{p}{m} n_I \; I(p,\vartheta)[\varphi(\vec{p}\,',t) - \varphi(\vec{p},t)] \ , \tag{2.5}$$

where ϑ is the <u>angle of deflection</u>, i.e. the angle between the initial and final momenta $\vec{p}$ and $\vec{p}\,'$, and $I(p,\vartheta)$ is the differential cross section. See Figure 13.1.a. In fact, the rate of collision, as discussed in §4.8, is given by (density of scatterers) x (speed) x (total cross section). If we apply this rule to the flux of particles with momentum $\vec{p}$, we can obtain the second integral of eq.(2.5), the integral with the minus sign. This integral corresponds to the <u>loss</u> of the flux due to the collision. The flux of particles with momentum $\vec{p}$ can <u>gain</u> by the inverse collision, which is shown in Figure 13.1.b. The contribution of the inverse collision is represented by the first integral.

So far, we have neglected the fact that electrons are fermions, and are therefore subject to Pauli's exclusion principle. We will now look at the effect of quantum statistics.

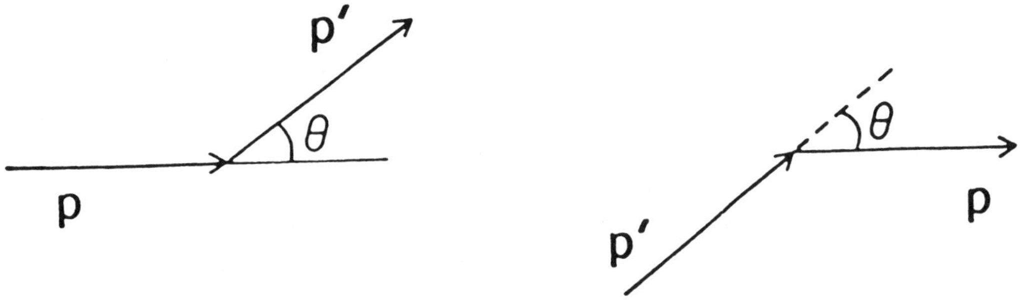

a. original collision b. inverse collision

Fig. 13.1 In a, the electron suffers the change in momentum from $\vec{p}$ to $\vec{p}'$ after scattering. The inverse collision is shown in b.

If the final momentum state $\vec{p}'$ was already occupied, the scattering from the state $\vec{p}$ to the state $\vec{p}'$ should not have occurred. The probability of this scattering therefore should be reduced by the factor $1-\varphi(\vec{p}',t)$, which represents the probability that the final state $\vec{p}'$ is unoccupied. Consideration of the exclusion principle thus modifies the Boltzmann collision term given in (2.5) to:

$$\left(\frac{d\varphi}{dt}\right)_{\text{collision}} = \int d\Omega \, \frac{p}{m} \, n_I \, I(p,\vartheta) \{ \varphi(\vec{p}',t)[1 - \varphi(\vec{p},t)] - \varphi(\vec{p},t)[1 - \varphi(\vec{p}',t)]\}. \qquad (2.6)$$

When expanded, the two terms proportional to $\varphi(\vec{p}',t)\,\varphi(\vec{p},t)$ in the curly brackets have opposite signs and cancel each other out. We then have the same collision term as given by (2.5).

Gathering the result from (2.3) - (2.4), we obtain

$$\frac{\partial\varphi(\vec{p},t)}{\partial t} + (e\,\vec{E}) \cdot \frac{\partial\varphi(\vec{p},t)}{\partial\vec{p}} = \frac{n_I}{m}\int d\Omega\; p\; I(p,\vartheta)\;[\varphi(\vec{p}\,',t) - \varphi(\vec{p},t)]\;. \tag{2.7}$$

This is the Boltzmann equation for the electron-impurity system. This equation is linear in φ, and much simpler than the Boltzmann equaiton for a dilute gas, eq.(4.12.12), which was discussed in Chapter 4. In particular, we can solve eq.(2.7) by elementary methods and calculate the conductivity σ. We will do this in the next section. For simple forms of the scattering cross section, we can also solve eq.(2.7) as an initial-value problem [Problems 2.1 and 2.2].

As mentioned in the beginning to this section, the Boltzmann equation is practically very important, but it is an approximate equation. If the impurity density is not low or if the range of the interaction is not short, we must then take account of simultaneous scatterings by two or more impurities. If we include the effect of the Coulomb interaction among electrons, which has hitherto been ignored in our consideration, the collision term should be further modified. It is, however, very difficult to estimate appropriate corrections arising from these various effects.

Problem 2.1* Obtain the Boltzmann equation for an electron-impurity system in two dimensions, which may be deduced by inspection, from eq.(2.7). Assume that all electrons have the same velocity at the initial time $t = 0$. Further, assuming no electric field ($E = 0$) and isotropic scattering (I = constant), solve the Boltzmann equation.

Problem 2.2* Solve the Boltzmann equation (2.7) for three dimensions with the same condition as in Problem 2.1. Define the Boltzmann H-function by $H(t) \equiv \frac{1}{n} \int d^3p \; \varphi(\vec{p},t) \; \ln \left[\varphi(\vec{p},t)/n\right]$. Evaluate dH/dt. Plot H(t) as a function of time.

13.3 The Current Relaxation Rate

Let us assume that a small constant electric field $\vec{E}$ is applied to the system, and that a stationary homogeneous current is established. We take the positive x-axis along the field $\vec{E}$. In the stationary and homogeneous state, the distribution function φ depends on momentum $\vec{p}$ only. From eq.(2.7) the Boltzmann equaiton for φ is then given by

$$-e\,E\,\frac{\partial\varphi(\vec{p})}{\partial p_x} = \frac{n_I}{m}\int d\Omega \; pI[\varphi(\vec{p}\,') - \varphi(\vec{p})] \; . \tag{3.1}$$

We wish to solve this equation and calculate the conductivity σ.

In the absence of the field $\vec{E}$, the system, by assumption, is characterized by the equilibrium distribution funciton, that is, the Fermi distribution function for free electrons:

$$\varphi_o(\vec{p}) = f(\varepsilon_p) \equiv \frac{1}{e^{\beta(\varepsilon_p-\mu)} + 1}. \tag{3.2}$$

With the small field E, the function φ deviates from φ_o. Let us regard φ as a function of E and expand it in powers of E:

$$\begin{aligned}\varphi(\vec{p}) &= \varphi_o + \varphi_1 + \cdots \\ &= f(\varepsilon_p) + \varphi_1(\vec{p}) + \cdots ,\end{aligned} \tag{3.3}$$

where the subscripts denote the orders in E. For the determination of the conductivity σ we need φ_1 only. Let us introduce expansion (3.3) in eq.(3.1), and compare terms of the same order in E. In zeroth order we have

$$\frac{n_I}{m} \int d\Omega \; p \; I \; [\varphi_o(\vec{p}\,') - \varphi_o(\vec{p})] = 0. \tag{3.4}$$

Since the energy is conserved:

$$\varepsilon_{p'} = \varepsilon_p \tag{3.5}$$

in the scattering, $\varphi_o(\vec{p}) \equiv f(\varepsilon_p)$ clearly satisfies eq.(3.4). In the first order in E, we have

$$\text{the l.h.s.} = -eE \frac{\partial \varphi_o}{\partial p_x} = -eE \frac{\partial f(\varepsilon_p)}{\partial p_x} \qquad \text{[use of (3.2)]}$$

$$= -eE \frac{\partial \varepsilon_p}{\partial p_x} \frac{df}{d\varepsilon_p} = -eE \frac{p_x}{m} \frac{df}{d\varepsilon_p} .$$

Therefore, we obtain from (3.1)

$$-eE \frac{p_x}{m} \frac{df}{d\varepsilon_p} = \frac{n_I}{m} \int d\Omega \; p \; I \; [\varphi_1(\vec{p}\,') - \varphi_1(\vec{p})] . \tag{3.6}$$

For the moment, let us neglect the first term on the r.h.s. In this case, the r.h.s. equals $-n_I \, m^{-1} p \, \varphi_1(\vec{p}) \int d\Omega \, I$. Then, the function $\varphi_1(\vec{p})$ is proportional to p_x. Noting this property, let us now try a solution of the form

$$\varphi_1(\vec{p}) = p_x \, \Phi(\varepsilon_p), \tag{3.7}$$

where $\Phi(\varepsilon_p)$ is a function of ε_p (no angular dependence). Substitution of (3.7) into eq.(3.6) yields

$$-eE\, p_x \frac{df}{d\varepsilon_p} = n_I \int d\Omega\, pI\, \Phi(\varepsilon_p)[p_x' - p_x] \ . \tag{3.8}$$

Let us look at the integral on the r.h.s. We introduce a new frame of reference with the polar axis (the Z-axis) pointing along the fixed vector p as shown in Figure 13.2. The old positive x-axis, which is parallel to the electric field $\vec{E}$, can be specified by the angles (ϑ, ϕ). From the diagram we have

$$p_x = p \cos \vartheta. \tag{3.9}$$

The vector $\vec{p}'$ can be represented by (p, χ, ϕ_1). If we denote the angle between $\vec{p}'$ and $\vec{E}$ by ψ, we have

$$p_x' = p \cos \psi. \tag{3.10}$$

We wish to express $\cos \psi$ in terms of the angles (ϑ, ϕ) and (χ, ϕ_1). This can be done by using the vector decomposition property as follows:

$$\begin{aligned} \cos \psi &= \frac{\vec{p}'}{p} \cdot \frac{\vec{E}}{E} \\ &= (\vec{i} \sin \chi \cos \phi_1 + \vec{j} \sin \chi \sin \phi_1 + \vec{k} \cos \chi) \cdot \\ &\quad (\vec{i} \sin \vartheta \cos \phi + \vec{j} \sin \vartheta \sin \phi + \vec{k} \cos \vartheta) \\ &= \sin \chi \sin \vartheta [\cos \phi_1 \cos \phi + \sin \phi_1 \sin \phi] + \cos \chi \cos \vartheta \end{aligned}$$

or

$$\cos \psi = \sin \chi \sin \vartheta \cos (\phi - \phi_1) + \cos \chi \cos \vartheta. \tag{3.11}$$

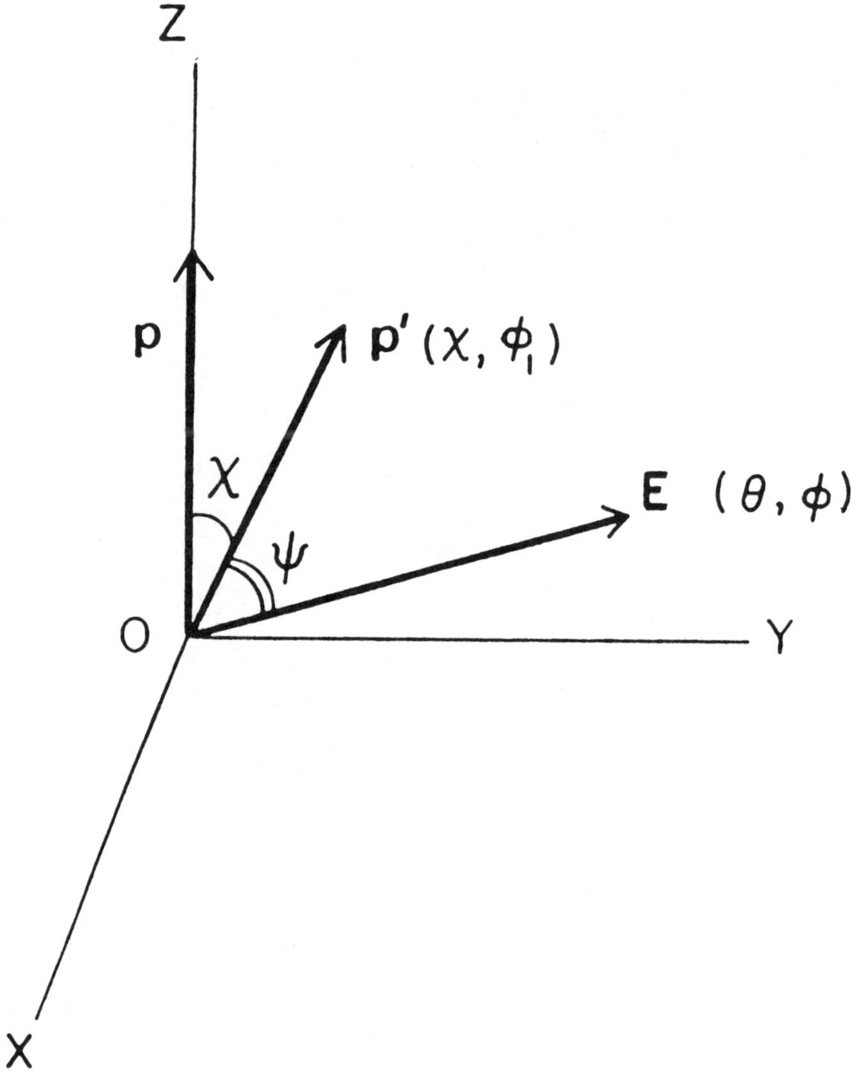

Fig. 13.2 The new frame of reference in which the positive Z-axis points in the direction of the fixed vector $\vec{p}$. In this frame, the direction of the electric field $\vec{E}$ is specified by (ϑ,ϕ) and that of the momentum $\vec{p}'$ by (χ,ϕ_1).

Let us now consider the first integral in eq.(3.8):

$$A \equiv n_I \int d\Omega\, p\, I(p,\chi)\, \Phi(\varepsilon_p)(p \cos\psi)$$

$$= n_I \int_0^{2\pi} d\,\phi_1 \int_0^{\pi} d\chi \;\; \sin\chi\, p^2\, I(p,\chi)\, \Phi(\varepsilon_p)$$

$$\times [\sin\chi \sin\vartheta \cos(\phi-\phi_1) + \cos\chi \cos\vartheta]. \qquad \text{[use of (3.11)]} \tag{3.12}$$

Since

$$\int_0^{2\pi} d\phi_1 \cos(\phi-\phi_1) = 0 , \tag{3.13}$$

the first part of the integral (3.12) can be dropped. We then obtain

$$A = n_I \int_0^{2\pi} d\phi_1 \int_0^{\pi} d\chi\, \sin\chi\, p\, I(p,\chi) \cos\chi\, [p \cos\vartheta\, \Phi(\varepsilon_p)] \qquad \text{[use of (3.12) and (3.13)]}$$

$$= \varphi_1(\vec{p})\, n_I \int d\Omega\, p\, I(p,\chi) \cos\chi . \qquad \text{[use of (3.7) and (3.9)]} \tag{3.14}$$

The second integral in eq.(3.8) therefore is proportional to $\varphi_1(\vec{p})$. We thus obtain the solution of eq.(3.8) as follows:

$$\varphi_1(\vec{p}) = eE \frac{p_x}{m} \frac{df}{d\varepsilon_p} \frac{1}{\Gamma(p)} \tag{3.15}$$

$$\Gamma(p) \equiv \frac{n_I}{m} \int d\Omega\, p\, I(p,\chi)[1-\cos\chi] > 0. \tag{3.16}$$

The Γ here is positive, and depends only on $p \equiv |p|$ (or equivalently on the energy ε_p); it has the dimension of frequency and is called the <u>energy-</u>

dependent current relaxation rate or simply the (energy-dependent) relaxation rate. Its inverse is called the relaxation time.

The electric current density j_x can be calculated from (2.2),

$$j_x = \frac{-e}{(2\pi\hbar)^3 m} \int d^3p \, p_x \, \varphi(\vec{p}) \, . \tag{3.17}$$

We introduce $\varphi = \varphi_0 + \varphi_1 + \ldots$ in this expression. The first term gives a vanishing contribution. (No current in equilibrium.) The second term yields, using (3.15),

$$j_x = -\frac{e^2 E}{(2\pi\hbar)^3 m^2} \int d^3p \, \frac{p_x^2}{\Gamma(p)} \frac{df}{d\varepsilon_p} \, . \tag{3.18}$$

Comparing this with Ohm's law (1.5), we obtain the following expression for the conductivity:

$$\sigma = \frac{e^2}{(2\pi\hbar)^3 m^2} \int d^3p \, \frac{p_x^2}{\Gamma(p)} \left(-\frac{df}{d\varepsilon_p}\right) \, . \tag{3.19}$$

A few applications of this formula will be discussed in the following section.

Problem 3.1 For a classical hard-sphere interaction, the scattering cross section I is given by $\frac{1}{2}a^2$, where a is the radius of the sphere. Evaluate the relaxation rate $\Gamma(p)$ given by (3.16). Using this result, calculate the conductivity σ through (3.19). Verify that the conductivity calculated is temperature-independent.

Problem 3.2 Formula (3.19) was obtained with the assumption that the equilibrium distribution function in the absence of the field is given by the Fermi distribution function (3.2).

(a) Verify that the same formula applies when we assume that the equilibrium distribution function is given by the Boltzmann distribution function.

(b) Show that the conductivity calculated by this formula does not depend on the temperature.

13.4 Applications to Semiconductors

The elements Ge, Si, ... in the fourth column of the atomic (or Mendeleev) periodic table are known to form crystals of diamond structure. The four electrons in the outermost shell of each atom are paired with the electrons of the nearest neighbor atoms and form the so-called covalent bond configuration as represented schematically in Figure 13.3 for the case of the square lattice. This lattice, which has the same number of nearest neighbors as the diamond lattice, was substituted here for visual convenience.

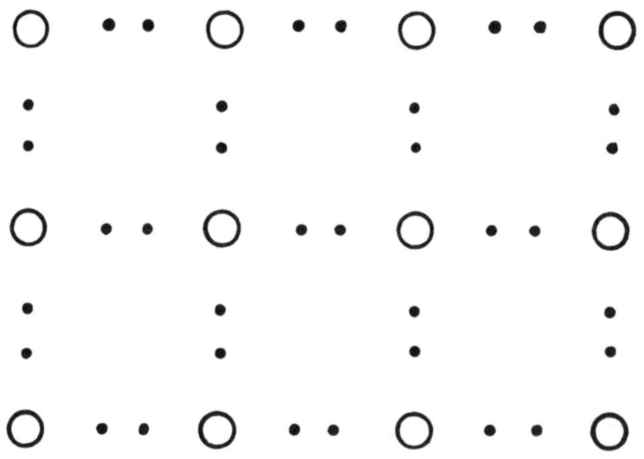

Fig. 13.3 The elemental semiconductor forms a crystal of diamond structure with the number of nearest neighbors equal to four, here schematically represented by a square lattice. Small black dots in pairs represent the covalent bonding.

Let us now replace one of the atoms by the phosphorus atom P, which lies in the fifth column of the periodic table. Four of the five electrons in the outer shell of this atom are available to form the covalent bonds, and the remaining electron will be loosely attached to the atom. See Figure 13.4. In fact, the state of the last electron may be represented by the

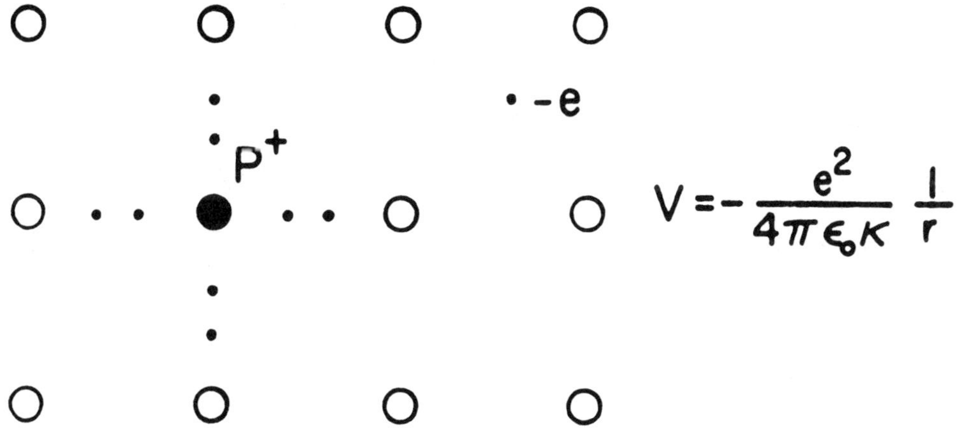

Fig. 13.4 The positively ionized impurity P^+ and the released electron interact with a Coulomb-like potential V reduced by the dielectric constant $\varkappa$.

hydrogen-like configuration in which the electron interacts with the ionized phosphorus atom P^+ through the Coulomb interaction:

$$V = -\frac{e^2}{4\pi\varepsilon_0\varkappa}\frac{1}{r} . \tag{4.1}$$

Here, $\varkappa$ is the dielectric constant, which represents the polarization effect of the host lattice. This $\varkappa$ has the numerical value of about 16 for Ge and about

12 for Si. Thus, the Coulomb interaction is reduced by one order of magnitude. The hydrogen-like binding energy calculated by use of (4.1) is reduced by the factor κ^2 compared with that for the true hydrogen atom. Because of such a reduction, the binding energy becomes of the order 0.01 eV or hundreds of degrees times the Boltzmann constant k_B. Then, a portion of phosphorus (impurity) atoms may be ionized at room temperature. The electrons released from foreign atoms in this fashion can move in the crystal and contribute to the transport of charge.

Let us now consider another case. We replace one of the host atoms by the aluminum atom Al, which lies in the third column of the periodic table. The three electrons in the outer shell of this atom are paired with the electrons

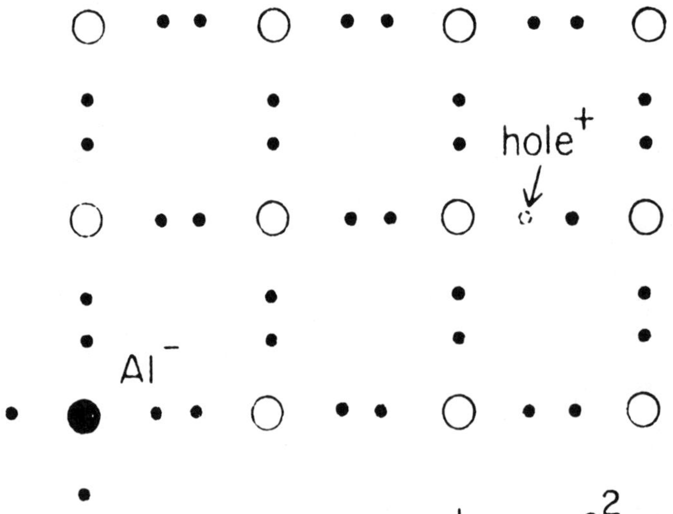

$$V = -\frac{1}{4\pi\epsilon_0\kappa}\frac{e^2}{r}$$

Fig. 13.5 The negatively ionized impurity Al^- and the hole interact with a reduced Coulomb-like potential V.

of the neighboring atoms to form covalent bonds. This leaves one unpaired electron "dangling" brtween the impurity atom and its four neighbors. See Figure 13.5. The site of the unpaired electron may be regarded as that of the vacancy in bonding.

This vacancy, also called the hole, may be filled by the transfer of an electron from the neighboring atoms. If this happens, the impurity atom will be ionized with the negative charge (-e), and the hole which was created in the process will carry the positive net charge e. Such a hole may move in the whole crystal and can also contribute to the transport of the electric charge.

In summary, at room temperature, then, there are certain numbers of charge carriers, that is, electrons and/or holes, which move about within the whole crystal. The density n of charge carriers is very small compared with that of the lattice atoms. This naturally leads to the fact that the electrical conductivity σ, which is proportional to the carrier density n, is much smaller in semiconductors than in metals. The term "semiconductor" principally arises from this transport property.

The density of charge carriers, n, depends on the temperature rather strongly. As the temperature is lowered, the ionized impurities may catch electrons or holes, and this recombination process reduces the number of mobile electrons. Therefore, the conductivity σ is lower at lower temperatures. This behavior is in contrast with the case of a metal, where the opposite temperature-dependence is observed.

Let us now introduce the mobility μ through the relation

$$\sigma \equiv z\, en\mu. \tag{4.2}$$

Comparing with the order-of-magnitude formula: $\sigma = z^2 e^2 n\, \tau_f / m^*$ [see (1.7)] we obtain

$$\mu = z\, e\, \tau_f / m^* . \tag{4.3}$$

We observe that the mobility μ is negative for electrons and positive for holes. The magnitude of the mobility is greater for a smaller effective mass. Physically, the mobility represents a measure of the facility with which the charge carrier moves in a material under a small applied field.

Figure 13.6 shows typical data for the mobility μ against the temperature T. As the temperature is lowered from room temperature, the mobility becomes greater in magnitude, reaches a maximum at a certain temperature, and tends to decrease at still lower temperatures. We may interpret this behavior as follows. At room temperature there are many phonons, which act as scatterers

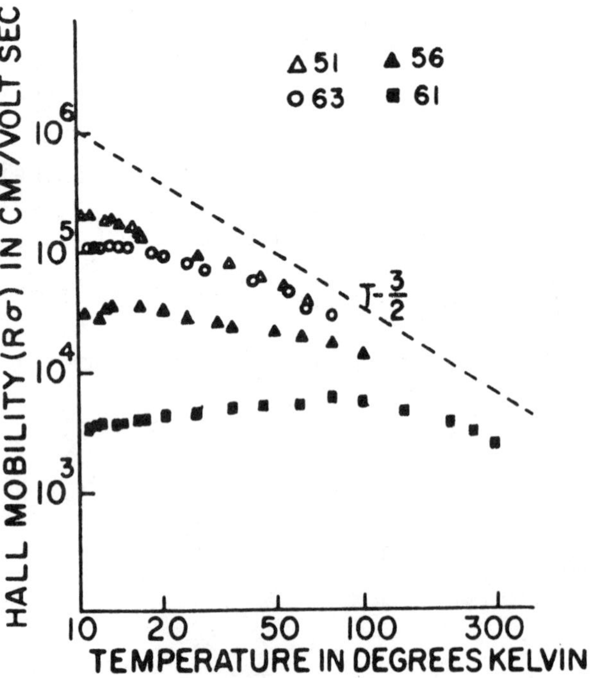

Fig. 13.6 The mobility against temperature for typical germanium samples with different concentrations of impurities. The broken line represents the ideal mobility expected from the phonon scattering. The data are due to Debye and Conwell [1].

of the conduction electrons. As the temperature is lowered, the number of phonons becomes smaller and therefore the scattering by phonons is less effective. The mobility should therefore rise. If there are no other scattering mechanisms, the mobility should rise indefinitely as indicated by the broken line in the diagram. The measured mobility indicates that there are other scatterers which are effective at low temperatures.

Conwell and Weisskopf [2] studied the mobility by taking account of the scattering by ionized impurities by means of the Boltzmann equation. In the last section we solved the Boltzmann equation and obtained a formula for the conductivity σ. By taking the Coulomb interaction (4.1), let us evaluate this formula.

The scattering cross section I (p,χ) calculated classical mechanically for the Coulomb force is known as the Rutherford scattering cross section. It is given by [3]

$$I\,(p,\chi) = \left(\frac{m\,e^2}{8\pi\varepsilon_0\,\varkappa\,p^2}\right)^2 \frac{1}{\sin^4\left(\frac{\chi}{2}\right)} \,. \tag{4.4}$$

If we introduce this in the formula (3.16) for the relaxation rate $\Gamma(p)$, the angular integral diverges! This means that something went wrong in setting up the Boltzmann collision term (2.5).

The divergence occurs for small-angle scatterings (small χ) or for scatterings with great impact parameters. Because of the special long-range nature of the Coulomb interaction, the electron is scattered even if it is far away from the ionized impurity, which by assumption is located at the origin. In a real semiconductor, however, there exist many impurities distributed at random with a certain number density n_I. The electron therefore may be scattered not only by the impurity at the origin but also by other impurities. If the electron was moving far away from the origin initially, it may be scattered by these other impurities even before it can approach the impurity at the origin. This effect, which may be called the scattering in medium, is not taken into account in the Boltzmann collision term (2.5).

Conwell and Weisskopf proposed to neglect the contribution of low-angle scatterings of impact parameters greater than half the average inter-impurity distance $r_I \equiv (n_I)^{-1/3}$. In this approximation, an electron is scattered by a particular ion only when it comes within the distance $\frac{1}{2} r_I$ of that ion. The limitation on impact parameter b $[b< \frac{1}{2} r_I]$ is equivalent to the limitation on the scattering angle χ, $[\chi > \chi_{min}]$ where

$$\chi_{min} = 2 \tan^{-1}[m^* e^2/2\pi\varepsilon_o \varkappa\, p^2\, r_I] \,. \tag{4.5}$$

Using χ_{min} as the lower limit of the χ-integration, we then obtain from (3.16) [Problem 4.1]

$$\Gamma_{C-W} = \frac{n_I\, p}{m^*} (2\pi) \int_{\chi_{min}}^{\pi} d\chi\, \sin\chi\;\; I(p,\chi)[1 - \cos\chi\,]$$

$$= (\hbar^{-2}\, m^*\, n_I\, e^4/8\pi\varepsilon_o^{\,2}\varkappa^2 p^3)\, \ell n\, [1 + (2\pi\varepsilon_o\varkappa\, p^2\, r_I/m^*\, e^2)^2]. \tag{4.6}$$

[use of (4.4-5)]

Since the number density of conduction electrons, n, in a semiconductor is low, the electrons are statistically non-degenerate [except for lowest temperatures]. We may therefore assume a Boltzmann distribution function for electrons:

$$f = n(2\pi\, m^* k_B\, T)^{-3/2} \exp\,(-\varepsilon_p/k_B\, T). \tag{4.7}$$

Using this and (4.6), we can compute the electrical conductivity σ from formula (3.19). The result translated for the mobility $\mu \equiv \sigma/en$ is given by [Problem 4.2]

$$\mu_{cw} = \frac{8}{\pi}\left(\frac{2}{\pi}\right)^{1/2} \frac{(4\pi\varepsilon_0 \varkappa)^2 (k_B\, T)^{3/2}}{e^3\, n_I\, m^{*1/2}\, F_{cw}(T)}\,, \tag{4.8}$$

where

$$F_{cw}(T) \equiv \ln\,[1 + (12\pi\varepsilon_0\, \varkappa\, r_I\, k_B\, T/e^2)^2]\,. \tag{4.9}$$

Generally speaking, the temperature behavior of μ_{cw} is more strongly influenced by $T^{3/2}$ in the numerator than by the T^2 terms in the logarithmic function $F_{cw}(T)$, and therefore the mobility rises as the temperature increases. This temperature dependence is compatible with a slight rise in the mobility on the low temperature side as seen in Figure 13.6.

Formula (4.8) was obtained without consideration of the electron-electron interaction. In most semiconductors, the densities of carriers and impurities, (n, n_I), are of comparable magnitude. Since the Coulomb interaction between electron and electron has the same magnitude as that between electron and charged impurity, we must take the electron-electron interaction into account in a realistic model. The most important effect of this interaction is the screening, which is brought about by moving electrons. Brooks and Herring [4] independently argued that by considering the screening, the low-angle scatterings can be effectively limited. This theory will now be reviewed.

The screening effect by electrons may be represented by the screened Coulomb potential [5]

$$V_D(r) = -\frac{e^2}{4\pi\varepsilon_0\,\varkappa}\,\frac{e^{-q_D r}}{r}, \tag{4.10}$$

where

$$q_D^{-1} \equiv (\varepsilon_0\,\varkappa\,k_B\,T/n\,e^2)^{1/2} \tag{4.11}$$

represents the Debye screening length. Notice that when the distance r exceeds the screening length q_D^{-1}, the potential $V_D(r)$ diminishes exponentially. This in turn limits the low-angle scattering associated with the long range nature of the Coulomb interaction. Brooks and Herring used the screened Coulomb potential (4.10) and arrived at a new formula for the mobility. This formula turns out to be very similar to the Conwell-Weisskopf formula (4.8); only the logarithmic factor $F_{CW}(T)$ is replaced by a new factor involving the Debye screening length.

The two model calculations presented indicate that the scattering by charged impurities can account for the low temperature behavior of the mobility. Note that the two mechanisms, which prevent the Coulomb divergence, are separate. Both corrections must be included in the realistic calculation of the mobility. Further discussions will be given in §13.10*.

Problem 4.1 Derive (4.6).

Problem 4.2* Derive (4.8).

13.5* Motion of a Charged Particle in an Electromagnetic Field

Let us consider a particle of mass m and charge q in a given electric and magnetic fields $(\vec{E},\vec{B})$. In this section we will study the motion of such a particle, first classically and then quantum mechanically. We will be interested mainly in those situations for which the electric field E is very small and the magnetic field B may be arbitrarily large but constant in space and time.

Let us first consider the case in which E = 0. Newton's equation of motion for a classical particle is

$$m \frac{d\vec{v}}{dt} = q \vec{v} \times \vec{B}. \tag{5.1}$$

We take the dot product of this equation with $\vec{v}$:

$$m \vec{v}\cdot\frac{d\vec{v}}{dt} = q \vec{v}\cdot(\vec{v} \times \vec{B}).$$

The r.h.s. vanishes since

$$\begin{aligned} &\vec{v}\cdot(\vec{v} \times \vec{B}) \\ &= (\vec{v} \times \vec{v})\cdot\vec{B} && \text{[use of (C.25)]} \\ &= 0. && \text{[use of (C.22)]} \end{aligned}$$

We therefore obtain

$$m \vec{v} \cdot \frac{d\vec{v}}{dt} = \frac{d}{dt} \left(\frac{1}{2} m v^2\right) = 0, \tag{5.2}$$

which means that <u>the kinetic energy is conserved.</u>

This result is valid regardless of how the magnetic field $\vec{B}$ varies in space. If the magnetic field $\vec{B}$ varies in time, an electric field is necessarily induced, and the above result will not hold in a strict manner.

In the case of a constant magnetic field, we can rewrite eq.(5.1) as

$$\frac{d\vec{v}}{dt} = \vec{v} \times \vec{\omega}_o, \tag{5.3}$$

where $\vec{\omega}_o$ is the constant vector pointing along the direction of the magnetic field and having the magnitude

$$\boxed{|\vec{\omega}_o| \equiv \omega_o \equiv \frac{qB}{m}.} \tag{5.4}$$

This quantity ω_o has the dimension of frequency and is called the cyclotron frequency. It is proportional to the magnetic-field strength and inversely proportional to the mass of the particle. For example, for an electron ($m = m_e$ and $q = -e$) in a field of 1000 Gauss, we have

$$\omega_o \sim 10^{10} \quad \text{sec}^{-1}.$$

From eq.(5.3) [or from the study in § 3.13] we can deduce that the motion of the electron consists of the uniform motion along the magnetic field with velocity v_z plus a circular motion with constant speed $v_\perp$ about the magnetic field. The radius R of this circular orbit about the magnetic field, called the cyclotron radius, can be determined from (centripetal force) = (magnetic force) or

$$m\, v_\perp^2 / R = e\, v_\perp B. \tag{5.5}$$

Solving this equation with respect to R, we obtain

$$R = \frac{m\,v_\perp}{e\,B} = \frac{v_\perp}{\omega_o} . \tag{5.6}$$

For the case : $B = 1000$ Gauss and $v_\perp = 10^5$ cm/sec, the cyclotron radius R is of the order 10^{-5}cm. Note that the radius is inversely proportional to B. Thus, as the magnetic field gets greater, the electron spirals around more rapidly in smaller orbits. As we saw eariler, the magnetic field $\vec{B}$ does not change the kinetic energy of the electron. The cyclotron motion therefore describes an orbit of a constant energy. This important feature is preserved in quantum mechanics. In fact, it can be used to explore the energy band structure for the conduction electrons in a metal or semiconductor [5].

The motion of an electron in a static, uniform electric and magnetic fields is found to be similar to the case we have just discussed.

First, if the electric field $\vec{E}$ is parallel to $\vec{B}$, the motion perpendicular to $\vec{B}$ is not affected so that the electron spirals around the magnetic field lines as before. The motion along $\vec{B}$ is subjected to a uniform acceleration equal to $-e\vec{E}/m$.

Let us now turn to the second and more interesting situation in which the electric and magnetic fields are perpendicular to each other. Let us introduce a quantity $\vec{V}_D$ defined by

$$\vec{V}_D \equiv \vec{E} \times \vec{B}/B^2, \tag{5.7}$$

which is constant in time and has the dimension of velocity. Let us decompose the velocity $\vec{v}$ in two parts :

$$\vec{v} \equiv \vec{v}' + \vec{V}_D . \tag{5.8}$$

Substituting this into the equation of motion

$$m \frac{d\vec{v}}{dt} = \text{Lorentz force}$$

$$= q [\vec{E} + \vec{v} \times \vec{B}], \qquad (5.9)$$

we obtain

$$\text{the l.h.s.} = m \frac{d}{dt} (\vec{v}' + \vec{V}_D) = m \frac{d\vec{v}'}{dt}$$

$$\text{the r.h.s.} = q (\vec{E} + \vec{V}_D \times \vec{B} + \vec{v}' \times \vec{B}) \qquad \text{[see of (5.8)]}$$

$$= q [\vec{E} + \frac{(\vec{E} \times \vec{B}) \times \vec{B}}{B^2}] + q \vec{v}' \times \vec{B} \qquad \text{[use of (5.7)]}$$

$$= q [E + \frac{\vec{B} (\vec{E} \cdot \vec{B}) - \vec{E} (\vec{B} \cdot \vec{B})}{B^2}] + q \vec{v}' \times \vec{B} \qquad \text{[use of (C.26)]}$$

$$= 0 + q \vec{v}' \times \vec{B}$$

or

$$m \frac{d\vec{v}'}{dt} = q \vec{v}' \times \vec{B}, \qquad (5.10)$$

which has same form as eq.[5.1]. The motion can then be regarded as the superposition of the motion in a uniform magnetic field <u>and</u> a drift of the cyclotron orbit with the constant velocity $\vec{V}_D$ as given in (5.7). Such a motion is indicated in Figure 13.7.

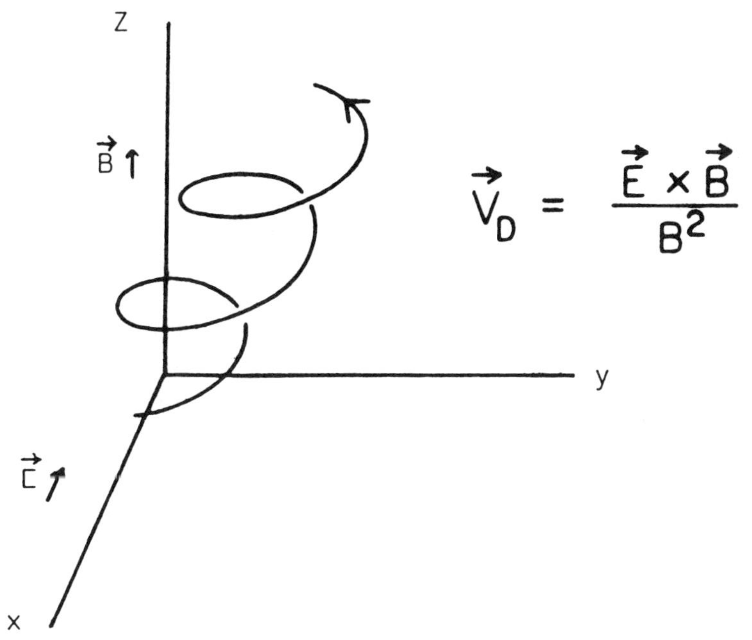

Fig. 13.7 A charged particle spirals around a slanted axis under the action of the electric and magnetic fields, $\vec{E}$ and $\vec{B}$ perpendicular to each other.

The drift velocity $\vec{V}_D$ in (5.7) is perpendicular to both $\vec{E}$ and $\vec{B}$. This implies that the weak electric field will induce a macroscopic current $\vec{j}$ in the direction perpendicular to both $\vec{E}$ and $\vec{B}$:

$$\vec{j} = - e n \frac{\vec{E} \times \vec{B}}{B^2} , \tag{5.11}$$

where n is the number density of the electrons. This current is often called the Hall current. We note that the drift velocity $\vec{V}_D$ is independent of charge and mass. This turns out to be a very important property. The measurement of this effect in a semi-conductors gives information about the sign of the charge (electron or hole) and the number density of carriers, uninfluenced by the complexity of the energy band structure [5].

We have so far discussed the motion of an electron, using classical mechanics. Most of the qualitative features hold also in quantum mechanics.

A most distinctive quantum feature is the quantization of the cyclotron orbit. To see this in more detail, let us calculate the energy levels of an electron in a constant magnetic field $\vec{B}$. We choose the vector potential

$$(A_x, A_y, A_z) = (0, Bx, 0), \tag{5.12}$$

which yields a constant field $\vec{B}$ in the z - direction as can be verified easily from $\vec{B} = \nabla \times \vec{A}$. [Problem 5.1]. The Hamiltonian H then is given by

$$H = \frac{1}{2m} (\vec{p} + e\vec{A})^2 = \frac{1}{2m} [p_x^2 + (p_y + eBx)^2 + p_z^2]. \tag{5.13}$$

The Schrödinger equation can now be written down from (8.8.4) as follows :

$$-\frac{\hbar^2}{2m} \left[\frac{\partial^2}{\partial x^2} \psi(x,y,z) + \left(\frac{\partial}{\partial y} + \frac{ieB}{\hbar} x \right)^2 \psi + \frac{\partial^2 \psi}{\partial z^2} \right] = E \psi. \tag{5.14}$$

Since the Hamiltonian H contains neither y nor z explicitly, we may try the wave function ψ of the form :

$$\psi(x,y,z) = e^{-i(k_y y + k_z z)} \varphi(x). \tag{5.15}$$

Substitution of this expression into eq.(5.14) yields the following equation for $\varphi(x)$: [Problem 5.2]

$$\left[-\frac{\hbar^2}{2m}\frac{d^2}{dx^2} + \frac{1}{2} m \omega_o^2 \left(x - \frac{\hbar k_y}{eB} \right)^2 \right] \varphi(x) = E_1 \varphi(x), \quad (5.16)$$

where

$$E_1 \equiv E - \frac{\hbar^2 k_z^2}{2m} . \quad (5.17)$$

Comparison with eq.(8.9.5) indicates that eq.(5.16) is the wave equation for a harmonic oscillator with the angular frequency

$$\omega_o \equiv \frac{eB}{m} ,$$

and the center of oscillation displaced from the origin by

$$X = \frac{\hbar k_y}{eB} . \quad (5.18)$$

According to the study in § 8.9, the energy eigenvalues are given by (8.9.13):

$$E_1 = (n + \frac{1}{2}) \hbar\omega_o , \quad n = 0,1,2, \ldots \quad (5.19)$$

Combining this with (5.17) we obtain

$$\boxed{E = (n + \frac{1}{2}) \hbar\omega_o + \frac{\hbar^2 k_z^2}{2m} .} \quad (5.20)$$

These energy eigenvalues are called Landau levels. The corresponding quantum states, called the Landau states, are characterized by the quantum number (n, k_y, k_z). We note that the energies do not depend on the quantum number k_y, and are therefore highly degenerate.

As we see here, the Landau states are quite different from the momentum eigenstates. This has a significant consequence on the quantum theory of transport phenomena. Let us recall that the Boltzmann collision term for a field-free case is expressed in terms of a scattering crose section, a concept which can be regarded as a quantum transition from one momentum state to another momentum state. With a finite magnetic field present, we cannot envision a scattering and therefore cannot write down the Boltzmann collision term in an obvious manner.

The electron in a Landau state can be viewed as in rotation with the angular frequency ω_0 around the magnetic field. If a radiation with the same frequency ω_0 is applied, the electron may jump up from one Landau state to another with absorption of the photon energy $\hbar\omega_0$. This generates a phenomenon of cyclotron resonance, which will be discussed briefly in § 13.11*.

Problem 5.1 Show that the vector potential given by (5.12) generates a magnetic field pointing in the positive z-direction.

Problem 5.2 Derive eq.(5.16) from (5.14) and (5.15).

13.6* Generalized Ohm's Law. Absorption Power

When a beam of radiation falls on a semiconductor, part of the radiation energy is absorbed. We will study such absorption loss in the present section.

Let us suppose that a semiconductor sample is irradiated by a monochromatic wave as shown in Figure 13.8.

Fig.13.8 A semiconductor sample is irradiated by a monochromatic microwave characterized by the wave vector $\vec{k}$ and angular frequency ω. This generates an oscillatory current in the sample, and some radiation energy is absorbed.

The electric field $\vec{E}$ associated with the radiation is a running wave whose space and time variations are characterized by the wave vector $\vec{k}$ and the frequency ω.

We may represent this field by

$$\vec{E}(\vec{r},t) = \vec{E}\, e^{i\omega t - i\vec{k}.\vec{r}}. \tag{6.1}$$

Responding to this field, a current wave characterized by the same set of the k-vector-frequency, $(\vec{k},\omega)$, may arise within the sample. Normally this running-wave part of the current is proportional to the field $\vec{E}$. Let us express this linear relation between part of the current density $\vec{j}$ and the field as follows:

$$\boxed{\text{the running-wave part of } j_\mu(\vec{r},t) = \sum_{\nu = x,y,z} \sigma_{\mu\nu}(\vec{k},\omega)\, E_\nu\, e^{i\omega t - i\vec{k}\cdot\vec{r}}.} \tag{6.2}$$

Here the proportionality factor $\sigma_{\mu\nu}$ must be a tensor since the two vectors $(\vec{j},\vec{E})$ are related linearly. Eq.(6.2) represents the generalized Ohm's Law, see (1.6), and the tensor $\sigma_{\mu\nu}(\vec{k},\omega)$ is called the dynamic conductivity tensor. This tensor $\vec{\vec{\sigma}}(\vec{k},\omega)$ in general depends on the material and on the temperature. It also depends on the frequency ω and the wave vector $\vec{k}$.

The field associated with the same running wave can equally well be represented by

$$\vec{E}(\vec{r},t) = \vec{E}\, e^{-i\omega t + i\vec{k}\cdot\vec{r}}. \tag{6.3}$$

The generalized Ohm's law in terms of this expression can be formulated as the complex conjugate of eq.(6.2). That is,

$$\text{the running-wave part of } j_\mu(\vec{r},t) = \sum_{\nu = x,y,z} \sigma^*_{\mu\nu}(\vec{k},\omega)\, E_\nu e^{-i\omega t + i\vec{k}\cdot\vec{r}}. \tag{6.4}$$

It is instructive to rewrite eq.(6.2) as

$$j_\mu(\vec{r},t) = \sum_\nu \sigma_{\mu\nu}(\vec{k},\omega)\, E_\nu\, e^{i\omega t - i\vec{k}\cdot\vec{r}} + j_\mu^{(tr)}(\vec{r},t), \tag{6.5}$$

where $\vec{j}^{(tr)}$ represents the transient current, which includes everything but the persistent current wave generated by the applied EM wave.

It is important to note that the generalized Ohm's law for an oscillatory field is expressed in terms of a complex field $\vec{E}(\vec{r},t) = \vec{E}\, e^{i\omega t - i\vec{k}\cdot\vec{r}}$. This is unavoidable if we postulate the linear relationship between the monochromatic running wave of the current and that of the field. A consequence of this is that the dynamic conductivity $\overleftrightarrow{\sigma}(\vec{k},\omega)$ is a complex tensor.

The wavelength $\lambda \equiv 2\pi/k$ of the radiation has a magnitude of the order of 4000 Å for blue light, and is much greater for infrared or microwave. Thus, the wavelength λ in normal experiments is far greater than the linear size of the electron wave packet, which is of the order of 1 Å. The electron therefore is subject to a field varying very slowly in space, as shown in Figure 13.9. In such a case, we expect the k-dependence of the dynamic conductivity $\overleftrightarrow{\sigma}(\vec{k},\omega)$ to be negligible. Omitting the $\vec{k}$-dependence in eq.(6.5), we have

$$\text{part of } j_\mu(t) = \sum_{\nu = x,y,z} \sigma_{\mu\nu}(\omega)\, E_\nu\, e^{i\omega t}. \tag{6.6}$$

This represents the generalized Ohm's law in the homogeneous-field approximation, and will be used in most applications of the theory.

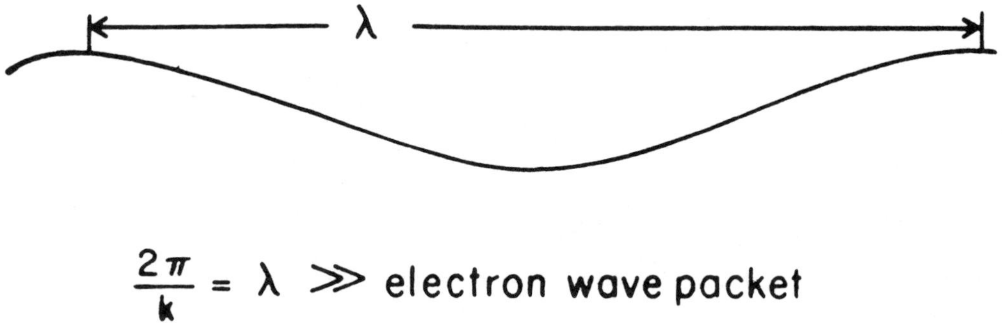

Fig. 13.9 The wavelength of a typical radiation used in the normal experiments, is much greater than the linear size of the electron wave packet.

Let us now look at the interaction between electrons and radiation. The work done per unit time to a typical conduction electron is $(-e\vec{E}) \cdot \frac{d\vec{r}}{dt}$ according to the usual rule that (force) times (velocity) = power . Considering all conduction electrons in the sample, the power per unit volume delivered by the electric field $\vec{E}(t)$ is given by

$$V^{-1} < \sum_k [-e\vec{E}(t)] \cdot \frac{d\vec{r}_k}{dt} > = \vec{E}(t) \cdot [V^{-1} < \sum_k (-e) \frac{d\vec{r}_k}{dt} >]. \quad (6.7)$$

The quantity in the angular brackets is just the current density of conduction electrons,

$$V^{-1} \quad < -e \sum_k \frac{d\vec{r}_k}{dt} > \equiv \vec{j}(t). \quad (6.8)$$

Therefore, we can rewrite eq.(6.7) in the form :

$$V^{-1} < \sum_k [-e\vec{E}(t)] \cdot \frac{d\vec{r}_k}{dt} > = \vec{E}(t) \cdot \vec{j}(t). \quad (6.7a)$$

In the normal stationary-state experiments, the power averaged over a time much greater than the period of oscillation $2\pi/\omega$, is measured. The average power p per unit volume is then given by

$$\boxed{p \equiv \lim_{t\to\infty} \frac{1}{t} \int_0^t dt' \, \vec{E}(t') \cdot \vec{j}(t').} \quad (6.9)$$

This equation means that the time integral,

$$W(t) \equiv \int_0^t dt' \, \vec{E}.\vec{j}, \quad (6.10)$$

representing the work accumulated up to the time t, must have a component proportional to t for a large t. In mathematical terms, we may write (6.9) in the form :

$$W(t) \to pt \text{ as } t \to \infty. \tag{6.11}$$

Other work done which may arise from the transient current $\vec{j}^{(tr)}$ and from other causes, will not grow with the time t, and then does not contribute to the average power p. See below for examples of such contributions.

We now wish to find an expression for the average power p with the assumption of generalized Ohm's law (6.6).

For definiteness, we assume that the applied EM wave is plane-polarized such that

$$E_x(t) = E \cos \omega t, \; E_y = E_z = 0, \tag{6.12}$$

where E is the field amplitude. Introducing these expressions in (6.10), we obtain

$$W(t) = \int_0^t dt' \, E_x(t') \, j_x(t').$$

Writing

$$E_x(t') = \frac{1}{2} E \, (e^{i\omega t'} + e^{-i\omega t'}), \tag{6.13}$$

we apply Ohm's law (6.6) and its complex conjugate to the first and second terms, respectively, of this expression, and obtain

$$W(t) = \int_0^t dt' \; [\tfrac{1}{2} E(e^{i\omega t'} + e^{-i\omega t'})] \; [\sigma_{xx}(\omega) \tfrac{1}{2} E \, e^{i\omega t'}$$

$$+ \sigma_{xx}^*(\omega) \tfrac{1}{2} E \, e^{-i\omega t'} + \text{(transient current)}]$$

$$= \frac{1}{4} E^2 t[\sigma_{xx}(\omega) + \sigma^*_{xx}(\omega)] + \text{oscillatory terms in } t. \qquad (6.14)$$

We observe that the first term is proportional to t. Comparing eq.(6.14) with (6.11), we then obtain

$$p = \frac{1}{4} E^2 [\sigma_{xx}(\omega) + \sigma^*_{xx}(\omega)]$$

or

$$\boxed{p = \frac{1}{2} E^2 \, \text{Re}\{ \sigma_{xx}(\omega) \}.} \qquad \text{(plane-polarized wave)} \qquad (6.15)$$

Note that the obtained power is real, and is proportional to the square of the field amplitude E. The power p is expressed concisely in terms of the conductivity tensor-element $\sigma_{xx}(\omega)$.

Formula (6.15) was obtained with the assumption that the EM wave is linearly polarized as given by (6.12). If the applied EM wave is circularly polarized such that

$$E_x(t) = E \cos \omega t, \; E_y(t) = E \sin \omega t, \; E_z(t) = 0, \qquad (6.16)$$

the average power p can be expressed by

$$p = \frac{1}{2} E^2 \, \text{Re} \{ \sigma_{xx}(\omega) + \sigma_{yy}(\omega) + i\sigma_{xy}(\omega) - i\sigma_{yx}(\omega) \}, \qquad (6.17)$$

whose derivation is left as an excercise for the reader [Problem 6.1].

Problem 6.1 Derive an expression for the absorption power for a circularly polarized microwave. The answer is given by (6.17).

13.7* Kubo's Formula for the Electrical Conductivity

Before the mid fifties, the quantitative calculations of transport coefficients had been done invariably in terms of the Boltzmann equation. In 1957 Kubo published a classic paper [6] in which the electrical conductivity is expressed in a closed form without going through the kinetic equation. We will discuss kubo's formula in this section.

Let us consider a system of many particles which is characterized by a time-independent Hamiltonian H. At the initial time : t=0, the system, by assumption, is in equilibrium, and its thermodynamic state is described in terms of the canonical density operator :

$$\rho \equiv e^{-\beta H}/\mathrm{Tr}\{e^{-\beta H}\} \equiv e^{-\beta H}/Z. \qquad (7.1)$$

We then apply a weak electric field $\vec{E}(t)$, which is associated with a monochromatic wave,

$$\vec{E}(t) \equiv \vec{E}\, e^{i\omega t}. \qquad (7.2)$$

Responding to this perturbation an electric current which oscillates with the same frequency ω will be generated in the system. After a long time t, the generalized Ohm's law as given in (6.6) may be postulated :

$$\text{part of } j_\mu(t)$$

$$= \sum_{\nu = x,y,z} \sigma_{\mu\nu}(\omega)\, E_\nu e^{i\omega t} \quad \text{for large } t, \qquad (7.3)$$

where $\sigma_{\mu\nu}(\omega)$ is the conductivity tensor. We note here that to determine the conductivity $\sigma_{\mu\nu}(\omega)$ we are required to calculate that component of the current density which is linear in the complex field $\vec{E}(t) = \vec{E}\, e^{i\omega t}$.

For a positive time $t > 0$, the quantum density (or statistical) operator $\rho'(t)$ changes with time, following the quantum Liouville equation [see (9.8.5)] :

$$i\hbar \frac{\partial \rho'(t)}{\partial t} = [H'(t),\ \rho'(t)], \tag{7.4}$$

where $H'(t)$ is the total Hamiltonian consisting of the internal Hamiltonian H and the potential $\Phi(t)$ representing the <u>interaction energy</u> with the probinq field :

$$H'(t) = H + \lambda\, \Phi(t). \tag{7.5}$$

Here, λ is the coupling constant (which may be set equal to unity at the end of the calculations). The interaction energy $\Phi(t)$ can explicitly be represented by

$$\Phi(t) = -\vec{P}\cdot\vec{E}(t), \tag{7.6}$$

where $\vec{P}$ is the <u>total polarization vector</u> defined by

$$\vec{P} \equiv \sum_j (-e)\vec{r}_j. \tag{7.7}$$

In fact, the j-th electron is subjected to the force given by [Problem 7.1]

$$-\frac{\partial \Phi}{\partial \vec{r}_j} = -e\,\vec{E}(t), \qquad \text{[use of (7.6)]} \tag{7.8}$$

as it should be.

We now wish to solve eq.(7.4) with the initial condition given by (7.1). As mentioned earlier, we only need a solution valid up to the first order in the coupling constant λ (or the field $\vec{E}$) . We can accomplish this goal in the following manner. As we can check by simple calculations [Problem 7.2], the solution of eq.(7.4) subject to the initial condition (7.1), satisfies the following integral equation :

$$\rho'(t) = \rho - \frac{i\lambda}{\hbar} \int_0^t d\tau \, e^{-i(t-\tau)H/\hbar} [\Phi(\tau), \rho'(\tau)] \, e^{i(t-\tau)H/\hbar} . \tag{7.9}$$

Since the last term is already of the first order in λ , after substituting ρ for $\rho'(\tau)$, we obtain the required solution in the form :

$$\rho'(t) = \rho - \frac{i\lambda}{\hbar} \int_0^t d\tau \, e^{-i(t-\tau)H/\hbar} [\Phi(\tau), \rho] \, e^{i(t-\tau)H/\hbar} . \tag{7.10}$$

We now set λ equal to unity, substitute the explicit form of $\Phi(\tau)$ from (7.6), and obtain

$$\rho'(t) = \rho - \frac{i}{\hbar} \int_0^t d\tau \, e^{-i(t-\tau)H/\hbar} [-\vec{P}\cdot\vec{E} \, e^{i\omega\tau}, \rho] \, e^{i(t-\tau)H/\hbar}$$

$$= \rho + \frac{i}{\hbar} \int_0^t d\tau' \, e^{-i\tau' H/\hbar} [\vec{P}\cdot\vec{E}, \rho] \, e^{i\tau' H/\hbar} \, e^{i\omega(t-\tau')} . \tag{7.11}$$

$[\tau' = t - \tau]$

Using this, we can calculate the current density. Comparing the result with Ohm's law (7.3), we can then find the conductivity tensor. Before actually performing this program, let us re-express (7.11) in a more compact form.

We first note the identity :

$$[A, e^{-\beta H}] = -i\hbar\, e^{-\beta H} \int_0^\beta d\beta_1\, e^{\beta_1 H} \dot{A}\, e^{-\beta_1 H}, \tag{7.12}$$

where

$$\dot{A} = (i/\hbar)\,[H,A] \tag{7.13}$$

denotes the time-derivative of a quantum Heisenberg operator A [see (8.14.20)]. Eq.(7.12) may be proved as follows. Differentiating $e^{\beta_1 H} A\, e^{-\beta_1 H}$ with respect to β_1, we have

$$\frac{d}{d\beta_1}[e^{\beta_1 H} A\, e^{-\beta_1 H}] = (e^{\beta_1 H} H)\, A\, e^{-\beta_1 H} + e^{\beta_1 H} A\, (-H\, e^{-\beta_1 H})$$

$$= e^{\beta_1 H} [H,A]\, e^{-\beta_1 H}.$$

Integrating both sides from 0 to β and multiplying the results by the factor $e^{-\beta H}$ from the left, we obtain

$$e^{-\beta H} \int_0^\beta d\beta_1 \frac{d}{d\beta_1}[\, e^{\beta_1 H} A\, e^{-\beta_1 H}] \qquad \text{[from } \ell\text{.h.s.]}$$

$$= e^{-\beta H} [e^{\beta_1 H} A\, e^{-\beta_1 H}]_0^\beta$$

$$= e^{-\beta H} [\, e^{\beta H} A\, e^{-\beta H} - A] \equiv [\, A, e^{-\beta H}]$$

$$= e^{-\beta H} \int_0^\beta d\beta_1\, e^{\beta_1 H} [H,A]\, e^{-\beta_1 H} \qquad \text{[from r.h.s.]}$$

$$= e^{-\beta H} \int_0^\beta d\beta_1\, e^{\beta_1 H} (\hbar/i)\, \dot{A}\, e^{-\beta_1 H}, \quad \text{[use of (7.13)]}$$

which establishes (7.12).

By applying the identity (7.12) with $A \equiv P_x = \sum_j (-e)\, x_j$ and $\dot{A} \equiv \dot{P}_x = \sum_j (-e)\, \dot{x}_j$ = the x-component of the total current $\equiv J_x$, we obtain

$$[\, P_x,\, e^{-\beta H}] = -\, i\hbar\, e^{-\beta H} \int_0^\beta d\beta_1\, e^{\beta_1 H}\, J_x\, e^{-\beta_1 H}. \qquad (7.14)$$

Using this and similar relations for y- and z-components, we can then re-express (7.11) as

$$\rho'(t) = \rho + \int_0^t d\tau\, e^{i\omega(t-\tau)} \int_0^\beta d\beta_1\, e^{-(\beta-\beta_1)H}\, e^{-i\tau H/\hbar}\, \vec{J}\cdot\vec{E}\, e^{-\beta_1 H}\, e^{i\tau H/\hbar}. \qquad (7.15)$$

Let us now calculate the average current through the standard formula : [see (9.8.3)]

$$\langle J_\mu \rangle_t \equiv \mathrm{Tr}\{ J_\mu\, \rho'(t) \} . \qquad (7.16)$$

Since there is no current in equilibrium, the first term $\rho = e^{-\beta H}/Z$ of (7.15) generates no current. The second term (the double integral) generates

$$j_\mu(t) \equiv \frac{1}{V} \langle J_\mu \rangle_t$$

$$= \sum_\nu \frac{1}{ZV} \int_0^t d\tau\, e^{-i\omega\tau} \int_0^\beta d\beta_1 \quad \mathrm{Tr}\{ J_\mu\, e^{-(\beta-\beta_1)H}\, e^{-i\tau H/\hbar}\, J_\nu\, e^{-\beta_1 H}\, e^{i\tau H/\hbar}\, E_\nu\, e^{i\omega t} \}$$

[use of (7.2) and (7.15)] (7.17)

Comparing this expression with (7.3), we then obtain

$$\sigma_{\mu\nu} = \frac{1}{ZV} \int_0^\infty d\tau\, e^{-i\omega\tau} \int_0^\beta d\beta_1\, \mathrm{Tr}\{ J_\mu e^{-(\beta-\beta_1)H} e^{-i\tau H/\hbar} J_\nu e^{-\beta_1 H} e^{i\tau H/\hbar}\}$$

$$= \frac{1}{V} \int_0^\infty d\tau\, e^{-i\omega\tau} \int_0^\beta d\beta_1\, \mathrm{Tr}\{ Z^{-1} e^{-\beta H} e^{\beta_1 H} J_\nu e^{-\beta_1 H} e^{i\tau H/\hbar} J_\mu e^{-i\tau H/\hbar}\}$$

$$[\mathrm{Tr}\{AB\} = \mathrm{Tr}\{BA\}]$$

or

$$\boxed{\sigma_{\mu\nu} = \int_0^\infty d\tau\, e^{-i\omega\tau} \int_0^\beta d\beta_1\, \mathrm{Lim}\, \frac{1}{V}\, \mathrm{Tr}\{\rho\, J_\nu(-i\hbar\beta_1)\, J_\mu(t)\},} \qquad (7.18)$$

where

$$\vec{J}(t) \equiv e^{itH/\hbar}\, \vec{J}\, e^{-itH/\hbar} \qquad (7.19)$$

is the total current operator in the Heisenberg picture. Also note that $\vec{J}(-i\hbar\beta_1) \equiv e^{\beta_1 H}\, \vec{J}\, e^{-\beta_1 H}$. The generalized Ohm's law is valid in a rigorous sense only in the bulk limit. Because of this, the bulk limit(Lim) is indicated explicitly in (7.18).

The formula (7.18) is referred to as Kubo's formula or the current correlation function formula. It gives a rigorous closed expression for the conductivity in microscopic terms. Once the Hamiltonian of the system is given, we can in principle compute the conductivity. This situation is similar to that of the calculation of the equilibrium properties of the system. The actual computation of formula (7.18) is much more involved than that of the partition function Z. A few simple applications will be considered in the following sections.

Problem 7.1 Verify (7.8), using (7.6) and (7.7).

Problem 7.2 By differentiating $\rho'(t)$ in (7.9) with respect to t, show that $\rho'(t)$ satisfies eq.(7.4).

13.8* More about Kubo's Formula

A. Reduction of Kubo's formula

Kubo's formula as given by (7.18) requires integration with respect to the reciprocal temperature β_1 . This process may be eliminated in the following manner.

Let us consider the operator exp ($-\beta H'$) with the modified Hamiltonian

$$H' \equiv H - \vec{J}\cdot\vec{u}, \tag{8.1}$$

where $\vec{u}$ is a constant vector independent of the dynamical variables. Its derivative with respect to u_x can be represented by

$$\frac{\partial}{\partial u_x} e^{-\beta H'}$$

$$\equiv \frac{\partial}{\partial u_x} e^{-\beta H + \beta \vec{J}\cdot\vec{u}} = \int_0^\beta d\beta_1 \, e^{-(\beta-\beta_1)H'} J_x e^{-\beta_1 H'}. \tag{8.2}$$

This formula may be proved as follows. We consider

$$\frac{\partial}{\partial \beta_1} [e^{\beta_1 H'} \frac{\partial}{\partial u_x} e^{-\beta_1 H'}]$$

$$= e^{\beta_1 H'} [(H - \vec{J}\cdot\vec{u}) \frac{\partial}{\partial u_x} - \frac{\partial}{\partial u_x} (H - \vec{J}\cdot\vec{u})] e^{-\beta_1 H'} \quad \text{[use of (8.1)]}$$

$$= e^{\beta_1 H'} J_x e^{-\beta_1 H'}.$$

Integrating both sides with respect to β_1 from 0 to β , and multiplying the results by $e^{-\beta H'}$ from the left, we obtain the desired equation (8.2).

Expanding $\mathrm{Tr}\,\{ e^{-\beta H + \beta \vec{J}\cdot\vec{u}} \}$ in $\vec{u}$ and noting that

$$\mathrm{Tr}\,\{ \vec{J}\, e^{-\beta H} \} = 0, \text{ (no current in equilibrium)} \tag{8.3}$$

we can show that [Problem 8.1]

$$\mathrm{Tr}\,\{ e^{-\beta H + \beta \vec{J}\cdot\vec{u}} \} = \mathrm{Tr}\{ e^{-\beta H} \} + \text{quadratic terms in } \vec{u}. \tag{8.4}$$

Using this and (8.2), we can rewrite (7.18) in the form :

$$\boxed{\sigma_{\mu\nu}(\omega) = \int_0^\infty dt\, e^{-i\omega t} \,\mathrm{Lim}\, \frac{1}{V} \frac{\partial}{\partial u_\nu} \,\mathrm{Tr}\,\{ \rho' J_\mu(t) \} ,} \tag{8.5}$$

where

$$\rho' \equiv e^{-\beta H + \beta \vec{J}\cdot\vec{u}} / \mathrm{Tr}\,\{ e^{-\beta H + \beta \vec{J}\cdot\vec{u}} \} \tag{8.6}$$

is the modified density operator ; and the u_ν- derivatives are to be calculated at $u = 0$.

The alternate form of Kubo's formula,(8.5), can be handled in a simpler manner than the original form (7.18). First there is no need for carrying out the β_1 - integration. Second and more importantly, we can work with the average current rather than with the current - current correlation function .

B. Classical limit

The derivation of Kubo's formula (7.18) and the subsquent transformation to (8.5) were carried out with the assumption of quantum mechanics. This is of course desirable and necessary when we deal with the motion of electrons. In certain cases including the case of the ionic conduction, we may use the classical mechanical limit of Kubo's formula (7.18). In this limit, the β_1 - integration simply yields the factor β . We then obtain

$$\sigma_{\mu\nu}(\omega) = \beta \int_0^\infty dt\, e^{-i\omega t} \operatorname{Lim} \frac{1}{V} \operatorname{Tr}\{\rho J_\nu J_\mu(t)\} \quad \text{(classical limit)} \tag{8.7}$$

This formula allows a simple interpretation : the conductivity can be obtained by evaluating the integral of the current-current correlation at different times.

We note that the transformed formula (8.5) can still be used with appropriate redefinitions of the quantities in terms of classical mechanics.

C. The true conductivity

In the derivation of Kubo's formula, we assumed that a small external field $\vec{E}$ generates a small perturbation in the system as explicitly represented by (7.11) for the change in the density operator. The mathematical treatment (perturbation theory) by means of the expansion in the coupling constant λ is fairly complicated but straightforward. In the real experimental condition in which we measure the conductivity the system in general undergoes a far more complicated change. In fact, let us take a semiconductor with impurities. When a small external field $\vec{E}$ is applied, the dynamical motion of the electrons creates an induced field $\vec{E}_i$ due to the Coulomb interaction among the electrons. Such induced fields are far from negligible in practice. The electrons in the system are then expected to move under the action of the total electric field

$$\vec{E}_T \equiv \vec{E} + \vec{E}_i. \tag{8.8}$$

This implies that the generalized Ohm's law must take the form

$$j_\mu = \sum_\nu \sigma'_{\mu\nu} E_{T\nu}, \tag{8.9}$$

which defines the true conductivity tensor $\sigma'_{\mu\nu}$. On the other hand, the formula (7.3) defines the tensor $\overleftrightarrow{\sigma}$ as the proportionality factor between the current density $\vec{j}$ and the external field $\vec{E}$. Kubo's formula (7.18) derived on the basis of (7.3), therefore does not correspond to the true conductivity $\overleftrightarrow{\sigma}'$. To relate the conductivities $\overleftrightarrow{\sigma}$ and $\overleftrightarrow{\sigma}'$, a more careful study is therefore necessary. The detailed calculations [7] however indicate that the true conductivity $\overleftrightarrow{\sigma}'$ can be obtained from a slight modification of (7.18).

D. Einstein relation.

When a system is not homegeneous, that is, its density changes from place to place, there will be a current. If the density gradient is small, the particle-current density $\vec{j}'$ will be proportional to the density gradient [see § 4.1] :

$$\vec{j}' = - D \nabla n, \tag{8.10}$$

where the proportionality constant D is called the diffusion coefficient. This empirical law (Fick's law) holds whether the particles constituting the system carry charge or not.

Let us now apply (8.10) to conduction electrons. The particle-current density $\vec{j}'$ will generate the electric current density $\vec{j}$. They are distinct from each other only by the constant factor -e :

$$\vec{j} = (-e)\,\vec{j}'. \tag{8.11}$$

This means that the electrical current can be generated by the density gradient as well as by the electric field as it is known in Ohm's law (1.6). The responses of the system to the electrodynamical force ($-e\,\vec{E}$) and the non-mechanical "force" ($-\nabla n$) are expected to be similar; and therefore the diffusion coefficient D will be proportional to the electric conductivity σ :

$$\boxed{D \propto \sigma,} \tag{8.12}$$

which is known as the Einstein relation or Nernst-Einstein relation. This relation is observed experimentally in a general manner.

The Einstein relation (8.12) suggests that the diffusion coefficient D may also be expressed in the form (8.7) of the current-current correlation integral. This turns out to be true [8] for the classical mechanical systems although its derivation is much more involved [since no simple perturbation theory can be applied]. A detailed quantum mechanical calculation indicates however [9] that the Einstein relation does not hold strictly in the presence of a high magnetic field.

Problem 8.1 Prove (8.4).

13.9*. The Dynamic Conductivity of Free Electrons.

As an application of Kubo's formula, we will calculate, in this section, the dynamic conductivity tensor $\sigma_{\mu\nu}(\omega)$ for free electrons.

A The case in which there is no magnetic field : B = 0.

We may start with formula (8.5). Let us compute this formula, using classical mechanics. The total current $\vec{J}$ can be written as

$$\vec{J} \equiv \sum_k (-e)\vec{v}_k \equiv \sum_k \vec{j}_k, \tag{9.1}$$

where

$$\vec{j} \equiv -e\,\vec{v} = -e\,m^{-1}\,\vec{p} \equiv \vec{j}(\vec{p}) \tag{9.2}$$

represents the <u>single-particle electric current.</u> For free electrons, the momentum of each electron is conserved. This means that the total current $\vec{J}$ also is conserved. That is,

$$\vec{J}(t) = \vec{J}. \tag{9.3}$$

Let us write the total current $\vec{J}$ in the form :

$$\vec{J} = \sum_k (-e\,m^{-1}) \int d^3p\ \delta^{(3)}(\vec{p}_k - \vec{p})\ \vec{p}_k \qquad \text{[use of (4.2.5)]}$$

$$= (-e\,m^{-1}) \int d^3p\ \vec{p} \sum_k \delta^{(3)}(\vec{p}_k - \vec{p}). \quad \text{[use of (4.2.8)]} \tag{9.4}$$

Using this, we obtain

$$\mathrm{Tr}\{\ \rho'\ \vec{J}(t)\} \ =\ \mathrm{Tr}\{\ \rho'\ \vec{J}\} \quad \text{[use of (9.3)]}$$

$$= \int d^3p\ \vec{j}(\vec{p})\ \mathrm{Tr}\ \{\ \rho'\ \sum_k \delta^{(3)}(\vec{p}_k - \vec{p})\} \quad \text{[use of (9.2) and (9.3)]}$$

$$= \int d^3p\ \vec{j}\ \ \varphi'(\vec{p}), \tag{9.5}$$

where

$$\varphi'(\vec{p})\ \equiv\ \mathrm{Tr}\{\ \rho'\ \sum_k\ \delta^{(3)}(\vec{p}_k - \vec{p})\} \tag{9.6}$$

represents the momentum distribution function. Using the explicit form (8.6) for the distribution function ρ', we obtain

$$\varphi'(\vec{p})\ =\ \frac{N}{(2\pi\ m/\beta)^{3/2}}\ e^{-\beta(\varepsilon_p - \vec{j}\cdot\vec{u})}. \tag{9.7}$$

Differentiation with respect to u_ν yields

$$\frac{\partial}{\partial u_\nu}\ \varphi'(\vec{p})\Big]_{u=0}\ =\ \frac{N}{(2\pi\ m/\beta)^{3/2}}\ \beta\ j_\nu\ e^{-\beta\varepsilon_p}. \tag{9.8}$$

We can now compute the conductivity $\sigma_{\mu\nu}$ as follows :

$$\sigma_{\mu\nu}$$

$$= \int_0^\infty dt\ e^{-i\omega t}\ \mathrm{Lim}\ \frac{1}{V}\frac{\partial}{\partial u_\nu}\ \mathrm{Tr}\{\ \rho'\ J_\mu(t)\} \quad \text{[use of (8.5)]}$$

$$= \lim \frac{1}{V}\frac{1}{i\omega}\frac{\partial}{\partial u_\nu}\int d^3p\; j_\mu \varphi'(\vec{p}) \qquad \text{[use of (9.3) and (9.5)]}$$

$$= \lim \frac{N}{i\omega V}\frac{\beta}{(2\pi\; m/\beta)^{3/2}}\int d^3p\; j_\mu\; j_\nu\; e^{-\beta\varepsilon_p} \quad \text{[use of (9.8)]}$$

$$= \lim \frac{n_o}{i\omega}\frac{\beta}{(2\pi\; m/\beta)^{3/2}}\left(\frac{e}{m}\right)^2\int d^3p\; p_\mu\; p_\nu\; e^{-\beta\varepsilon_p}\; .$$

$$\text{[use of (9.2)}; n_o \equiv \frac{N}{V}\text{]} \tag{9.9}$$

The integrand is odd unless $\mu = \nu$. We therefore obtain

$$\sigma_{\mu\nu} = 0 \quad \text{if} \quad \mu \neq \nu\; . \tag{9.10}$$

From the symmetry consideration, we also have

$$\sigma_{xx} = \sigma_{yy} = \sigma_{zz} \equiv \sigma\; . \tag{9.11}$$

After carrying out the $\vec{p}$ - integration, we obtain [Problem 9.1]

$$\boxed{\sigma = -\, i\,\frac{e^2 n_o}{m\omega}\; .} \tag{9.12}$$

This expression was first obtained by Drude, and is often called the Drude formula. This simple expression should be contrasted with the kinetic-theoretical formula for the static conductivity, (1.7):

$$\sigma = \frac{e^2}{m}\, n_o \tau_f .$$

Only the relaxation time τ_f is replaced by $-i\,\omega^{-1}$.

In summary, we can exhibit the conductivity tensor $\sigma_{\mu\nu}$ in the matrix form :

$$\overleftrightarrow{\sigma} = -i\,\frac{e^2 n_o}{\omega\, m} \begin{bmatrix} 1 & 0 & 0 \\ 0 & 1 & 0 \\ 0 & 0 & 1 \end{bmatrix}. \tag{9.13}$$

All elements are pure imaginary or zero. According to formula (6.15) or (6.17) this means that there will be no absorption of the electromagnetic energy by the system.

The Drude formula (9.12) can also be derived from formula (8.7). The derivation is left as an excercise for the reader [Problem 9.2]. The same result can be obtained for the quantum statistical system of free electrons. This derivation again is left as an excercise [Problem 9.3].

B. The case in which there is a constant magnetic field $\vec{B}$ pointed in the z-direction.

In the presence of a constant magnetic field, a classical electron spirals around the field with the cyclotron frequency [see §§3.13* and 13.5]

$$\omega_o \equiv \frac{eB}{m}, \tag{9.14}$$

which is independent of the electron energy. We may decompose the velocity $\vec{v}$ of the electron in two parts :

$$\vec{v}(t) \equiv v_z\,\vec{k} + \vec{v}_\perp(t); \tag{9.15}$$

The longitudinal component v_z is time-independent and the transversal velocity $\vec{v}_\perp$ change its direction in the x - y plane with the angular speed ω_0. The single-particle electric current $\vec{j}$ associated with the electron motion is defined by

$$\vec{j}(t) \equiv -e\,\vec{v}(t). \tag{9.16}$$

This time-dependent current is quite different from that of the field-free case as given by (9.2). Using (9.15) and (9.16), we can calculate the dynamic conductivity $\overleftrightarrow{\sigma}$ from formula (8.5) or (8.7) [Problem 9.4]. The result is given by

$$\overleftrightarrow{\sigma} = \frac{e^2 n_0}{m} \begin{pmatrix} -i\dfrac{\omega}{\omega^2 - \omega_0^2} & \dfrac{\omega_0}{\omega^2 - \omega_0^2} & 0 \\ -\dfrac{\omega_0}{\omega^2 - \omega_0^2} & -i\dfrac{\omega}{\omega^2 - \omega_0^2} & 0 \\ 0 & 0 & -\dfrac{i}{\omega} \end{pmatrix}. \tag{9.17}$$

In the no-field limit: $\omega_0 \to 0$ expressions (9.17) reduce to expressions (9.13). The non-vanishing off-diagonal elements $\sigma_{xy}(= -\sigma_{yx})$describe the Hall current, discussed in §13.5. The dynamic conductivity diverges at $\omega = \omega_0$, that is, at the cyclotron frequency. A more careful study with consideration of scattering by impurities, phonons, etc., reveals that energy absorption can occur, and this has a maximum at the cyclotron frequency $\omega = \omega_0$. This is called the <u>cyclotron resonance.</u> Formula (9.17) is valid also for free quantum statistical electrons. [Problem 9.5]

Problem 9.1 Carry out the p-integration in (9.9) and obtain (9.12).

Problem 9.2 Derive the Drude formula, starting with (8.7).

Problem 9.3* Derive the Drude formula (9.3) for a quantum statistical system of free electrons.

Problem 9.4* Derive (9.17) for free classical electrons.

Problem 9.5* Derive (9.17) for a quantum statistical system of free electrons.

13.10* More about the Mobility. Quasi-Particle Effect

Earlier in § 13.4, we discussed the mobility in a semiconductor by the method of the Boltzmann equation. If we apply this method to the case of the Coulomb interaction between electron and charged impurity, we have faced with infinite relaxation rate and zero conductivity. Conwell and Weisskopf overcame this difficulty by restricting the contribution of the low-angle scatterings. The reason for doing so is quite convincing but the actual procedure for limiting low-angle scattering is artificial. In general, it is quite difficult (and almost impossible) to modify the Boltzmann collision term in consideration of the effects such as the scattering in medium and three-body collision. The method of Kubo's formula, which has been developed in the last few sections, can be applied without having to initially set up the correct kinetic equation. In the present section, we will study the problem by this method briefly without going into details.

To get some physical insight, let us compute Kubo's formula in a simple-minded manner. For definiteness, we will consider an isotropic system. By symmetry consideration, we then have

$$\sigma_{xx} = \sigma_{yy} = \sigma_{zz} \equiv \sigma . \tag{10.1}$$

Introducing

$$\vec{J} = \sum_k (-e/m)\, \vec{p}_k \tag{10.2}$$

twice in the classical formula (8.7), we have

$$\sigma = \beta \int_0^\infty dt \ \operatorname{Lim} \frac{1}{V} \ \mathrm{Tr}\, \{\rho\, J_x\, J_x(t)\}$$

$$= \beta \int_0^\infty dt \operatorname{Lim} \frac{1}{V} (-e/m)^2 \sum_k \sum_\ell \operatorname{Tr} \{\rho\, p_{kx}\, p_{\ell x}(t)\} . \tag{10.3}$$

The (classical) trace $\operatorname{Tr}\{\rho\, p_{kx}\, p_{\ell x}(t)\}$ vanishes on account of the parity unless $k = \ell$. After using the identity

$$\sum_k p_{kx}\, p_{kx}(t) = \int d^3p \sum_k p_x\, p_x(t)\ \delta^{(3)}(\vec{p}_k - \vec{p}), \tag{10.4}$$

and re-arranging terms, we can reduce (10.3) to [Problem 10.1]

$$\sigma = \beta\, (-e/m)^2\, n \int_0^\infty dt\ \langle p_x\, p_x(t) \rangle , \tag{10.5}$$

where

$$\langle p_x\, p_x(t) \rangle \equiv \int d^3p\ p_x\, p_x(t)\ \frac{e^{-\beta\varepsilon_p}}{(2\pi\, m/\beta)^{3/2}} \tag{10.6}$$

represents the time-dependent momentum correlation function averaged with the equilibrium (Boltzmann) distribution.

For a system of free electrons, the momentum of each particle does not change with time. Then the momentum correlation function $\langle p_x\, p_x(t) \rangle$ is constant in time, and is given by

$$\langle p_x\, p_x(t) \rangle = \langle p_x^2 \rangle = m\, k_B\, T. \quad \text{(free electrons)} \tag{10.7}$$

In this case, the conductivity σ calculated from (10.5), diverges as it should. If a system contains impurities and other scatterers, then the momenta do change in time and the momentum correlation function depends on the time t. Let us assume that this function decays exponentially :

$$\langle p_x \, p_x(t) \rangle = \langle p_x^2 \rangle \, e^{-\Gamma t} = m \, k_B T \, e^{-\Gamma t} , \tag{10.8}$$

where Γ is a decay constant of the dimension of frequency. Substitution of (10.8) into (10.5) yields

$$\sigma = \frac{e^2 n}{m} \, \frac{1}{\Gamma} , \tag{10.9}$$

which is in agreement with the order-of-magnitude formula : $\sigma = e^2 n \, m^{-1} \tau_f$ [see (1.7)].

Our assumption (10.8) is equivalent to the assumption of a single relaxation rate for all momentum states. In the case of the elastic scattering by impurities, each group of electrons with a definite energy will suffer a change in momentum independently of any other group. We may then assume an exponential decay of the momentum correlation function with a certain energy-dependent decay rate $\Gamma(\varepsilon_p)$:

$$\langle p_x \, p_x(t) \rangle$$

$$= \int d^3p \, p_x^2 \, e^{-\Gamma(\varepsilon_p)t} \, \frac{e^{-\beta\varepsilon_p}}{(2\pi \, m/\beta)^{3/2}} . \tag{10.10}$$

In this case, we obtain

$$\sigma = \left(\frac{e}{m}\right)^2 n \int d^3p \, p_x^2 \, \frac{e^{-\beta\varepsilon_p}}{(2\pi \, m/\beta)^{3/2}} \, \frac{1}{\Gamma(\varepsilon_p)} . \tag{10.11}$$

Note that this is similar to formula (3.19), obtained earlier by means of the Boltzmann equation ; only here we used classical mechanics. The rate $\Gamma(\varepsilon_p)$

is a function of the impurity density n_I. We see from (10.10) and (10.11) that

$$\sigma \to \infty \quad \text{as} \quad n_I \to 0 . \tag{10.12}$$

This means that we cannot simply expand the conductivity σ in powers of the impurity density n_I. The discussions in § 13.4 indicate that the relaxation rate as given in (3.16) diverges for the case of a pure Coulomb scattering. This means that the relaxation rate $\Gamma(p)$ may not be expanded in powers of n_I. Thus, computation of Kubo's formula cannot be carried out in a straightforward manner although this formula provides a sound starting point for the calculation.

In recent years, a number of techniques for the computation of Kubo's formulas and again for the treatment of more general non-equilibrium problems have been developed. These techniques are generally quite sophisticated. Interested readers are recommended to look at Zubarev's book, Non-equilibrium Statistical Thermodynamics [10], which gives an excellent overview before plunging into a particular line of approach.

In the remainder of this section, we will discuss the effect of the scattering in medium based on the correlation function formula and its expansion in terms of proper connected diagrams [11].

Let us recall that formula (3.16) for the relaxation rate involves the scattering cross section $I(p,\chi)$. The latter calculated by using clasical mechanics for the Coulomb interaction, is given by (4.4). We may, of course, calculate the cross section quantum mechanically. After using a quantum mechanical perturbation theory, we obtain [12]

$$I\,(p,\chi)\,d\Omega \;=\; \frac{2\pi}{\hbar}\,(2\pi\hbar)^3\,m^2\;|\,\langle\,\vec{p}\,'\,|\,V\,|\,\vec{p}\,\rangle\;|^2\,d\Omega\,. \tag{10.13}$$

This approximate result, which is of second order in the perturbation potential V, is known to be reasonably good for the case of the Coulomb interaction. Using (10.13), we can re-express the current relaxation rate (3.16) in the form [Problem 10.1] :

$$\Gamma^{(0)} \;=\; (2\pi)^4\,n_I\,\hbar^{-1}\int d^3q\;|v(q)|^2\;\delta[\varepsilon(|\vec{p}+\hbar\vec{q}|)-\varepsilon(p)]\,[1-\cos\chi]\,, \tag{10.14}$$

where

$$v(q)\;\equiv\;(2\pi)^{-3}\int d^3r\;e^{-i\vec{q}\cdot\vec{r}}\;V(r) \tag{10.15}$$

represents the Fourier transform of the potential V, and χ is the angle between the two vectors $(\vec{p},\,\vec{p}+\hbar\,\vec{q})$. The delta-function $\delta[\varepsilon(|\vec{p}+\hbar\vec{q}|)-\varepsilon(p)]$ in (10.14) insures that the electron energy is conserved before and after scattering. For the Coulomb interaction $V = -e^2\,(4\pi\,\varepsilon_0\,\kappa\,)^{-1}\,r^{-1}$, the Fourier transform v(q) is given by [Problem 10.2]

$$v(q)\;=\;-\,\frac{1}{4\pi\varepsilon_0}\;\frac{e^2}{2\pi^2\kappa}\;\frac{1}{q^2}\,. \tag{10.16}$$

If we introduce this expression in (10.14), the integral diverges in the small q region which corresponds to the region of small scattering angles.

The relaxation of the electric current is generated by the scattering by impurities. It is therefore desirable to express the exact correlation function formula (7.18) for the conductivity in terms of <u>collision operators</u>.

This can be done in a systematic manner by means of a proper connected diagram expansion [11]. Applying this method, we can obtain an equation for the relaxation rate Γ as follows : [13]

$$\Gamma = 2(2\pi)^3 n_I \int d^3q \, |v(q)|^2 \frac{\hbar\Gamma}{[\varepsilon(|\vec{p} + \hbar\vec{q}|) - \varepsilon(p)]^2 + \hbar^2\Gamma^2} [1 - \cos\chi] \, . \tag{10.17}$$

Note that this equation has a great similarity with (10.14).

As mentioned earlier, an electron in a crystal must move in the potential field of many impurities and therefore cannot stay in the same momentum state forever unlike an electron moving in a vacuum. In other words, the impurities generate a quasi-particle with a finite stay-time (lifetime) for each momentum state. A result of this many-body (impurity) effect called the quasi-particle effect is a weakening of the energy conservation restraint as represented by the Lorentzian function :

$$D(q,p) \equiv \frac{\hbar\Gamma}{[\varepsilon(|\vec{p} + \hbar\vec{q}|) - \varepsilon(p)]^2 + \hbar^2\Gamma^2} \, . \tag{10.18}$$

In (10.17), if other factors in the integrand vary slowly near $\varepsilon(|\vec{p} + \hbar\vec{q}|) = \varepsilon(p)$, then the integral can be evaluated by replacing $D(q,p)$ by $\pi\delta[\varepsilon(|\vec{p} + \hbar\vec{q}|) - \varepsilon(p)]$. See below for further explanation. In this case, the solution of eq.(10.17) approaches $\Gamma^{(o)}$ as given by (10.14). In the case of the Coulomb interaction, however, $v(q)^2$ ($\propto q^{-4}$) has an extremely sharp maximum at $q = 0$. The Lorentzian function $D(q,p)$ also has a maximum at $q = 0$. The maximum of $v(q)^2$ is much stronger, and should overshadow the maximum due to $D(q,p)$.

We may then take

$$D(q=0,\ p) \ = \ \frac{\hbar\Gamma}{[\varepsilon(p) - \varepsilon(p)]^2 + \hbar^2\Gamma^2} \ = \frac{1}{\hbar\Gamma} \tag{10.19}$$

out of the integral sign, and obtain from (10.17)

$$\Gamma = 2\ (2\pi)^3\ n_I \frac{1}{\hbar\Gamma} \int d^3q\ \left[- \frac{1}{4\pi\varepsilon_0} \frac{e^2}{2\pi^2\varkappa} \frac{1}{q^2} \right]^2 [1 - \cos X], \tag{10.20}$$

which immediately yields a solution [Problem 10.3]

$$\Gamma = \frac{1}{\pi\varepsilon_0}\ \hbar^{-1/2}\ e^2\ \varkappa^{-1}\ n_I^{\ 1/2}\ p^{-1/2}. \tag{10.21}$$

This approximate solution is greater than the exact solution of eq.(10.17) because $D(q = 0,\ p) > D(q,p)$. Therefore, the exact solution, which may be obtained numerically by computer, is finite for the Coulomb interaction without ever introducing an artificial cut-off. The approximate formula (10.21) differs significantly from the Conwell-Weisskopf formula Γ_{C-W} in (4.6). In particular, its square-root n_I-dependence is noteworthy. We will comment further on this dependence in the following section, where we discuss the cyclotron resonance lineshape of very pure Ge.

As mentioned earlier in §13.4, the mobility is influenced by the Coulomb interaction among the moving electrons. If the screening, the main effect of this interaction, is taken into consideration, the relaxation rate Γ satisfies eq. (10.17) with the Fourier transform of the screened Coulomb potential [Problem 10.2] :

$$v_D(q) \equiv (2\pi)^{-3} \int d^3r \, e^{-i\vec{q}\cdot\vec{r}} \, V_D(r)$$

$$= -\frac{1}{4\pi\varepsilon_0} \frac{e^2}{2\pi^2 \varkappa} \frac{1}{q^2 + q_D^2} . \quad \text{[use of (4.10)]} \qquad (10.22)$$

Note that this $v_D(q)$ has a q-dependence : $(q^2 + q_D^2)^{-1}$ different from that of $v(q)$ [$\propto q^{-2}$] for the pure Coulomb potential ($q_D = 0$). Depending on the size of q_D, which depends on the number density n of electrons, the maximum of $v_D(q)^2$ may or may not be broader than the maximum of the Lorentzian function $D(q,p)$. Simple calculations show that the maximum of $v_D(q)^2$ is generally broader if the number density n exceeds 10^{15} cm^{-3} for Ge with $n \sim n_I$. If the density is significantly higher than this value, then the rate Γ can be calculated from the standard formula (10.14) after the "delta-function approximation". This calculation leads to the so-called Brooks-Herring formula [4] for the mobility [Problem 13.4] :

$$\mu_{BH} = \frac{8}{\pi} \left(\frac{2}{\pi}\right)^{1/2} \frac{(4\pi\varepsilon_0 \varkappa)^2 (k_B T)^{3/2}}{e^3 n_I m^{*1/2} F_{BH}(T)} , \qquad (10.23)$$

where

$$F_{BH}(T) \equiv \ln(1 + b) - b(1 + b)^{-1}; \qquad (10.24)$$

$$b \equiv 4\varepsilon_0 \varkappa k_B T p^2 / n e^2 . \qquad (10.25)$$

The formula (10.23) was refered to earlier in § 13.4.

We now look at the "delta-function approximation". Let us consider the integral

$$I \equiv \int_{-\infty}^{\infty} d\varepsilon \frac{\hbar\Gamma}{(\varepsilon - \varepsilon_0)^2 + \hbar^2 \Gamma^2} F(\varepsilon). \tag{10.26}$$

If the function $F(\varepsilon)$ changes smoothly, we may Taylor-expand it at $\varepsilon = \varepsilon_0$, and obtain

$$I = F(\varepsilon_0) \int_{-\infty}^{\infty} d\varepsilon \frac{\hbar\Gamma}{(\varepsilon - \varepsilon_0)^2 + \hbar^2\Gamma^2} + \,..\, = \pi F(\varepsilon_0) + \ldots \tag{10.27}$$

Since $\int_{-\infty}^{\infty} d\varepsilon \, [\pi \, \delta(\varepsilon - \varepsilon_0)] \, F(\varepsilon) = \pi F(\varepsilon_0)$, the leading term in (10.27) can be obtained effectively by the replacement

$$\frac{\hbar\Gamma}{(\varepsilon - \varepsilon_0)^2 + \hbar^2\Gamma^2} \rightarrow \pi \, \delta(\varepsilon - \varepsilon_0). \tag{10.28}$$

Note that the "delta-function approximation" depends on the smoothness of the integrand $F(\varepsilon)$.

Problem 10.1 Using (10.13), establish the equivalence between (3.16) and (10.14).

Problem 10.2 Verify (10.16) and (10.22).

Problem 10.3 Derive (10.21) from (10.20).

Problem 10.4* Derive the Brooks-Herring formula (10.23).

13.11* The Cyclotron Resonance

In the presence of a magnetic field $\vec{B}$, a free electron spirals around the field with the cyclotron frequency $\omega_o \equiv e\,B/m_e$ independent of its kinetic energy. Quantum mechanically, the electron is in one of the Landau states with energies $(n + \frac{1}{2})\,\hbar\omega_o + \hbar^2 k_z^2\,/2m_e$ [see §13.5]. If an E M radiation of the matching frequency $\omega = \omega_o$ is applied, the electron may absorb a photon of energy $\hbar\,\omega_o$, and make an upward quantum transition $(n \to n + 1)$. This is called a cyclotron resonance. Much the same phenomenon is expected to occur to the conduction electron in a crystal. An important difference is that the cyclotron frequency for the conduction electron is given by

$$\omega_o \equiv e\,B\,/m^* \tag{11.1}$$

with m* representing the effective mass characteristic of the crystal. Measurements of the resonance frequency give immediately the value of the effective mass m* and therefore represent an important method for exploring the electronic band structure. In the present section we will discuss the cyclotron resonance in a heuristic manner.

Let us first look at the data for the cyclotron resonance as reported in a classic paper by Dresselhaus, Kip and Kittel for germanium samples [14]. Fig. 13.10 represents typical experimental results for the power absorption against static magnetic field at $\nu = 24{,}000$ M cycles $\sec^{-1}$ and $T = 4$ K. [Note that in the abscissa, the magnetic field, rather than the angular frequency, is taken. In normal experimental runs, the magnetic field is varied.] A pure Ge [with a diamond-like crystal] has a quite complicated electronic band structure. This is reflected by a number of peaks in the figure, which are associated with different effective masses. To get a physical picture, let us find the order of magnitude of several physical quantities relevant to such experiments. We choose : ν = 24,000 M cycles sec^{-1} , $\omega = 2\pi\nu = 1.5 \times 10^{11}$ sec^{-1} , $m^* = 0.1\ m_e$. From (11.1), we then obtain B = 860 Gauss at resonance. In order to obtain a distinctive resonace, the linewidth Γ_c must be smaller than the resonance frequency :

$$\Gamma_c < \omega_o. \tag{11.2}$$

The linewidth Γ_c is determined by the relaxation rate caused by scatterers such as phonons and impurities. The relaxation rates at room temperature are typically in the range 10^{13} to 10^{15} sec^{-1}. The condition (11.2) that Γ_c <

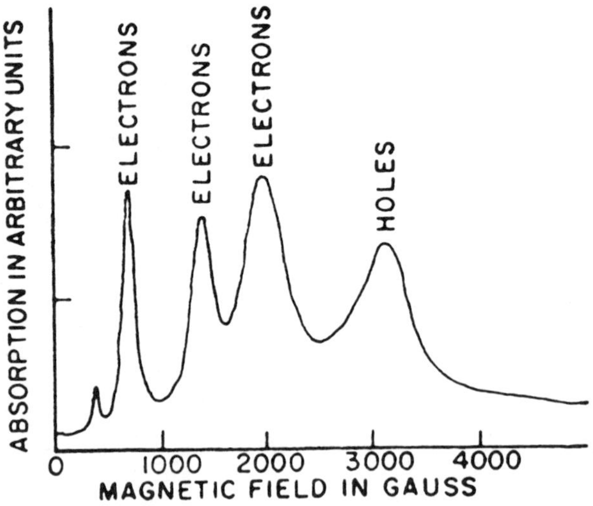

Fig. 13.10 Cyclotron resonance absorption for Ge at ν = 24,000 M cycles sec^{-1} and T = 4 K. After Dresselhaus, Kip and Kittel [14].

1.5×10^{11} sec^{-1}, normally requires to work with high-purity crystals and in the liquid helium temperature range.

Incidentally, observation of the cyclotron resonance for a metal is quite difficult for several reasons. The effective mass of a typical monovalent metal like K is close to the electron mass. This makes the resonance frequency ($\omega \sim \omega_0$) lower by one or more orders of magnitude compared with the case of an elemental semiconductor like Ge or that of a compound simiconductor like InSb. This in turn makes it more difficult to satisfy the required inequality (11.2). The relaxation rate Γ is high for a metal because high purity metals are harder to produce and because intrinsically there are more conduction electrons, which tend to enhance the scattering effect. Besides, a probing E M wave is mostly reflected at the metal surface as the common phrase : "a shiny metal" indicates this phenomenon. This will necessitate a more refined and therefore more difficult experimental arrangement.

As we just saw, the center position of the resonance maximum yields information about the effective mass m*. Let us now look at the linewidth Γ_c. This quantity is normally defined as half the width at the half-height after suppressing the background absorption. The width will be zero if the conduction electrons are free. The scatterings by impurities and phonons make Γ_c finite. In this respect, the width Γ_c is quite similar to the current relaxation rate Γ associated with the static conductivity. Both are expected to be of same order of magnitude, but are different from each other since the quantum states of the electrons, that is, the momentum and Landau states, are quite different.

Kawamura et al [15] made an extensive study of the cyclotron resonance for very pure Ge ($n \sim n_I$: 10^{12} cm^{-3}) at very low temperatures (T : 1.5 to 15 K) and at high resonance frequency (ω : 50 G cycles sec^{-1} = 5×10^{13} cycles s^{-1}

They reported that the width Γ_c becomes temperature-independent below 2.6 K and approximately follows the following formula :

$$\Gamma_c = 0.9 \quad \hbar^{-3/4} (e^2/4\pi\varepsilon_0\kappa) (m*)^{-1/4} \omega_0^{-1/4} n_I^{1/2}. \tag{11.3}$$

Besides the T - independence, they checked exprimentally the n_I-dependence (range : 10^{11} - 10^{13} cm^{-3}), the m* - dependence and also the absolute magnitude. This is a remarkable result. In particular, the square-root n_I-dependence is noteworthy ; a linear dependence is expected from the semiclassical theory.

As mentioned repeatedly, the Landau states for electrons are quite different from the momentum states, and the method of the Boltzmann equation cannot be applied to the computation of the cyclotron resonance width. However, we know that (a) the absorption power p can be expressed in terms of the dynamic conductivity tensor $\sigma_{\mu\nu}(\omega)$ as given by (6.15) or (6.17), and (b) the latter $\sigma_{\mu\nu}(\omega)$ can be obtained from the correlation function formula (7.18). We can therefore compute the cyclotron resonance linewidth from this route. It is possible to perform this program in terms of proper connected diagrams in a systematic manner [16]. Applying then this theory to the case of an extreme high field :

$$\hbar\omega_0 \sim \hbar\omega \gg k_B T, \tag{11.4}$$

we can obtain the following equation for the energy-dependent width $\Gamma' \equiv \Gamma'(k_z)$ [17] :

$$\Gamma' = (2\pi)^3 n_I \int d^3q \, v(q)^2 \, t^2 e^{-t} \frac{\hbar\Gamma'}{[\varepsilon(k_z) - \varepsilon(k_z - q_z)]^2 + (\hbar\Gamma')^2}, \tag{11.5}$$

where

$$t \equiv \frac{1}{2}\left(\frac{\hbar}{eB}\right)(q_x{}^2 + q_y{}^2) \; ; \tag{11.6}$$

and

$$\varepsilon\,(k_z) \equiv \hbar^2 k_z^2 /2m^* \tag{11.7}$$

represents the kinetic energy along the magnetic field. Note that eq.(11.5) has a great similarity with eq.(10.17) for the energy-dependent current relaxation rate. The factor $t^2\,e^{-t}$ arises from the fact that the transitions between the Landau states of the two lowest oscillator quantum numbers ($n = 0,1$), which are important for the extreme high field region ($\hbar\,\omega_0 \gg k_B\,T$), are taken into account.

At the lowest temperatures, the most important scatterers are ionized impurities. Assuming the Coulomb interaction, we now seek a solution of eq. (11.5). First, let us apply the "delta-function approximation" to the q - integral. We then obtain

$$\Gamma' = (2\pi)^3\, n_I \int d^3q \left[-\frac{1}{4\pi\varepsilon_0}\,\frac{e^2}{2\pi^2 \varkappa\, q^2} \right]^2 t^2 e^{-t}\, \pi\delta\,[\varepsilon(k_z) - \varepsilon(k_z - q_z)]. \tag{11.8}$$

After carrying out the above integration, we obtain [Problem 11.1]

$$\Gamma'(k_z) = n_I \left(\frac{e^2}{4\pi\varepsilon_0 \varkappa} \right)^2 \frac{\pi m^*}{\hbar e B}\, \frac{1}{|k_z|}\, F(\eta), \tag{11.9}$$

where

$$F(\eta) \equiv (2+\eta)\,[\,1-\eta\, e^{\eta}\, Ei\,(\eta)\,], \qquad (11.10)$$

$$\eta \equiv \frac{2\hbar}{eB}\, k_z^2\,; \qquad (11.11)$$

and Ei(x) is the exponential integral defined by

$$Ei\,(x) \equiv \int_x^{\infty} du\, \frac{e^{-u}}{u}\,. \qquad (11.12)$$

The function $F(\eta)$ is a monotonically decreasing function of η starting with the value 2 at $\eta = 0$ and reaching unity at $\eta = \infty$.

It is interesting to note that the energy-dependent width Γ' obtained here is finite in contrast with the case of the current relaxation rate computed with the same delta-function approximation [see (10.14)], which diverges for the Coulomb interaction. The width Γ' grows indefinitely for small $|k_z|$. Yet, the behavior of $\Gamma'(|k_z|)$ for small $|k_z|$ is needed for the low-temperature behavior of the resonance width. A better estimate of $\Gamma'(k_z)$ near $k_z = 0$ can be obtained from eq.(11.5) by replacing the Lorentzian function with $1/\hbar\Gamma'$, yielding [Problem 11.2]

$$\Gamma'^2$$

$$= (2\pi)^3\, n_I \int d^3q \left(-\frac{1}{4\pi\varepsilon_0}\,\frac{e^2}{2\pi^2\varkappa}\,\frac{1}{q^2}\right)^2 t^2\, e^{-t}$$

$$= \left(\frac{1}{2}\pi\right)^{3/2} \left(\frac{\hbar}{eB}\right)^{1/2} n_I \left(\frac{e^2}{4\pi\varepsilon_0\varkappa}\right)^2$$

or

$$\Gamma' = 1.403\, \hbar^{-3/4}\, (e^2/4\pi\varepsilon_0\, \varkappa)\, (m^*)^{-1/4}\, \omega_0^{-1/4}\, n_I^{1/2}$$

$$\text{for small } |k_z| . \qquad (11.13)$$

Note that the width Γ' in this approximation approaches a constant in the small k_z - limit. Further note that expression (11.13) is almost identical with formula (11. 3) experimentally verified in the low temperature limit; only the numerical factor is different.

Eq.(11.5) can, of course, be solved numerically by computer. Using this result and assuming the Boltzmann distribution for the thermal average, we can calculate the temperature - dependent width $\Gamma_c(T)$. The result of this calculation [18] is shown in solid line in Figure 13.11, where the small circles indicate the experimental data due to Kawamura et al. The agreement between theory and experiment appears quite good.

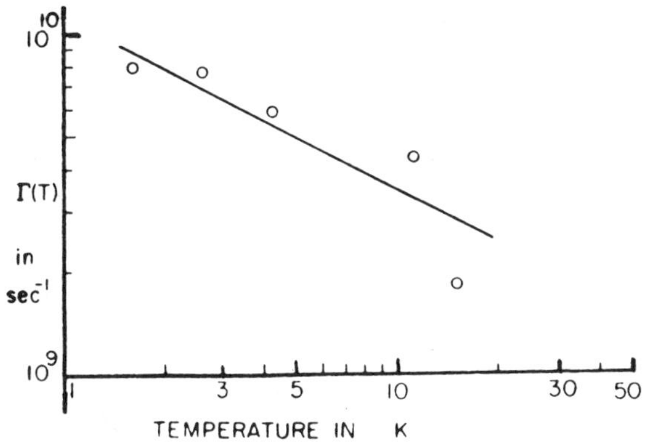

Fig. 13.11 The half-width Γ for Ge at $n \simeq n_I = 10^{12} cm^{-3}$ as a function of temperature T, observed by Kawamura et al [15] are shown in circles. The solid line represents the numerical solution from eq.(11.5), [18].

When the width Γ does not change linearly with the scatterer density n_I, Matthiessen's rule of the usual form (1.10) : $\Gamma = \Gamma_1 + \Gamma_2$ is violated in general (see below). In particular if there are two kinds of scatterers which generate the square-root-density-dependent widths :

$$\Gamma_1 = a_1 \sqrt{n_1} \quad , \qquad \Gamma_2 = a_2 \sqrt{n_2}, \tag{11.14}$$

the total width Γ must be given by

$$\boxed{\Gamma = \sqrt{\Gamma_1^2 + \Gamma_2^2}} \; . \tag{11.15}$$

This can be proved as follows. Let us assume that the system contains a single kind of scatterer which generates the square-root-density-dependent width : $\Gamma = a\sqrt{n}$. Imagine now that a portion of the scatterers are "colored" in red. The sum of the densities of colored and uncolored scatterers (n_1, n_2) equals the total scatterer density n. Artificial coloring does not change the scattering effect so that $\Gamma_j = a\sqrt{n_j}$. If we apply the modified form of Matthiessen's rule (11.15) to colored and uncolored impurities, we can recover the correct answer $\Gamma = a\sqrt{n}$. Since n_1 and n_2 can be changed arbitrarily, no other form of combination rule is possible. The same method can be used to prove that the usual form of Matthiessen's rule : $\Gamma = \Gamma_1 + \Gamma_2 + \ldots$ strictly holds only if the component widths vary linearly with the scatterer densities.

Problem 11.1 Derive (11.9)

Problem 11.2 Derive (11.13)

13.12* The Diffusion

Earlier in § 4.1 we discussed the elementary notion of diffusion. The diffusion of ink in water will serve as a good example for the following discussions.

In Figure 13.12 we show an inhomogeneous particle distribution such that the particle density $n(x,t)$ changes in the x-direction only. The diffusion of the particles normally follows Fick's law:

$$j_x(x,t) = - D \frac{\partial n\ (x,t)}{\partial x} , \tag{12.1}$$

where D is the diffusion coefficient. The change in the density obeys the equation of continuity : [see (4.8.19)]

$$\frac{\partial n}{\partial t} + \nabla \cdot \vec{j} = \frac{\partial n\ (x,t)}{\partial t} + \frac{\partial}{\partial x} j_x(x,t) = 0. \tag{12.2}$$

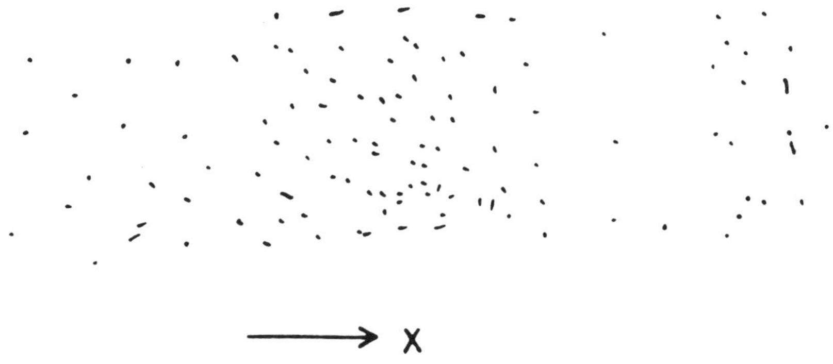

Fig. 13.12 An inhomogeneous particle distribution

Introducing eq.(12.1) here and re-arranging the terms, we obtain

$$\boxed{\frac{\partial n}{\partial t} = D\,\frac{\partial^2 n}{\partial x^2}\ .} \tag{12.3}$$

This **is called the <u>diffusion equation</u>.**

The diffusion process can visually be illustrated from this equation in a simple manner. In Figure 13.13, we plot the density $n(x,t)$ against the location x. The curve in general may comprise convex and concave parts with respect to the x-axis. The convex parts have negative second

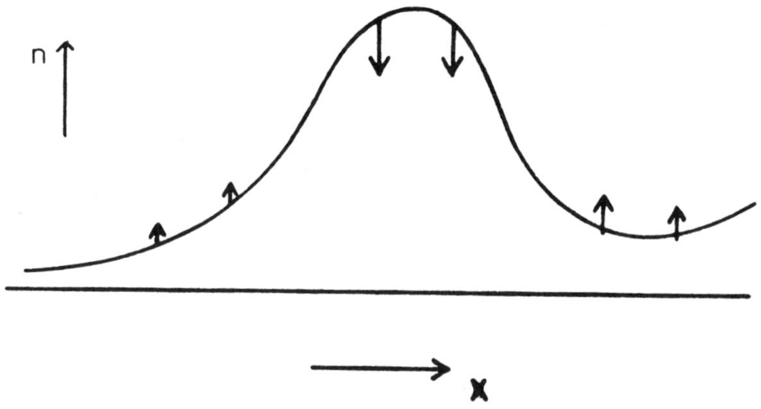

Fig. 13.13 The number density $n(x,t)$ against the location x. The density increases if $\partial^2 n / \partial x^2 > 0$ or if the curve $n(x,t)$ is concave ; such parts are indicated by upward arrows.

derivatives: $\frac{\partial^2 n}{\partial x^2} < 0$, and therefore correspond to those parts where the density should decrease in time ($\frac{\partial n}{\partial t} < 0$). The concave parts, on the contrary, have positive second derivatives and correspond to those parts where the density tends to increase. The change in density will cease when the density gradient $\frac{\partial n}{\partial x}$ becomes constant everywhere.

Let us now look at the diffusion from a different angle. By definition,

$n\ (x,t)\ dx$

= the number of particles in the range $(x, x + dx)$ at time t

= the relative probability of finding a particle in the range $(x, x + dx)$ at time t. (12.4)

We also have the normalization condition :

$$\int_{-\infty}^{\infty} n\ (x,t)\ dx = N, \qquad (12.5)$$

where N is the total number of particles.

Let us assume that all the particles move independently of each other. This condition is usually satisfied if the density of particles is low and if the interaction between particles is negligible. The motion of the particles can then be studied by following a single particle. In particular, the density n and the probability distribution for a single particle change in the same manner.

We now consider a hypothetical situation. At the initial time $t = 0$, all of the particles are concentrated at the origin 0, and are set to diffuse subsequently. The mean square displacement $\langle x^2 \rangle_t$ is defined by

$$\langle x^2 \rangle_t \equiv \frac{1}{N} \int_{-\infty}^{\infty} dx\ x^2\ n(x,t). \qquad (12.6)$$

Differentiating it with respect to t, we obtain

$$\frac{d}{dt}\langle x^2\rangle_t = \frac{1}{N}\int_{-\infty}^{\infty} dx\, x^2 \,\frac{\partial n(x,t)}{\partial t}$$

$$= \frac{1}{N}\int_{-\infty}^{\infty} dx\, x^2 \left[\, D \frac{\partial^2 n}{\partial x^2} \,\right]. \qquad \text{[use of (12.3)]} \qquad (12.7)$$

The x-integration can be carried out as follows :

$$\int_{-\infty}^{\infty} dx\, x^2 \frac{\partial^2 n}{\partial x^2}$$

$$= \left[x^2 \frac{\partial n}{\partial x}\right]_{-\infty}^{\infty} - 2\int_{-\infty}^{\infty} dx\, x \frac{\partial n}{\partial x} \quad \text{[integration by parts]}$$

$$= [0 - 0] - 2\,[xn]_{-\infty}^{\infty} + 2\int_{-\infty}^{\infty} dx\, n \quad \text{[integration by parts]}$$

$$= 2N. \qquad \text{[use of (12.5)]}$$

Introducing this result in (12.7), we obtain

$$\frac{d}{dt}\langle x^2\rangle_t = 2D, \qquad (12.8)$$

which, upon integration, yields

$$\boxed{\langle x^2\rangle_t = 2Dt.} \qquad (12.9)$$

This means that the spread of the particles as characterized by the mean square displacement increases linearly with the time t, and the rate of the spread is given by the diffusion coefficient D, up to a constant. It is stressed that this result was obtained in a very general manner based only on Fick's law (12.1) and the equation of continuity (12.2).

The above treatment can be extended in three dimensions, [Problem 12.1]. The result is as follows :

$$\boxed{\langle r^2 \rangle_t \equiv \langle x^2 + y^2 + z^2 \rangle_t = 6Dt.} \tag{12.10}$$

We note that this expression is different from (12.9) only by the numerical factor 3, the factor equal to the ratio of the numbers of dimensions.

Problem 12.1 Derive (12.10).

13.13* Simulation of the Lorentz Gas

Earlier in § 13.2, we discussed a system of (non-interacting) electrons in a potential field of fixed scatterers. This model of an electron-impurity system is often called the Lorentz gas model of a metal, named after Lorentz's classic work on the kinetic theory of electrons [19]. In this model, any given particle (electron) will be scattered elastically by scatterers (impurities). Since only elastic scatterings enter here, each group of particles with a definite speed (and therefore energy) will move independently of any other group. In the present section, we will discuss a simulation of the dynamics of the Lorentz gas in terms of correlated walks.

Let us set up a simple cubic (sc) lattice of lattice constant a_0. A particle is allowed to move on the lattice sites with speed v and only along the cubic axes. Scatterers (impurities) by assumption are distributed over the lattice sites at random; that is, each site is occupied by an

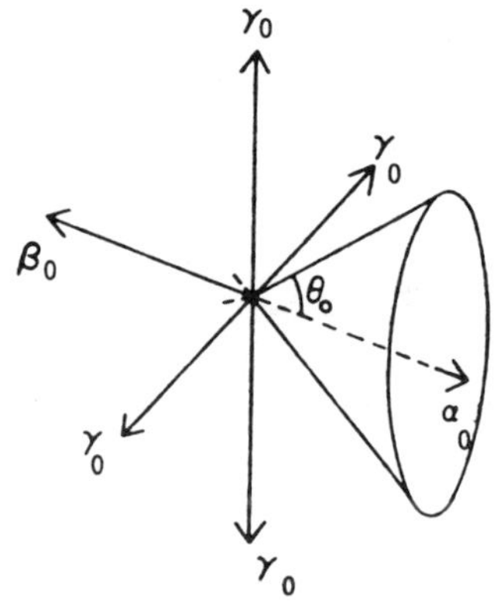

Fig. 13.14

The particle, after hitting a scatterer, will move forward, sideways, or backward with the probabilities α_0, γ_0, or β_0, respectively.

impurity with the same probability. If the particle hits a scatterer, which happens with the probability q, it may, by assumption, move forward, reverse or turn at right angles with probabilities α_0, β_0 or γ_0 with the normalization

$$\alpha_0 + \beta_0 + 4\gamma_0 = 1. \tag{13.1}$$

These probabilities may be related with the scattering cross section $I(v,\vartheta)$ in the following manner. Consider first the case of forward scattering which corresponds to one sixth (1/6) of the total solid angle 4π, as indicated in Figure 13.14. The limiting scattering angle ϑ_0 is defined by

$$\frac{1}{6}(4\pi) = \int_0^{2\pi} d\phi \int_0^{\vartheta_0} d\vartheta \sin\vartheta = 2\pi(1 - \cos\vartheta_0)$$

or

$$\vartheta_0 = \cos^{-1}\left(\frac{2}{3}\right) = 48.19^\circ . \tag{13.2}$$

We will choose the probability α_0 to be the ratio of the forward scattering cross section within the angle ϑ_0 to the total cross section:

$$\alpha_0(\text{forward}) \equiv \frac{\int_0^{2\pi} d\phi \int_0^{\vartheta_0} d\vartheta \sin\vartheta \; I(v,\vartheta)}{\int_0^{2\pi} d\phi \int_0^{\pi} d\vartheta \sin\vartheta \; I(v,\vartheta)}$$

$$= \int_0^{\vartheta_0} d\vartheta \sin\vartheta \; I(v,\vartheta) \Big/ \int_0^{\pi} d\vartheta \sin\vartheta \; I(v,\vartheta). \tag{13.3a}$$

By similar arguments, we may choose the probabilities (β_0, γ_0) as follows :

$$\beta_0 \text{ (reverse)} = \int_{\pi-\vartheta_0}^{\pi} d\vartheta \sin\vartheta \; I(v,\vartheta) \Big/ \int_0^{\pi} d\vartheta \sin\vartheta \; I(v,\vartheta)$$

$$\gamma_0 \text{ (turn)} = \frac{1}{4}(1-\alpha_0 - \beta_0). \tag{13.3b}$$

We note here that for an isotropic scattering, all of the probabilities (α_0, β_0, γ_0) are equal to one sixth.

If the particle does not hit any scatterer, which should happen with the probability 1-q, it should continue to move in the same direction.

Combining these arguments, we can choose the step probabilities (α, β, γ) as follows :

$$\alpha = q\,\alpha_0 + 1 - q$$

$$\beta = q\,\beta_0$$

$$\gamma = q\,\gamma_0. \qquad (13.4)$$

By assumption, the particle moves on the lattice one step per unit time :

$$\tau \equiv a_0/v\ . \qquad (13.5)$$

Let $P_a(x,y,z,N)$ be the probability that the particle arrives at the site $(xa_0,\ ya_0,\ za_0)$ with the direction a at the time $N\tau$. For brevity, the six general directions will be named as

$$(x_+,\ x_-,\ y_+,\ y_-,\ z_+,\ z_-) \equiv (1,\ 2,\ 3,\ 4,\ 5,\ 6). \qquad (13.6)$$

The probabilities of arrival with direction a, P_a, will satisfy the following equations :

$$P_1(x,\ y,\ z,\ N)$$

$$= \alpha P_1(x-1,\ y,\ z,\ N-1) + \beta\ P_2(x-1,\ y,\ z,\ N-1)$$

$$+ \sum_{b=3,4,5,6} \gamma\ P_b(x-1,\ y,\ z,\ N-1), \qquad (13.7)$$

and similar equations for $P_2, \ldots, P_6$. These difference equations may be solved subject to given initial conditions.

Let us consider a special case in which, at the initial time N = 0, the particle arrives at the origin with the direction a = 1; this can mathematically be expressed by

$$P_a(x, y, z, 0) = \delta_{x,0}\ \delta_{y,0}\ \delta_{z,0}\ \delta_{a,1}. \tag{13.8}$$

For small N, the solution may be worked out by drawing diagrams [Problem 13.1]. The analytic solution for $P_a(x,y,z,N)$ for a general N is difficult to obtain. However, the <u>mean square displacement</u> defined by

$$\langle r^2\rangle \equiv \sum_a \sum_x \sum_y \sum_z (x^2 + y^2 + z^2)\ P_a(x, y, z, N) \tag{13.9}$$

can be obtained as outlined earlier in §12.12*, and is given by, [see (12.12.3)]

$$\frac{\langle r^2\rangle}{a_0^2} = \frac{1 + \delta}{1 - \delta} N - \frac{2\delta}{(1 - \delta)^2} (1 - \delta^N), \tag{13.10}$$

where

$$\delta \equiv \alpha - \beta. \tag{13.11}$$

represents the <u>degree of correlation</u>.

If there were no impurities and therefore no scatterings, then q = 0, and $\alpha = 1$, $\beta = \gamma = 0$. In this free-particle limit ($\delta \to 1$) the r.h.s. of eq. (13.10) approaches N^2 so that [see (2.7.24)]

$$\langle r^2\rangle \to N^2 a_0^2 \qquad \text{as} \quad \delta \to 1. \tag{13.12}$$

With a finite amount of scattering, the first term on the r.h.s. of eq.(13.10), which is proportional to N, is predominant :

$$\langle r^2 \rangle \cong \frac{1+\delta}{1-\delta} N a_o^2 \qquad \text{for large N and } \delta < 1. \tag{13.13}$$

In the last section we saw that the mean square displacement $\langle r^2 \rangle$ is closely related to the diffusion coefficient D. In the derivation of the relation (12.10), we assumed that the particles move independently in the continuous space. In the present model, each particle can move only in the six directions along the orthogonal axes. In order for us to apply a fluid-dynamical description, we may look at the motion of the particle in a <u>macroscopic scale</u> in which distance and time are measured in units of length and time much greater than the lattice constant a_o and the jump time τ, respectively. We can then define the density and velocity fields as functions of space and time, and apply fluid-dynamical concepts such as Fick's law and the diffusion coefficient D.

Identifying the time $N\tau$ by t :

$$N\tau = t, \tag{13.14}$$

we get from (13.13)

$$\langle r^2 \rangle = \frac{1+\delta}{1-\delta} \frac{a_o^2}{\tau} t \qquad \text{for large t.} \tag{13.15}$$

Comparing this expression with eq.(12.10), we obtain

$$\boxed{D = \frac{1}{6} \frac{1+\delta}{1-\delta} \frac{a_o^2}{\tau}.} \tag{13.16}$$

This represents the diffusion coefficient for the lattice characterized by the lattice constant a_o and the unit step time τ.

To gain some insight , let us apply (13.16) to a simple case. The ratio a_o/τ is equal to the speed of the particle as seen from (13.5). If we choose

$$a_o = \text{mean free path} = \ell$$

and

$$\delta \equiv \alpha - \beta = 0 \quad \text{(isotropic scattering)}, \tag{13.17}$$

we can reduce (13.16) to

$$\frac{1}{6} v \, \ell \equiv D_{\text{random walk}}. \tag{13.18}$$

Note that the diffusion constant is proportional to the mean free path and the particle-speed v. The result (13.18) can be obtained within the framework of simple kinetic theory [Problem 13.3]. It also represents one of the principal results obtained in the random walk theory [20].

Eq. (13.16) obtained for the correlated walk model, contains two model parameters (a_o, τ). By eliminating these parameters more carefully, we can obtain essentially exact results as follows [21].

In the original Lorentz gas model there exists a definite collision rate $1/\tau_c$ given by [see (4.10.1)]

$$1/\tau_c = n_I \, v \, \sigma, \tag{13.19}$$

where

$$\sigma \equiv 2\pi \int_0^{\pi} d\vartheta \sin\vartheta \, I(v,\vartheta) \tag{13.20}$$

is the <u>total cross section</u>, and n_I the density of impurities. The mean free path defined by

$$\ell \equiv v\tau_c = (n_I \sigma)^{-1} \tag{13.21}$$

does not depend on the speed of the moving particle. The probability of hitting a scatterer for the first time is proportional to the distance L traveled, and should be given by L/ℓ for small L. We may therefore assume that the probability q of hitting a scatterer after the length of travel a_0 is given

$$q = a_0/\ell = a_0 n_I \sigma, \tag{13.22}$$

which is valid for small a_0. Let us now define the <u>kinetic-theoretical limit</u> as

$$a_0 \to 0, \quad \tau \to 0 \tag{13.23}$$

while

$$a_0/\tau = v(\text{speed}) = \text{finite},$$

$$(n_I \sigma)^{-1} = \ell(\text{mean free path}) = \text{finite}. \tag{13.24}$$

In this limit, we can show [Problem 13.2] that the diffusion coefficient D as given by (13.16) approaches

$$\boxed{\frac{1}{3}\ \frac{v^2}{\Gamma^*} = D,} \tag{13.25}$$

where

$$\boxed{\Gamma^* \equiv n_I\ v\sigma\ (1 - \alpha_0 + \beta_0)} \tag{13.26}$$

represents the current relaxation rate for the lattice model.

The last expression for the relaxation rate is in essential agreement with expression (3.16), obtained earlier in § 13.3 by solving the Boltzmann equation for the electron-impurity system. In fact, if we assume that the particle is scattered in one of the six general directions, that is, forward, backward, or sideways, the Boltzmann formula (3.16) reduces to (13.26). This can be seen from the following calculations :

$$\int d\Omega \quad I(v,\vartheta)\ [1 - \cos\vartheta] \qquad \text{[see (3.16)]}$$

$$= \sigma - \int d\Omega \quad I(v,\vartheta)\ \cos\vartheta \qquad \text{[use of (13.20)]}$$

$$\rightarrow \sigma - \sum_{a=1}^{6} [I(v,\vartheta)\ \cos\vartheta]_a\ \Delta\Omega_a \qquad \text{[discrete directions]}$$

$$= \sigma - \sigma\,[\alpha_0 \cos 0 + 4\ \gamma_0 \cos(\tfrac{\pi}{2}) + \beta_0 \cos(\pi)] \qquad \text{[use of (13.3a and 3b)]}$$

$$= \sigma - \sigma(\alpha_0 - \beta_0)$$

$$= \sigma\ (1 - \alpha_0 + \beta_0)\ ,$$

which indicates explicitly that the contribution of the gain term, that is, the integral with $\cos\vartheta$, is represented by terms involving α_0 and β_0.

Problem 13.1 Exhibit correlated walks after one and two units of time explicitly by drawing diagrams. Do this first for the two-dimensional lattice, and then for the three-dimensional lattice. Compute the mean square displacement by this method, and see if the obtained result is in agreement with the analytical formula given in (13.10).

Problem 13.2 Show that the formula (13.16) for the diffusion coefficient approaches (13.25) in the kinetic-theoretical limit defined by (13.22-24).

Problem 13.3 Let us consider a hypothetical cube of a side-length equal to the mean free path ℓ . By assumption, any molecules within the cube leave it without scattering. Suppose that we have two such cubes adjacent to each other and containing slightly different numbers of molecules. Compute the molecular flux passing through the common face with the assumption that all molecules move in all directions with the same speed v. By assuming Fick's law, find the diffusion coefficient D, and compare it with formula (13.18). [Answer : $D = (1/3)\, v\ell$]

13.14* Atomic Diffusion in Metals with Impurities

It is experimentally known that in a metallic crystal like niobium hydrogen atoms can move about quite freely. If impurities such as nitrogens are present in the crystal, these impurities attract, and trap, the hydrogen atoms, and thereby slow down the diffusion. In the present section we will discuss the atomic diffusion based on the correlated walk model.

Let us take a simple cubic (s.c.) lattice. A walker (model atom) is allowed to jump between nearest-neighbor lattice sites in unit time τ, or stay at the same site with the following rules: If it should arrive at any site, it may move in the same direction as that of the previous step with probability α, turn at right angles with probability γ, reverse with probability β, or remain at that site with probability σ. These possibilities are indicated in Figure 13.15a.

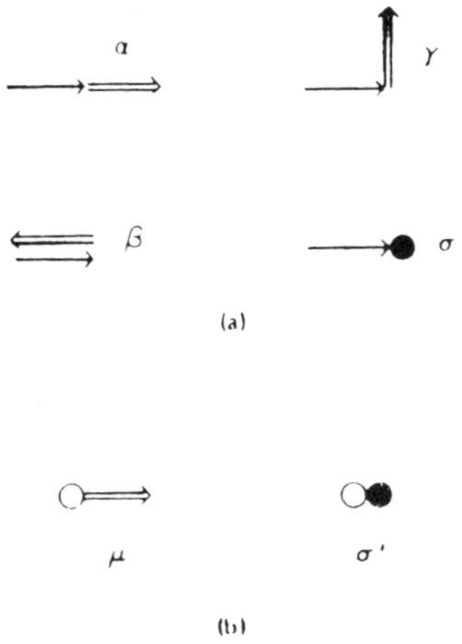

Fig. 13.15 The step probabilities with correlations for the s.c. lattice. They are normalized such that $\alpha + 4\gamma + \beta + \sigma = 1$ for (a) and $6\mu + \sigma' = 1$ for (b).

The step probabilities are normalized such that

$$\alpha + 4\gamma + \beta + \sigma = 1, \tag{14.1}$$

where the numerical factor 4 corresponds to the number of possibilities of turning at right angles. If the walker were at rest it may move in any direction with the same probability μ or remain with probability σ' with the normalization condition :

$$6\mu + \sigma' = 1. \tag{14.2}$$

See Figure 13.15 b.

The connection between the present model and the atomic diffusion in a metallic crystal is rather obvious. In many cases, the actual metallic ions form a body-centered cubic lattice. For simplicity however, we assumed a s.c. lattice for our model. The atoms are known to migrate on interstitial sites, which also form a s.c.lattice with the same lattice constant a_0. The latter can be used as the reference lattice in the model. If the concentration of diffusing atoms is small, atoms should move independently of each other. This allows us to describe the diffusion in terms of a single particle moving about on the lattice. Suppose that we have a perfect lattice without impurities. The jump rate should be the same everywhere on the lattice, and the average jump time from one site to the next will be chosen as the unit time τ in the model. In order to jump, the atom must overcome a certain energy barrier which is formed by the surrounding lattice molecules. Once the atom starts to move, it may jump over several sites in succession. Colloquially speaking, the atom acquires a "momentum", and this carries it over several sites. Such correlated motion can, to some degree, be described in terms of the step probabilities α , β and γ . If, on the other hand, the atom was stationary, it might move in any direction with no preference ; this possibility is represented by the start-to-move probability μ. The atom may also remain at the same site, or it may come to a stop temporarily. To account for these possibilities, we can use the probabilities σ and σ'.

The mean square displacement $\langle r^2 \rangle_N$ after N units of time can be computed by extending the method outlined in §12.13. After a considerable amount of computation, we can express the result as follows : [22]

$$\langle r^2 \rangle / a_0^2 = \frac{(1-\sigma')(1+\delta)}{(1-\sigma'+\sigma)(1-\delta)} N - (1-\sigma')(1-\sigma'+\sigma)^{-2}(1-\delta)^{-2}(\delta-\sigma'+\sigma)^{-1}$$

$$\times \{ 2\delta \, (1-\sigma'+\sigma) \, [\delta-\sigma'+\sigma-(1-\sigma'+\sigma) \, \delta^{N+1}]$$

$$+ (\sigma'-\sigma)(1-\delta)[(1+\delta)(\delta-\sigma'+\sigma)+(1-\delta)(\delta+\sigma'-\sigma)(\sigma'-\sigma)^N]\} \, .$$

$$(\, \delta \equiv \alpha - \beta \,) \qquad (14.3)$$

In the limit $\alpha \to 1$ (or $\delta \to 1$), which corresponds to the particle moving always in the same direction, the mean-square displacement $\langle r^2 \rangle$ approaches $N^2 \, a_0^2$ [as it should]. Except for this case, the first term on the r.h.s. of (14.3), which is proportional to N, dominates for large N:

$$\langle r^2 \rangle \simeq \frac{(1-\sigma')(1+\delta)}{(1-\sigma'+\sigma)(1-\delta)} N \, a_0^2 \qquad \text{for large N} \, . \qquad (14.4)$$

Comparing this with eq.(12.10), we obtain the diffusion coefficient D as follows :

$$D = \frac{(1-\sigma')(1+\delta)}{6(1-\sigma'+\sigma)(1-\delta)} \, v \, a_0, \qquad (14.5)$$

where

$$v \equiv a_0/\tau \qquad (14.6)$$

is the speed of the particle.

Using this result, we can discuss the atomic diffusion in the following manner.

(a) perfect crystals

Let us consider the case in which the particle moves on the lattice without stopping. In mathematical terms, $\sigma = \sigma' = 0$. The diffusion coefficient D is then reduced to

$$D = \frac{1}{6} \frac{1+\delta}{1-\delta} \, v \, a_0, \qquad (14.7)$$

which is in agreement with (13.16), the main result for the Lorentz gas simulation discussed in §13.13.

Let us consider another extreme case where the particle is allowed to jump just one step at a time. This case is characterized by

$$\alpha = \beta = \gamma = 0, \qquad (14.8)$$

as the probabilities α, β and γ all refer to the processes involving successive jumps. We then have $\delta \equiv \alpha - \beta = 0$, which simplifies expression (14.5) to

$$D = \frac{1}{6[1 + \sigma/(1-\sigma')]} v a_0 . \qquad (14.9)$$

We compare this result with that of the usual random walks: [see(13.18)]

$$D = \frac{1}{6} v a_0 . \qquad (14.10)$$

The difference arises from the fact that the walker in our model, may stop, stay and restart again. This allows the particle to leave the site immediately or a number of rest times after arrival with the probability weights: $(1, \sigma, \sigma\sigma', \sigma\sigma'^2 \ldots)$. This accounts for the correction factors :

$$1 + \sigma + \sigma\sigma' + \sigma\sigma'^2 + \ldots = 1 + \sigma(1 - \sigma')^{-1} . \qquad (14.11)$$

The come-to-stop probability σ and the remain-stationary probability σ' should have different values for different crystals. Equation (14.9) therefore represents the effect of the difference in the staying time.

The case of actual atomic diffusion should fall between these two extremes. The full expression (14.5) should yield a reasonably good description of the diffusion coefficient D.

(b) crystals with impurities

We assume that (interstitial) impurities are distributed uniformly over the lattice with a concentration c. These impurities in general should cause the change in jump probabilities not only at the particular interstitial sites where they are located, but also at the surrounding sites. But the most significant change should occur in the come-to-stop probability σ and the stay-stationary probability σ', both at the very same impurity sites. If the site is occupied by an impurity, which happens with the probability equal to the concentration c, the probabilites are denoted by (σ_1, σ_1') ; if the site is unoccupied, the probabilities are denoted by (σ_0, σ_0'). The successive jump probabilites (α, β, γ) will be chosen to be the same. Under these conditions, we may then replace the correction factor (14.9) by

$$1 + c \frac{\sigma_1}{1 - \sigma_1'} + (1 - c) \frac{\sigma_0}{1 - \sigma_0'} . \tag{14.12}$$

The three terms here correspond to the following processes : (i) The particle arrives and leaves with no intermediate stop, (ii) it arrives at an impurity site, which occurs with probability c, and eventually leaves, and (iii) it arrives at a regular site and leaves. Introducing (14.12) in (14.5) and re-arranging terms, we obtain

$$D = D_0 / \left[1 - c + c \frac{1 + \sigma_1/(1-\sigma_1')}{1 + \sigma_0/(1-\sigma_0')} \right], \tag{14.13}$$

where

$$D_0 \equiv \frac{(1 + \delta)\, v\, a_0}{6(1-\delta)\, [1+\sigma_0/(1-\sigma_0')]} \tag{14.14}$$

represents the diffusion coefficient for the pure crystal. The quantity appearing in the denominator in (14.13):

$$\frac{1 + \sigma_1/(1-\sigma_1')}{1 + \sigma_0/(1-\sigma_0')} = \frac{\tau_1}{\tau_0} \tag{14.15}$$

represents <u>the ratio of the mean staying time</u> at the impurity-occupied site, τ_1, to that at the unoccupied site, τ_0. In terms of the ratio of these staying (trap) times (τ_0, τ_1), we may re-express eq.(14.13) as

$$\boxed{D = \frac{D_0}{1 - c + c(\tau_1/\tau_0)}} . \qquad (14.16)$$

In a typical experimental condition, an interstitial nitrogen atom is estimated to make the trap time several hundred times greater. If the nitrogen concentration is 0.01 the formula (14.16) indicates that the diffusion coefficient could be a fraction of that of a pure crystal. This is in good agreement with the experimental evidence [23].

We can extend our theory to the case in which there exist impurities of several kinds. If the concentrations of different sites are denoted by c_I, c_{II},..., with the normalization

$$c_I + c_{II} + \dots = 1 , \qquad (14.17)$$

the correction factor should take the form

$$1 + c_I \frac{\sigma_I}{1 - \sigma_I'} + c_{II} \frac{\sigma_{II}}{1 - \sigma_{II}'} + \dots, \qquad (14.18)$$

in place of (14.12), where σ_α and σ_α' denote, respectively, the come-to-stop and stay-stationary probabilities at the site of type α, α = I, II, ...

The present theory is developed with the assumption of the s.c. lattice. The correlated walks on other cubic lattices can be formulated and solved in a similar manner [22]. The results for the mean square displacement and the diffusion coefficient are found to be expressed in a unified form. In particular, expressions (14.3) and (14.5) are valid with the appropriate redefinition of δ.

The treatment based on the correlated walk model is elementary, but can represent well the main feature of the atomic diffusion.

References

[1] Debye, P.P. and Conwell, E.M., Phys. Rev. 93, 693 (1954)

[2] Conwell, E. and Weisskopf, V.F., Phys. Rev. 77, 388 (1950)

[3] e.g. Symon, K., Mechanics, Addison-Wesley, Reading, Mass. Third Edition (1971)

[4] Brooks, H., Phys. Rev. 83, 879 (1954)

[5] e.g. Kittel, C., Introduction to Solid State Physics, Wiley, New York, Fifth Edition (1976)

[6] Kubo, R., J. Phys.Soc. Japan 12, 570 (1957) ; Nakano, H., Progr. Theoret. Phys. (Japan) 15, 77 (1956)

[7] Izuyama, T., Progr. Theoret.Phys. (Japan) 25, 964 (1961); Fujita, S., J. Phys.Soc. Japan 26, 505 (1969)

[8] Green,M.S., J. Chem. Phys.22, 898 (1954); Mori, H., J. Phys.Soc. Japan 11, 1029 (1956), Kubo, R., Yokota, M. and Nakajima, S., J.Phys.Soc. Japan 12, 1203 (1957); Luttinger, J.M., Phys. Rev. 135, A1505 (1964)

[9] Nakajima, S., Progr. Theoret.Phys. (Japan) 20, 448 (1958); see also, Chapter 7, in S. Fujita, Introduction to Non-Equilibrium Quantum Statistical Mechanics, Saunders, Philadelphia (1966)

[10] Zubarev, D.N., Nonequilibrium Statistical Thermodynamics, Consultant's Bureau, New York (1974)

[11] Fujita, S. and Chen, C. C., J. Theoret.Phys. 2, 59 (1969)

[12] e.g. Dirac, P.A.M., Principles of Quantum Mechanics, Oxford University Press, London, 4th Edition (1958), pp. 188-193

[13] Fujita, S., Ko, C.L. and Chi, J.Y., J. Phys.Chem. Solids, 37, 227 (1976)

[14] Dresselhaus, G., Kip, A.F. and Kittel, C., Phys. Rev. 98, 368 (1955)

[15] Kawamura, H., Saji, H., Fukai, M., Sekido K. and Imai, I., J.Phys. Soc. Japan 19, 288 (1964)

[16] Lodder, A. and Fujita, S., J. Phys.Soc. Japan 25, 774 (1968)

[17] Fujita, S. and Lodder, A., Physica 83 B, 117 (1976)

[18] Srinivas, T. K., Chaudhury, S. and Fujita, S., Solid State Commun. 37, 919 (1981)

[19] Lorentz, H.A., The Theory of Electrons, Dover, New York, Second Edition (1952), pp. 47-67, 267-74

[20] Chandrasekhar, S., Rev. Mod. Phys. 15, 1(1943)

[21] Fujita, S., Okamura, Y., Blaisten, E. and Godoy, S.V., J. Chem. Phys. 73, 4569 (1980)

[22] Okamura, Y., Blaisten-Barojas,E,Fujita, S. and Godoy, S.V., Phys. Rev. B 22, 1638 (1980)

[23] Richter, D., Topler, J. and Springer, T., J. Phys. F 6, L93 (1976)

1. Write down methematical expressions, figures, etc., and explain briefly.

(a) the order-of-magnitude formula for the electrical conductivity; the relative importance of the electronic and ionic conduction

(b) Matthiessen's rule

(c) diffuse scattering

(d) the Boltzmann equation for an electron-impurity system

(e) ionized impurities in a semiconductor

(f) holes as the charge carriers in a semiconductor

(g) mobility ; its connection with the conductivity

(h) screened Coulomb potential

(i) cyclotron frequency and radius

(j)* Hall current

(k)* Landau states and levels

(l)* absorption power; its connection with the probing electric field and the dynamic conductivity

(m)* Kubo's formula (current correlation function formula)

(n)* Einstein relation

(o)* Drude's formula for the dynamic conductivity

(p)* quasi-particle effect

(q)* delta-function approximation

(r)* cyclotron resonance

(s)* diffusion equation

(t)* diffusion coefficient and mean square displacement

General Problems

1. An ion of mass m and electric charge Ze is moving in a dilute gas of neutral molecules with which it collides. The mean time between collisions suffered by the ion is τ . Suppose that a weak uniform electric field $\vec{E}$ is applied in the x-direction.

(a) What is the mean distance $\overline{x}$ (in the direction of $\vec{E}$) which the ion travels between collisions if it starts out with zero mean velocity after each collision ?

(b) What is the current density $\vec{j}$ generated by a collection of ions ? Neglect the interaction among ions.

(c) Find the conductivity σ from Ohm's law : $\vec{j} = \sigma \vec{E}$.

2. Let us consider a hypothetical cube of a side-length equal to the mean free path ℓ . By assumption, any classical particles within the cube leave it without scattering. Suppose that we have two such cubes adjacent to each other and containing slightly different average energies. Compute the energy flux passing through the common face with the assumption that the net particle flux through the same face equals zero. By assuming Fourier's law : $\vec{q} = - K\nabla T$, find an expression for the thermal conductivity K.

[Answer $K = \frac{1}{3} n\, c v \ell$, $c \equiv$ heat capacity per particle]

3. Refer to problem 2. Instead of classical particles, we take phonons which carry the energy $\hbar\omega_k$ and which run with a constant speed v. Assuming the same situation, find an expression for the thermal conductivity K and investigate its temperature dependence. Note that this model may be used to describe the qualitative behavior of the thermal conductivity of a non-conducting solid. Look for experimental data for the inert-gas-atom solids and find the order of magnitude for the mean free path.

4. By using the simple kinetic theory, find the ratio of thermal to electrical conductivities, K/σ . Note that this ratic called the Wiedemann-Franz ratio is proportional to the absolute temperature T , and does not depend on the specific nature of a metal.

5. When a system is anisotropic because of its crystal structure and /or the presence of a magnetic field, the conductivity becomes a tensor. So does the resistivity $\rho_{\mu\nu}$, which is defined by

$$E_\mu = \Sigma_\nu \rho_{\mu\nu} j_\nu ,$$

and therefore is the inverse of the conductivity tensor. In the normal static experiments, the resistivity is measured while the conductivity can directly be analyzed in theory.

Given

$$\overleftrightarrow{\sigma} = \begin{bmatrix} \sigma_{xx} & \sigma_{xy} & 0 \\ \sigma_{yx} & \sigma_{xy} & 0 \\ 0 & 0 & \sigma_{zz} \end{bmatrix},$$

find the corresponding resistivity tensor $\rho_{\mu\nu}$.

BIBLIOGRAPHY

General textbooks at about the same level

Kittel, C. and Kroemer, H., Thermal Physics, W.H. Freeman, San Francisco, Ca., (1980)

Mandl, F., Statistical Physics, J. Wiley, London, (1971)

Morse, P.M., Thermal Physics, W.A. Benjamin, New York, N.Y. 10016, Second Edition (1969)

Reif, F., Fundamentals of Statistical and Thermal Physics, McGraw - Hill, New York, (1965)

Rosser, W. G. V., Introduction to Statistical Physics, E. Horwood, Chichester, England, (1982)

Terletskii, Ya. P., Statistical Physics, translated from the Russian by N. Froman, North - Holland, Amsterdam (1971)

Sears, F. W. amd Salinger, G.L., Thermodynamics, Kinetic Theory, and Statistical Thermodynamics, Addison-Wesley, Reading, Mass., (1975)

Zemansky, M.W. , Heat and Thermodynamics, McGraw-Hill, New York, N.Y. 5th ed., (1957)

General Textbooks at more advanced levels

Davidson, N., Statistical Mechanics, McGraw - Hill, N.Y. (1962)

Finkelstein, R. J., Thermodynamics and Statistical Physics, Freeman, SanFrancisco, California, (1969)

Heer, C. V., Statistical Mechanics, Kinetic Theory and Stochastic Processes, Academic Press, New York, N.Y. 10003 (1972)

Huang K., Statistical Mechanics, J. Wiley, New York, N.Y. (1967)

Isihara, A., *Statistical Physics*, Academic Press, New York, N.Y. 10003(1971)

Kestin, J. and Dorfman, J.R., *A Course in Statistical Thermodynamics*, Academic Press, New York, N.Y. 10003 (1971)

Landau, L. D. and Lifshitz, E. M., *Statistical Physics*, Addison - Wesley, Reading, Massachusetts, Second Edition (1969)

McQuarrie, D. A., *Statistical Mechanics*, Harper and Row, New York, N.Y. 10022 (1976)

Pathria, R. K., *Statistical Mechanics*, Pergamon, Oxford, England (1972)

Wannier, G. H., *Statistical Physics*, J. Wiley, New York (1966)

Probability Theory. *Stochastic Processes*

Chandrasekhar, S., *Rev. Mod. Phys.* 15, 1 (1943)

Cramer, H., *The Elements of Probability Theory*, J. Wiley, New York, N.Y. 10016 (1969)

Feller, W., *An Introduction to Probability Theory and Its Applications*, J. Wiley, New York, N.Y. 10016, Second Edition (1957)

Wang, M. C. and Uhlenbeck, G.E., *Rev. Mod. Phys*. *17*, 90 (1945)

Thermodynamics

Andrews, F. C., Thermodynamics : Priciples and Applications, J. Wiley, New York, N.Y. 10016 (1971)

Callen, H. B., Thermodynamics, J. Wiley, New York, N.Y. 10016 (1960)

Fermi, E., Thermodynamics, Dover Publications, New York, (1957)

Pippard, A. B., Thermodynamics : Applications, Cambridge University Press, Cambridge, England (1957)

Mechanics

Kibble, T. W. B., Classical Mechanics, McGraw - Hill, London (1966)

Symon, K. R., Mechanics, Addison - Wesley, Reading, Mass. Third Edition (1971)

Marion, J.B., Classical Dynamics, Academic, New York, NY 10003 (1965)

Elasticity

Landau, L. D. and Lifshitz, E. M., Theory of Elasticity, Addison - Wesley, Reading, Mass. (1959)

Electricity and Magnetism

Lorrain, P. and Corson, D. R., Electromagnetism, Freeman, San Francisco (1978)

Wangsness, R. K., Electromagnetic fields , J. Wiley, New York (1979)

Reitz, J.R. and Milford, F.J., Foundations of Electromagnetic Theory,

Quantum Mechanics

Alonso, M. and Finn, E.J., Fundamental University Physics, III Quantum and Statistical Physics, Addison-Wesley, Reading, Mass. (1968)

Dirac, P.A.M., Priciples of Quantum Mechanics, Oxford University Press, London, 4th Edition (1958) A true classic.

Gasiorowitz, S., Quantum Physics, Wiley, New York NY 10015 (1974)

McGervey, J.D., Modern Physics, Academic Press, New York (1971)

Pauling, L. and Wilson, E.B., Introduction to Quantum Mechanics, McGraw Hill, New York (1935)

Powell, J. L. and Grasmann, B., Quantum Mechanics, Addison-Wesley, Reading Mass. (1961)

Special Topics in Statistical Mechanics

Jancel R., Foundations of Classical and Quantum Statistical Mechanics, Pergamon, Fairview Park, Elmsford, New York 10523 (1969)

Thompson, C. J., Mathematical Statistical Mechanics, Macmillan, New York, N.Y. 10022 (1972)

Uhlenbeck, G. E. and Ford, G.E., Lectures in Statistical Mechanics, American Mathematical Society, Providence, R. I., (1963) (critical look at foundations of statistical mechanics)

Yvon, J., Correlations and Entropy in Classical Statistical Mechanics, Pergamon, Fairview Park, Elmsford, New York 10523 (1969) (authoritative and careful treatment of correlations and entropy in terms of a grand canonical ensemble)

Zubarev, D. N., Non-Equilibrium Statistical Thermodynamics, Consultants Bureau, New York, N.Y. 10011 (1974)

History

Brush, S.G., Kinetic Theory, I. and II., Pergamon, Oxford, England (1966)

Fermi, E., Thermodynamics and Statistics, University of Chicago Press, Chicago, (1966)

Klein, M.J., Paul Ehrenfest, North-Holland, Amsterdam (1970)

Ter Haar, D., L.D. Landau, I. and II., Pergamon, Oxford, England (1965) and (1969)

Mathematics

Arfken, G., Mathematical Methods for Physicists, Academic Press, New York, NY 10003 (1970)

Boyce, W.E. and DiPrima, R.C., Elementary Differential Equations and Boundary Value Problems, Wiley, New York, N.Y., Third Edition (1977)

Butkov, E., Mathematical Physics, Addison-Wesley, Reading Mass. (1968)

Kaplan, W., Advanced Calculus, Addison-Wesley, Reading , Mass., Second Edition (1973)

Kreyszig, E., Advanced Engineering Mathematics, Wiley, New York, Fourth Edition (1979)

Sokolnikoff, I.S. and Redhefler, R. M., Mathematics of Physics and Modern Engineering, McGraw-Hill, New York, Second Edition (1966)

Solid State Physics

Kittel, C., Introduction to Solid State Physics, Wiley, New York, NY 10016, Fifth Edition (1976)

Solid State Physics at more advanced levels

Haug, A., Theoretical Solid State Physics, Pergamon Press, Elmsford, NY 10523 (1978), vols. 1 and 2.

USEFUL PHYSICAL CONSTANTS

Quantity	Symbol	Value
Absolute zero on Celsius scale		-273.16°C
Avogadro's number	N_o	6.02×10^{23} molecules mol^{-1}
Boltzmann constant	k_B	$1.38 \times 10^{-16} erg\ K^{-1} = 1.38 \times 10^{-23} JK^{-1}$
Bohr magneton	μ_B	$9.22 \times 10^{-21} erg\ gauss^{-1}$
Bohr radius	a_o	$5.29 \times 10^{-9}\ cm = 5.29 \times 10^{-11} m$
Electron mass	m_e	$0.911 \times 10^{-27} g = 9.11 \times 10^{-31} kg$
Electron charge (absolute value)	e	4.80×10^{-10} esu= $1.6 \times 10^{-19} C$
Gas constant	R	$8.314\ J\ mol^{-1}\ K^{-1}$
Molar volume (gas at STP)		$2.24 \times 10^{4}\ cm^{3}$ = 22.4 liter
Mechanical equivalent of heat		$4.186\ J\ cal^{-1}$
Permittivity of free space	ε_o	$8.85 \times 10^{-12}\ C^{2}\ N^{-1}\ m^{-2}$
Permeability of free space	μ_o	$4 \times 10^{-7}\ Wb\ A^{-1}\ m^{-1}$
Planck's constant	h	6.63×10^{-27} erg sec = 6.63×10^{-34} J s
Planck's constant divided by 2π	$\frac{h}{2\pi} \equiv \hbar$	$1.05 \times 10^{-27} = 1.05 \times 10^{-34}$ J s
Proton mass	m_p	$1.67 \times 10^{-24} g = 1.67 \times 10^{-27} kg$
Speed of light	c	$3.00 \times 10^{10}\ cm\ sec^{-1} = 3.00 \times 10^{8}\ m\ s^{-1}$
Standard atmospheric pressure	1 atm	$1.013 \times 10^{5}\ N\ m^{-2} = 1.013 \times 10^{5}$ Pa

LIST OF SYMBOLS

The following list is not intended to be exhaustive.

It includes symbols of special importance.

Å	Angstrom (= 10^{-10}m)
$\vec{A}$	vector potential
$\vec{B}$	magnetic field
b	impact parameter
C	heat capacity
c	velocity of light
c	specific heat
E	total energy
E	internal energy
$\vec{E}$	electric field
$\vec{\vec{E}}$	unit tensor, see Appendix D.
e	base of natural logarithm
e	electronic charge (absolute value)
e	internal free energy density
F	Helmholtz free energy
f	Helmholtz free energy density
f	One-body distribution function in the μ-space
G	Gibbs free energy
g	Gibbs free energy density
g	gracitational acceleration
H	Hamiltonian
H	enthalpy
h	Planck's constant

h	single-particle Hamiltonian
$\hbar$	Planck's constant divided by 2π
$i \equiv \sqrt{-1}$	imaginary unit
$\vec{i}, \vec{j}, \vec{k}$	unit vectors for the Cartesian frame of reference
J	Jacobian of transformation
$\vec{J}$	total current
$\vec{j}$	single-particle current
$\vec{j}$	current density
$\vec{k}$	angular wave vector
k_B	Boltzmann constant
L	Lagrangian function
L, L_o	normalization length
$\hat{L}$	Lagrangian density
Lim	bulk limit, see (7.8.1)
l	mean free path
ln	natural logarithm
M	total mass
m	mass
N	number of particles
n	particle-number density
P	pressure
$\vec{P}$	total momentum
$\vec{p}$	momentum
p	(absorption) power
Q	quantity of heat
$\vec{R}$	position of the center of mass
r	radial coordinate

S	entropy
T	kinetic energy
T	absolute temperature
$\overleftrightarrow{T}$	stress tensor
t	time
Tr	trace [classical trace, see (4.6.10); quantum trace, see (8.16.10)]
TR	sum of N - particle traces, see (7.11.13).
V	potential energy
V	volume
$\vec{v}$	velocity field
W	work
Z	partition function
$e^{\alpha} \equiv z$	fugacity
λ	wavelength
μ	chemical potential
μ_B	Bohr magneton
$\mu_o \equiv \epsilon_F$	Fermi energy
$\nu = \omega/2\pi$	frequency
Ξ	grand partition function
ρ	density (statistical) operator
σ	electrical conductivity
σ	total cross section
τ_c	average time between collisions
ω	angular frequency

INDEX